Microbial Subversion of Immunity
Current Topics

Edited by

Peter J. Lachmann
Michael B.A. Oldstone

Caister Academic Press

Caister Academic Press
32 Hewitts Lane
Wymondham
Norfolk NR18 0JA
England

www.caister.com

British Library Cataloguing-in-Publication Data

A catalogue record for this book is available from the British Library

ISBN: 1-904455-05-0
ISBN-13: 978-1-904455-05-9

Printed and bound in Great Britain

Contents

Contributors

Antonio Alcami
Centro Nacional de Biotecnología (CSIC)
Campus Universidad Autónoma
Cantoblanco
28049 Madrid
Spain

Dennis R. Burton
Departments of Molecular Biology and Immunology
10550 North Torrey Pines Road
The Scripps Research Institute
La Jolla
California 92037
USA

Michael J. Doenhoff
School of Biological Sciences
University of Wales
Deiniol Road
Bangor
Gwynedd LL57 2UW
UK

Kurt Edelmann
The Scripps Research Institute
Department of Neuropharmacology
(IMM-6)
10550 N. Torrey Pines Road
La Jolla CA 92037
USA

Barbara Fernie-King
Centre for Veterinary Science
University of Cambridge
Madingley Road
Cambridge CB3 0ES
UK

Michael M. Frank
Division of Allergy/Immunology
Rm. 5416 Duke North
Duke University Medical Center
Durham NC 27710
USA

Denis Gerlier
Virologie & Pathogénèse Virale
CNRS-UCBL UMR 5537
IFR 62 Laennec
Rue Paradin
69372 Lyon Cedex 08
France

Stipan Jonjic
Dept. of Histology and Embryology
University of Rijeka
B. Branchetta 20
51000 Rijeka
Croatia

Ulrich H. Koszinowski
Max von Pettenkofer-Institut
University of Munich
Pettenkoferstrasse 9a
80336 Munich
Germany

David Laine
INSERM U503
CERVI
IFR 128 Bioscience
Lyon
France

Peter Lachmann
Centre for Veterinary Science
Madingley Road
Cambridge CB3 0ES
UK

Mansun Law
The Scripps Research Institute
Department of Immunology
10550 North Torrey Pines Road
La Jolla CA 92037
USA

Michael B. A. Oldstone
The Scripps Research Institute
Dept Neuropharmacology (IMM-6)
10550 N. Torrey Pines Road
La Jolla CA 92037
USA

Chantal Rabourdin-Combe
INSERM U503
CERVI
IFR 128 Bioscience
Lyon
France

Pietro Paolo Sanna
Department of Neuropharmacology
The Scripps Research Institute
10550 North Torrey Pines Road
La Jolla
CA 92037
USA

Christine Servet-Delprat
INSERM U503
CERVI
IFR 128 Bioscience
Lyon
France

Hélène Valentin
INSERM U503
CERVI
IFR 128 Bioscience
Lyon
France

Reinhard Würzner
Department for Hygiene
Microbiology and Social Medicine
Innsbruck Medical University
& Ludwig-Boltzmann-Institute for AIDS-Research
Fritz-Pregl-Str. 3
A-6020 Innsbruck
Austria

Elina Zuniga
The Scripps Research Institute
Department of Neuropharmacology (IMM-6)
10550 N. Torrey Pines Road
La Jolla CA 92037
USA

Foreword

All pathogenic microorganisms achieve pathogen status by virtue of succeeding, in one way or another, in subverting the immune responses of their host. The mechanisms for so doing are just as wide ranging as the immune responses themselves and the study of microbial subversion techniques is a good method of exploring the manifold complexities of the immune response. The scope of this book is therefore, potentially, extremely wide ranging and only a proportion of the field is covered in any detail. The choice of what has been covered was not wholly made by the editors; the inability of some authors to deliver manuscripts has played its part! Nevertheless the book does have some shape. Besides providing an overview of the field it has concentrated, in the mechanisms section, on the innate immune response and, in the organisms sections, on viruses.

The first section consists of overview chapters on important components of the innate immune response: complement, NK cells and the mucosal innate immune response. The trouble that microorganisms go to in order to subvert these aspects of immunity, that come into play before the adaptive immune response is recruited, provides strong evidence of the importance of innate immunity in resistance against infection.

The second section contains overview chapters on three major classes of human pathogens: viruses, bacteria and nematodes. This section approaches the subject by considering the subverting organism rather than its target and illustrates the versatility of the mechanisms deployed.

The final section is devoted to dealing with some aspects of viral subversion in much greater depth. It includes a detailed account of how viruses subvert the humoral immune response; an analysis of the interaction between viruses and dendritic cells, now recognised as being of central importance in the generation of the adaptive immune response; and concludes with a detailed case study of the measles virus - the virus first recognised as being able to subvert immunity (von Pirquet noticed that patients with measles lose their tuberculin reactivity in 1904) and one of the most successful human pathogens ever, in that, in the days before anti-measles vaccination, it infected almost everyone on the planet!

Peter Lachmann
Michael Oldstone

1

Subversion of Complement

*Reinhard Würzner**

Abstract

Complement is probably the most important arm of the innate humoral immune system and virtually all pathogens invading the human host are attacked by it directly following entry and usually also during consecutive stages of disease, especially when they are in contact with blood. In order to escape or subvert the destructive action of complement, both its direct action and the interplay with the phagocytic system, most microorganisms have developed an effective battery of specific strategies in order to survive the hostile environment within the host. The measures to avoid recognition and destruction by complement via complement-mediated attraction, opsonisation and activation of phagocytic cells, and lysis involves microbial molecules which are expressed on the surface or secreted into the near vicinity and can therefore be considered as virulence factors. Of all these highly sophisticated mechanisms imitation (molecular mimicry) and employment of host complement proteins appear to be the most evolutionary elaborated.

1. Complement – Structure and control

The complement system was first identified as a heat labile humoral component of the immune system that 'complements' the heat-stabile action of antibodies more than a century ago. It comprises more than thirty plasma or membrane proteins, including important regulatory proteins. Its activation relies initially on a cascade of proteolytic steps, performed by serine protease domains in the components involved, and proceeds via three different pathways: the phylogenetically oldest, alternative pathway; the first discovered but phylogenetically most recent, classical pathway; and the last discovered, lectin pathway (Lachmann, 1992; Morgan and Harris, 1998; Rother et al., 1998, Volanakis and Frank, 1998; Walport, 2001; Prodinger et al., 2003).

Activation of the complement system by the classical pathway is initiated by binding of complement factor C1q to an antigen-antibody complex or to a target surface. The consequence of C1q binding is activation of the associated subunits

*For correspondence email Reinhard.Wuerzner@uibk.ac.at

C1r and C1s to serine-proteases which cleave the peptide C4a from the N-terminus of the soluble C4, leaving C4b as the main fragment. C4b interacts via a very short-lived thioester bond with nearby acceptor sites, e.g. on the membrane of the pathogen, and thereafter allows attachment of soluble C2. C4b-bound C2 is also efficiently cleaved by C1s and the generated membrane-bound complex C4b2a forms the active C3 convertase of the classical pathway.

The same C3 convertase can also be generated antibody-independently by adherence of mannan-binding lectin (MBL) to carbohydrate structures on the surface of a variety of microorganisms. MBL is a collectin, i.e. a protein containing collagenous structures as well as globular C-type lectin domains. Binding of MBL to carbohydrates activates the MBL-associated serine proteases (MASPs) which cleave, in a manner similar to C1r and C1s, the complement factors C4 and C2 to generate C4b2a.

In the alternative pathway of complement activation low amounts of C3b are formed by other inflammatory proteases, or by C3 which is inactivated by H_2O (C3 H_2O), forming a complex with Factor B. This complex C3bB has only weak intrinsic C3-converting activity, but the serine protease Factor D cleaves Factor B proteolytically, generating C3bBb, a complex which is inherently labile but is stabilized by interaction with properdin. This is the alternative pathway C3 convertase and cleaves C3. The resulting C3b amplifies this loop of complement activation. In the absence of inhibitors this constant activation would result in total consumption of C3.

All three pathways thus merge into activation of C3, and, via the then generated C5 convertases C4b2a3b or C3bBb3b, into cleavage of C5. During activation, potent small proteolytic peptides, the anaphylatoxins (C3a and C5a), as well as opsonins (C3b and iC3b), are generated.

The membrane attached C5b is the initiator of the common terminal pathway, where the complement components C6, C7, C8 and C9 are activated in a non-proteolytic manner and assembled into the terminal complement complex (TCC), either as potentially membranolytic membrane attack complex (MAC), or, in extracellular fluids, as non-lytic SC5b-9 in the presence of S protein (vitronectin).

The precise mechanism of terminal complement-mediated cytotoxicity after insertion of C9, however, is still unclear. According to one model the polar domains of inserted complement proteins, particularly C9, cause local distortion of the phospholipid bilayer resulting in 'leaky patches' (Esser, 1991). The other theory postulates that the terminal complement proteins form a hydrophilic channel ('pore') through the membrane with consequent disruption of the cell (Bhakdi, 1991).

Such powerful machinery needs safe and redundant control mechanisms. Therefore, about half of the complement components serve as fluid-phase or membrane inhibitors for controlling the critical steps in activation, especially those dealing with C3b or C5b generation. Others function as inhibitors of membrane attack complex formation.

Inhibition of C3/C5 convertases occurs via different mechanisms: inhibition of complex formation, acceleration of decay of the enzyme complexes or promotion of C3b and C4b cleavage by Factor I. Fluid-phase inhibitors of C3/C5 convertases are Factor H, Factor I, C1q inhibitor and C4b binding protein; Properdin is the only fluid-phase positive regulator stabilizing alternative C3 convertase. The C3/C5 negative regulators MCP (CD46), DAF (CD55) and CR1 (CD35) are associated with the cell membrane.

Control of the terminal pathway is executed both before the integration of the nascent MAC (by S-protein, clusterin/SP-40,40, lipoproteins, antithrombin III, proteoglycans and by C7 and C8, which represent not only essential components of the lytic complex but, paradoxically, also modulate attack locally and preclude their firm insertion into the membrane (Nemerow, 1979; Würzner, 2000) and at the stage of pore formation (by 'homologous restriction factors', as they show some degree of species restriction, i.e. they prevent lysis by autologous complement attack (Lachmann, 1991): (i) a shadowy 65-68 kDa molecule (C8bp, HRF, MIP); and (ii) a well characterized glycolipid-anchored 18 -20 kDa molecule (CD59), which interferes with the C9 interaction site on the C8 α-chain that is needed for membrane insertion.

Apart from the functions of complement in host defense against microorganisms (see below), complement has another major function. It plays an important role in the processing and disposal of immune complexes and outdated host proteins, and in the clearance of apoptotic cells from the circulation. This is facilitated mainly by C1q, a fragment of the initial component of the classical pathway, in detail reviewed elsewhere (Lachmann, 1992; Morgan and Harris, 1998; Rother et al., 1998; Volanakis and Frank, 1998; Walport, 2001; Prodinger et al., 2003).

2. Complement: Protecting the host

Upon initial invasion into the human body microbes activate predominantly the alternative and lectin pathways of complement (triggered by their surface structures, such as lipopolysaccharides, sialic acids and peptidoglycans). Complement may be therefore considered as important first-line defence system of higher vertebrates against invading microorganisms. Following induction of the adaptive immune response, antibodies bind to the surface of the intruder. C1 will bind to the Fc portion of the antibody activating the classical pathway. Certain surface structures, e.g. the gp120 molecule of HIV, are able to bind to C1q directly, and thus activate the classical pathway independent of antibody formation.

2.1. Complement: Three main direct functions in defence against microorganisms

Complement proteins and functions and regulatory proteins are identified early in evolution e.g. in the teleost fish, barred sand bass, which in evolution is separated from humans by 100 million years. Three important direct functions can be distinguished for complement in immune defense against microorganisms.

i) CHEMOTAXIS: the attraction of phagocytic cells via newly generated anaphylatoxins to the site of infection. Activation of all three pathways eventually results into an activation of C3 and C5 and thus to a marked release of the anaphylatoxic peptides C3a and C5a. Using their C3a- and C5a-receptors, phagocytic cells, especially polymorphonuclear phagocytes (PMNs), sense concentration gradients and migrate towards the area of highest concentration, which is normally the site of inflammation.

ii) OPSONISATION: the generation of C3 fragments, mainly C3b and iC3b on the microbial surface facilitating the following phagocytosis. Upon arrival at the site of infection PMNs are unable to identify the invader on its own and require complement as a sort of signaller. Thus, the deposition of complement fragments (mainly C3b and iC3b) generated on the surface of the invader is a prerequisite for a sufficient engulfment and uptake via specific complement receptors (CR1 and especially CR3) displayed on the surface of the PMNs. Phagocytosis is most efficient when both C3 fragments and antibody are attached to the microorganism, suggesting enhancement via a dual-signal-procedure.

iii) LYSIS: the generation of a membrane attack complex, consisting of activated components C5b, C6, C7, C8 and C9, generating membrane leakage. Complement, as a phylogenetically old immune defence system, is able to act on its own as a potent first-line defence, independently of other parts of the immune system. Not only the number of C3-fragment molecules deposited on the invader but also the amount of C5b-9 inserted into the target membrane is decisive for host defence. The latter strongly correlates with bactericidal activity and inversely with complement resistance. The debate, how lysis is executed, i.e. by forming a physical hole or by generating membrane perturbations (see above), is especially interesting in the light of immune defence against *Neisseria meningitidis*. The particularly frequent occurrence of terminal complement deficiencies in patients with meningococcal infections confirms that the cytolytic activity of complement is important in resistance to gram-negative bacteria. It is established that the terminal pathway damages the LPS-containing cell wall. However, it is still unclear how the 'inner' cell membrane is destroyed: possibly via induction of injurious processes or directly by bacteria-derived lysozyme or bacterially generated ATP; alternatively, further terminal components, entering the periplasmic space via the outer membrane attack leaks, may insert into the inner membrane as MACs or even as polymerized or single C9 alone. While lysis plays a predominant role in immune defence against *Neisseria spp.*, it appears to be less important against other microorganisms, especially against those having a compound cell membrane or even a thick cell wall.

2.2. Complement: Fourth function in defence against microorganisms - as coordinator of immune responses

In addition to the three direct effects of complement towards the invader (see above), complement also possesses an only recently recognised and rather indirect but very powerful function by strongly interacting with both parts of the cellular immune system, the innate part, comprising macrophages, NK cells and granulocytes and also the adaptive (or acquired) part, consisting of B- and T-lymphocytes:

i) complement is involved in the initiation of adaptive immune responses, since antibody-independent, complement-mediated opsonization of microbes leads to an uptake and presentation of antigens via complement-receptor bearing antigen-presenting cells which then triggers the adaptive immune response;

ii) antigen-bound C3d augments antibody response by facilitating B-cell activation via crosslinking membrane IgM to the CR2/CD19/TAPA-1 complex;
iii) CR1 and CR2 expressed on follicular dendritic cells are essential for the formation of memory B cells by localizing C3b- and/or C4b-bearing immune complexes in the germinal centres;
iv) non-lytic MAC induces protection to the host cell, as upon single (and erroneous) attacks it provides resistance to subsequent, otherwise lytic doses;
v) non-lytic MAC activates the host cell in a procoagulant, which limits the area of attack, and proinflammatory manner, e.g. by augmenting arachidonic acid metabolism, intracellular Ca++ flux and protein kinase C activation; and
vi) activation of human CD4+ cells with CD3 and CD46 induces a T-regulatory cell 1 phenotype, defined by a suppression of T-helper cells and production of interleukin IL 10 (Kemper et al., 2003), indicating that complement has also a previously unappreciated role in T-cell-mediated immunity and tolerance.

3. Microorganisms: Subverting complement protection

Evolution during millions of years of both host and microorganisms has created a commensal relationship between man and microbes, so that in many cases potentially infectious microorganisms are not attacked and live in symbiosis with the host. Most of them cause disease only when the host defence is considerably weakened.

A different type of relationship is medically very important and scientifically most interesting: microorganisms which are highly pathogenic but nevertheless either evade appropriate recognition or constrain suitable attack and destruction. These pathogens invading the human body are attacked by the complement system directly following entry and usually also during most stages of the disease, especially when they are in contact with human blood. However, during co-evolution alongside their obligate hosts several pathogenic microorganisms have developed an effective battery of specific strategies to overcome the destructive human immune defence in general and complement in particular. These include both biochemical and biophysical measures to resist C3b deposition, opsonophagocytosis or complement-mediated cytolytic damage. These measures increase the likelihood of survival in a hostile environment and can be considered as virulence factors.

Microorganisms evade complement first by avoiding appropriate recognition and, as this is not always efficient, by preventing eradication by complement (see next two chapters; Cooper, 1991; Fishelson, 1994; Jokiranta et al., 1995; Kotwal, 1997; Würzner, 1999).

3.1. Avoidance of recognition: Disguise

The easiest principle to establish survival in the host is simply the one of disguise, i.e. the host does not recognise the invasion of the pathogen (Würzner, 1999): this is achieved by displaying non- or poorly-activating surface moieties. Such molecules which lead to an insufficient complement activation include sialic acids, e.g. of encapsulated bacteria or parasites as well as surface glycoproteins and lipophosphoglycans. Some gram-negative bacteria mask or change their surface-expressed lipopolysaccharides to avoid recognition. This leads to an absent or non-opsonic attachment of C3 fragments on the pathogen cell surface, resulting in an inappropriate recognition by PMNs. *Campylobacter fetus* even generates an entire protective paracrystalline protein S-layer (Thompson, 2002). Likewise, subversion is also accomplished by displaying the complement-activating properties somewhat hidden on the surface, so that because of sterical hindrance they are inaccessible by PMNs. A further clever way of subversion is to modulate the location of C3 deposition in a less-opsonic way, as described for the capsule of *Cryptococcus neoformans* (Zaragoza et al., 2003). The shedding of strongly activating surface molecules or initial immune complexes (*Fasciola hepatica*) is another way of disguise (Duffus and Franks, 1980). Not surprisingly, the virulence of a strain usually correlates well with its ability to prevent recognition.

3.2. Avoidance of eradication

Once recognised by the host's complement system the invader has to introduce measures to constrain suitable attack to avoid eradication by complement. A detailed list of pathogens and their ways to avoid eradication by complement (Würzner, 1999) and further figures detailing this interaction (Würzner, 2003) are published elsewhere. Literature concerning subversion mechanisms mentioned, but not referenced, herein will be likely found in those reviews.

Complement can be removed by shedding off, consuming, or destroying complement. Microorganisms may shed C3 (by *Trypanosoma cruzi*) or even assembling complement complexes, in particular at the membrane attack stage (*Naegleria fowleri*).

Removal of complement is also accomplished by consumption of complement, usually following the shedding of complement activating molecules, as consumption is much safer at a secure distance well away from the invader. However, this also leads to "reactive lysis" of bystander cells. Microorganisms exploiting this subversive procedure, such as *Streptococcus pyogenes*, must have a highly sophisticated regulation to ensure that not too much detrimental activation occurs. This is obviously quite efficient at an early stage, especially when an important inhibitor is knocked out: cleavage of C1 inhibitor by proteases (*E. coli*) results in an uninhibited consumption of C1 and consecutively of other complement proteins. A similar mechanism is employed by *Loxosceles* spider venom: by inducing metalloproteinases that can cleave MCP/CD46, host cells become more vulnerable to complement attack (Van den Berg et al., 2002).

Blocking both classical and alternative complement activation by simply binding to C3 is facilitated by the 19-kDa *Staphylococcus aureus* Efb protein thereby significantly reducing opsonophagocytosis (Lee et al., 2004).

Destruction by proteolytic cleavage and phosphorylation of complement proteins or inactivation of anaphylatoxins by microorganisms also represent ways to confer protection from opsonisation and lysis. Good examples are phosphorylating proteinkinases (*Leishmania spp.*), anaphylatoxin-inactivating proteases (*Serratia marcescens*) and C3-cleaving ureases (*Helicobacter pylori*). Recently, staphylokinase has also been found to degrade C3b thereby reducing opsonic activity of human serum (Rooijakkers et al., 2005).

4. Molecular mimicry of complement by microorganisms

Molecular mimicry, the imitation of complement-like proteins plays a dominant role in inhibition and employment of complement (see below). During co-evolution alongside their obligate hosts several pathogens have evolved functional or structural properties identical to those of host cells to prevent their own destruction by complement (convergent evolution). Thus, a number of distinct microbial proteins has been identified which share functional or structural similarities with complement proteins or complement receptors.

Some viruses, like Herpes Virus Saimiri, share genetic similarities (antigenic cross-reactivity, sequence homology) and even functional activities with complement proteins or receptors. Most likely they have, in evolutionary terms recently, incorporated complement gene segments into their genome (Rother et al., 1994).The same applies to kaposica, the complement control homolog of Kaposi sarcoma associated herpesvirus (HHV8 or KSHV), which inactivates C3 convertases by accelerating their decay (Spiller et al., 2003), especially via its SCRs 1-4 (Mullick et al., 2005).

For others a combination of convergent evolution and gene capture appears to be responsible. In the case of vaccinia virus the DNA encoding VCP, a functionally CR1-like and structurally C4bp-like complement control protein, was presumably originally acquired from the host, as it has sequence homologies to both. Over an evolutionary period the captured gene was constantly manipulated to retain only the most essential domains as any further manipulation of the small viral protein results in loss of function, indicating that the gene has achieved a maximum efficiency to encode a protein with the minimum number of amino acids (Kotwal, 1997).

5. Inhibition of complement

A variety of mechanisms are used by pathogens to inhibit complement, such as the display of C1 binding moieties which directly interfere with classical pathway activation, as the *Salmonella minnesota* 39 kDa porin, or inhibitory molecules interfering with the potentially lytic terminal pathway as the multipotent (as it also inhibits the antibacterial action of defensins, Fernie-King et al. (2004)) SIC protein of *Streptococcus pyogenes* (Fernie-King et al., 2001).

In order to inhibit complement, some pathogens can also take advantage of their endogenously expressed complement-like molecules mimicking host complement. Well characterised examples are the parasitic (*Trypanosoma*) or viral (Vaccinia virus, Herpes simplex virus) proteins related to mammalian CR1, DAF, MCP or C4BP, which all target C4b and C3b, limiting complement attack and confirm the importance of C3- and C4-binding molecules in host defence. The *Schistosoma* protein Sh-Tor exploits a short sequence identity with human C4b to interact with C2a thereby inhibiting the classical pathway (Inal and Sim, 2000). The different examples of molecular mimicry of the terminal cascade inhibitory protein CD59 are also quite remarkable, comprising functional, structural and/or antigenic similarities of CD59, inhibiting lysis by interacting with C8 and especially with C9, to viruses (Herpes Virus Saimiri ORF-15 (Albrecht et al., 1992)), parasites (the 170 kDa *Entamoeba histolytica* adhesin (Braga et al., 1992) and the 94 kDa

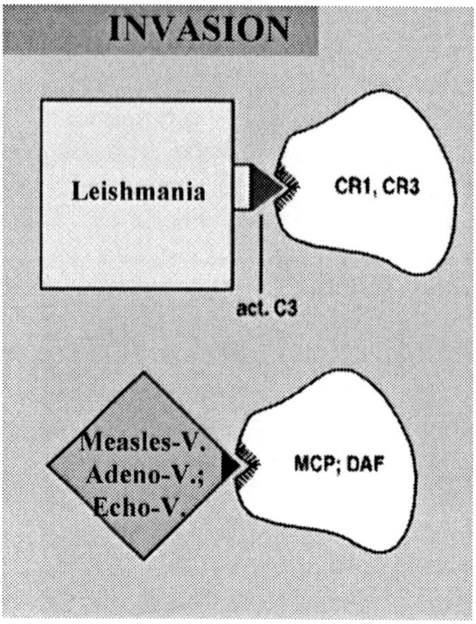

Figure 1. Aggressive employment of complement cell surface receptors to invade the host. Parasites (*Leishmania spp.*, square) or viruses (diamond), such as measles virus, adenovirus or echovirus may use complement receptors, e.g. CR1, CR3 or MCP or DAF, respectively, to adhere to (and subsequently invade) host cells, in an opsonisation-dependent (top) or independent (bottom) manner.

Schistosoma mansoni SCIP-1 protein (Parizade et al., 1994)) and to bacteria (the CD59-like molecule of *Borrelia burgdorferi* (Pausa et al., 2003).

6. Employment of complement

Some microorganisms evade complement in another very sophisticated way, namely by using host complement proteins or receptors for their own survival. Two main principles, one aggressive (facilitating adhesion and invasion) and one defensive one can be discerned (see next two sections): microorganisms may employ complement to adhere to and invade into host cells or to avert their own destruction, respectively, in part by an imitation of complement components or receptors (molecular mimicry).

6.1. Adhesion and invasion

Opsonisation of pathogens is not always harmful for the invader, and is sometimes even desired and initiated by microorganisms, as surface-deposited complement (in particular C3b or iC3b) can act as opsonin, facilitating entry of the invader into host cells via the complement receptors on the latter. This resembles a 'Trojan horse' principle as it can lead to a milder intracellular response of the phagocytes (lower respiratory burst) and thus intracellular survival and can rescue pathogens, such as Leishmania major, from the hostile extracellular environment. Opsonisation or cleavage, however, has to be very accurate so that most of the surface-deposited C3 is present in the optimum conformation (C3b or iC3b) for receptor binding (Da Silva et al., 1989, Figure 1). The same effect can be reached by cleaving C3, e.g. by the 56 kDa cysteine protease of Entamoeba histolytica (Reed et al., 1989).

Even more sophisticated, molecular mimicry of complement receptor ligands also facilitates entry into host cells, employing the complement receptors on them, and does not rely on the potentially dangerous opsonisation of the pathogen. The proteins involved on the invader are antigenically and/or functionally mimicking C3, as most convincingly shown for Epstein-Barr virus gp350/220 (Mold et al., 1988).

Other viruses are able to use the complement receptors MCP (CD46) and DAF (CD55) displayed on the surface of the nucleated cell. MCP is not only the cellular receptor for measles virus (Dorig et al., 1993), but also for other viruses, such as species B adenoviruses (Gaggar et al., 2003), in particular types 11 (Segerman et al., 2003) and 3 (Sirena et al., 2004), or meningococci (Johansson et al., 2003). Interestingly, the quite homologous bovine CD46 is also a receptor for a bovine diarrhea virus (Maurer et al., 2004). DAF on the other hand is a receptor for several echoviruses (Bergelson et al., 1994). Attachment to these cellular receptor allows cell entry directly without risking the for most pathogens extremely dangerous opsonisation procedure (Figure 1).

6.2. Protection

Pathogens do not only employ complement for invasion but also for their own defence. Both membrane bound as well as fluid phase complement inhibitors are utilized by microorganisms.

Exploitation of membrane inhibitors is a powerful way to subvert complement. *Neisseria gonorrhoeae* can give rise to downregulation of MCP (Gill et al., 2003) and ligation of this molecule plays a role in the induction of Treg cells by streptococcal M protein (Price et al., 2005). Intracellular pathogens, like CMV, can in particular harness the cellular machinery for their own benefit, e.g. by inducing expression of complement regulator DAF on the surface of the infected cell, preventing its destruction by the host immune system (Spiller et al. 1996).

It was recently shown that, in addition, CMV downmodulates CR3 and CR4 receptors on infected macrophages which then in turn leads to an impaired phagocytic capacity towards *Candida albicans* (Gafa et al., 2005), in part explaining the well-known immunosuppression accompanying anCMV infection. *Coxiella burnetii* also modulates CR3 expression, but here it is the local CR3 distribution (outside the pseudopodial extensions) which leads to immune subversion (Capo et al., 2003). A third way of exploiting CR3 is facilitated by *Blastomyces dermatitidis* BAD1 which binds to CR3 on macrophages and thereby suppresses TNF-alpha which is required for resistance (Brandhorst et al., 2004).

Both adhesion and possibly also inhibition can be achieved by mimicking integrins: a CR3-like molecule of the yeast *Candida albicans* facilitates not only host cell adhesion and cell invasion, but may also bind C3 in a non-opsonising way (Gale et al., 1998).

Some viruses employ complement proteins directly from the host cell membrane, e.g. upon budding HIV, coats itself with a lipid bilayer obtained from the host cell membrane and as a consequence carries, in addition to viral, also host cell, membrane proteins. Of the latter DAF and CD59 are of particular importance, protecting the virus from complement lysis.

Bacteria, such as *E. coli* (Rautemaa et al., 1998) or *Helicobacter pylori* (Rautemaa et al., 2001) can even acquire surface complement inhibitors, like DAF or CD59, respectively, especially as they are glycosylphosphoinositol-linked (Figure 2).

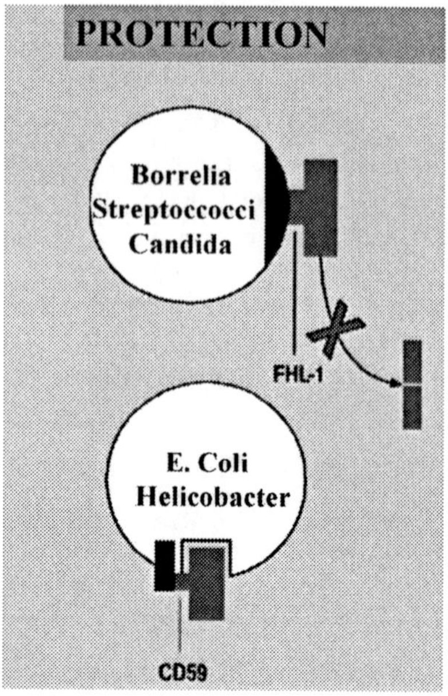

Figure 2. Defensive employment of complement inhibitors for own protection. Bacteria (*Borrelia spp., Streptococci*) or fungi (*Candida*) may employ complement inhibitors, e.g. factor H, FHL-1 (top) or CD59 (bottom), to inhibit complement activation, evading their own destruction.

Adsorption of fluid phase complement regulators/inhibitors to their cell surface is executed by a steadily increasing number of microorganisms, in part by mimicking the surface structure of host cells representing the physiological target. This acquisition results in a downregulation of the complement attack and in an inhibition of the generation of toxic complement activation products directly at the microbial surface and allows microbial growth and disease progression even in an immunocompetent host. Other benefits for the microbe include adhesion to, as well as invasion into, host cells.

Acquisition of host complement regulators has been shown not only for gram-positive or gram-negative bacteria (Jarva et al., 2003, Kraiczy and Würzner, 2006, Figure 2), but also for viruses, fungi (Figure 2) and parasites (Würzner and Zipfel, 2004), and recently also for worms (Baker et al., 2004). Soluble host complement inhibitor proteins which regulate the alternative pathway, (Factor H and the Factor H like protein 1 (FHL-1)), and proteins which control the classical pathway of complement, (C4 binding protein (C4BP)) are the components bound and this phenomenon has been extensively reviewed elsewhere (Würzner and Zipfel, 2004). It is interesting that in most cases the host regulators (Factor H and FHL-1) are attached via their C-terminal ends, i.e. domain 20 of Factor H and domain 7 of FHL-1, which both harbour heparin-binding-domains. This interaction results in a physical coat which efficiently separates the microbial cell membrane or capsule from the site of host attack and thus creates a real physical barrier which limits immune attack. Recent examples are the binding of factor H to *Borrelia recurrentes* (McDowell et al., 2003) and to *Aspergillus spp.* (Vogl et al., 2006), and new details concerning the binding to *streptococcus pyogenes* (Perez-Caballero et al., 2004; Wei et al., 2005), and the binding of C4BP to *streptococcus pyogenes* (Perez-Caballero et al., 2004), serogroup B *Neisseria meningitides* (Jarva et al., 2005) and *Candida albicans* (Meri et al., 2004).

7. General features

It is quite evident that the majority of microbial proteins and carbohydrates involved in subverting complement yet awaits discovery, but a lot of different principles have already been elucidated. Most microorganisms employ a whole arsenal of subversion strategies as *Streptococcus pyogenes* (Jarva *et al.*, 2003) or *Schistosoma mansoni* (Deng *et al.*, 2003), interacting at different stages within the complement cascade. For acquiring host factor H, e.g., *Borrelia burgdorferi* strain ZS7 uses at least five distinct surface proteins, termed BbCRASP-1 to BbCRASP-5, and one single microbial surface protein (e.g. the streptococcal M-protein) may bind multiple host regulators (the two alternative pathway regulators Factor H and FHL-1 and also the classical pathway regulator C4BP) by distinct sites (Kraiczy and Würzner, 2006).

Some microorganisms possess phage-encoded sets of complement inhibitors: the *Staphylococcus aureus* pathogenicity island-5 (SaPI5), comprising - among other virulence factors which are all not found in other non-aureus (less virulent) strains - not only the C5a receptor-blocking chemotaxis inhibitory protein CHIPS (De Haas et al., 2004), but also a staphylokinase, cleaving C3 (Rooijakkers et al., 2005), and a unique lectin pathway inhibitor (LPI) interacting with MASP-2 and thereby blocking phagocytosis (Rooijakkers et al., 2004). Not surprisingly, most pathogenic molecules involved in subversion are abundantly expressed on the cell surface at least at one stage of their development. Most are furthermore highly conserved within each species or indicate the virulent subtype.

It is likely important to characterise these principles in order to develop therapeutic or preventive approaches. Especially the sites of molecular mimicry have been considered as useful targets for vaccine development (Fishelson, 1991). However, taking their multiple as yet unrevealed interactions and, more importantly, their similarity to human molecules into account, a detrimental effect of such a vaccine, cannot be excluded as immunisation with a microbial antigen, mimicking a host protein, may induce an uncontrolled autoimmune reaction (Schwimmbeck *et al.*, 1987; Zhao *et al.*, 1998).

Recently, promising steps have been taken to utilize soluble (Lucas and McFadden, 2004) or membrane-associated immunomodulatory viral proteins (Chang et al., 2005) for the development of novel biotherapeutics.

References

Albrecht, J.C., Nichlas, J., Camerun, K.R., Newman, C., Fleckenstein, B., and Honess, R.W. (1992). Herpes virus saimiri has a gene specifying a homologue of the cellular membrane glycoprotein CD59. Virology *190*, 527-530.

Baker, S., Wang, H., Giacomin, P.R., Dent, L.A., and Gordon, D.L. (2004). Binding of complement factor H to the nematode *Nippostrongylus brasiliensis*. Mol. Immunol. *41*, 205.

Bergelson, J.M., Chan, M., Solomon, K.R., Stjohn, N.F., Lin, H.M., and Finberg, R.W. (1994). Decay-accelerating factor (CD55), a glycosylphosphatidyl-inositol-anchored complement regulatory protein, is a receptor for several echoviruses. Proc. Natl. Acad. Sci. USA *91*, 6245-6248.

Bhakdi, S., and Tranum Jensen, J. (1991). Complement lysis: A hole is a hole. Immunol. Today *12*, 318-320.

Braga, L.L., Ninomiya, H., McCoy, J.J., Eacker, S., Wiedmer, T., Pham, C., Wood, S., Sims, P.J., and Petri, W.A. Jr. (1992). Inhibition of the complement membrane attack complex by the galactose-specific adhesin of *Entamoeba histolytica*. J. Clin. Invest. *90*, 1131-1137.

Brandhorst, T.T., Wuthrich, M., Finkel-Jimenez, B., Warner, T., and Klein, B.S. (2004). Exploiting type 3 complement receptor for TNF-alpha suppression, immune evasion, and progressive pulmonary fungal infection. J. Immunol. *173*, 7444-7453.

Capo, C., Moynault, A., Collette, Y., Olive, D., Brown, E.J., Raoult, D., and Mege, J.L. (2003). *Coxiella burnetii* avoids macrophage phagocytosis by interfering with spatial distribution of complement receptor 3. J. Immunol. *170*, 4217-4225.

Chang, Y.J., Jiang, M., Lubinski, J.M., King, R.D., and Friedman, H.M. (2005). Implications for herpes simplex virus vaccine strategies based on antibodies produced to herpes simplex virus type 1 glycoprotein gC immune evasion domains. Vaccine *23*, 4658-4665.

Cooper, N.R. (1991). Complement evasion strategies of microorganisms. Immunol. Today *12*, 327-331.

Da Silva, R.P., Hall, B.F., Joiner, K.A., and Sacks, D.L. (1989). CR1, the C3b receptor, mediates binding of infective leishmania major metacyclic promastigotes to human macrophages. J. Immunol. *143*, 617-622.

De Haas, C.J., Veldkamp, K.E., Peschel, A., Weerkamp, F., Van Wamel, W.J., Heezius, E.C., Poppelier, M.J., Van Kessel, K.P., and van Strijp, J.A. (2004). Chemotaxis inhibitory protein of Staphylococcus aureus, a bacterial antiinflammatory agent. J. Exp. Med. *199*, 687-695.

Deng, J., Gold, D., LoVerde, P.T., and Fishelson, Z. (2003). Inhibition of the complement membrane attack complex by *Schistosoma mansoni paramyosin*. Infect. Immun. *71*, 6402-6410.

Dorig, R.E., Marcil, A., Chopra, A., and Richardson, C.D. (1993). The human CD46 molecule is a receptor for measles virus (Edmonston Strain). Cell *75*, 295-305.

Duffus, W.P.H., and Franks, D. (1980). *In vitro* effect of immune serum and bovine granulocytes on juvenile *Fasciola hepatica*. Clin. Exp. Immunol. *41*, 430-440.

Esser, A.F. (1991). Big MAC attack: complement proteins cause leaky patches. Immunol. Today *12*, 316-318.

Fernie-King, B.A., Seilly, D., Willers, C., Würzner, R., Davies, A., and Lachmann, P.J. (2001). Streptococcal inhibitor of complement inhibits the membrane attack complex by preventing uptake of C567 onto cell membranes. Immunology *103*, 390-398.

Fernie-King, B.A., Seilly, D.J., and Lachmann, P.J. (2004). The interaction of streptococcal inhibitor of complement (SIC) and its proteolytic fragments with the human beta defensins. Immunology *111*, 444-452.

Fishelson, Z. (1991). Complement evasion by parasites: search for "Achilles' heel". Clin. Exp. Immunol. *86* (Suppl.1), 47-52.

Fishelson, Z. (1994). Complement-related proteins in pathogenic organisms. Springer Sem. Immunopathol. *15*, 345-368.

Gafa, V., Manches, O., Pastor, A., Drouet, E., Ambroise-Thomas, P., Grillot, R., and Aldebert, D. (2005). Human cytomegalovirus downregulates complement receptors (CR3, CR4) and decreases phagocytosis by macrophages. J. Med. Virol. *76*, 361-366.

Gaggar, A., Shayakhmetov, D.M., and Lieber, A. (2003). CD46 is a cellular receptor for group B adenoviruses. Nat. Med. *9*, 1408-1412.

Gale, C.A., Bendel, C.M., McClellan, M., Hauser, M., Becker, J.M., Berman, J., and Hostetter, M.K. (1998). Linkage of adhesion, filamentous growth, and virulence in Candida albicans to a single gene, INT1. Science *279*, 1355-1358.

Gill, D.B., Koomey, M., Cannon, J.G., and Atkinson, J.P. (2003). Down-regulation of CD46 by piliated *Neisseria gonorrhoeae*. J. Exp. Med. *198*, 1313-1322.

Inal, J.M., and Sim, R.B. (2000). A Schistosoma protein, Sh-Tor, is a novel inhibitor of complement which binds human C2. Febs Lett. *470*, 131-134.

Jarva, H., Jokiranta, T.S., Würzner, R., and Meri, S. (2003). Complement resistance mechanisms of streptococci. Mol. Immunol. *40*, 95-107.

Jarva, H., Ram, S., Vogel, U., Blom, A.M., and Meri, S. (2005). Binding of the complement inhibitor C4bp to serogroup B *Neisseria meningitidis*. J. Immunol. *174*, 6299-6307.

Johansson, L., Rytkonen, A., Bergman, P., Albiger, B., Kallstrom, H., Hokfelt, T., Agerberth, B., Cattaneo, R., and Jonsson, A.B. (2003). CD46 in meningococcal disease. Science *301*, 373-375.

Jokiranta, T.S., Jokipii, L., and Meri, S. (1995). Complement resistance of parasites. Scand. J. Immunol. 42, 5-20.

Kemper, C., Chan, A.C., Green, J.M., Brett, K.A., Murphy, K.M., and Atkinson, J.P. (2003) Activation of human CD4+ cells with CD3 and CD46 induces a T-regulatory cell 1 phenotype. Nature *421*, 388-392.

Kotwal, G.J. (1997). The great escape: immune evasion by pathogens. Immunologist *4*, 157-164.

Kraiczy P, and Würzner R. (2006). Complement escape of human pathogenic bacteria by acquisition of complement regulators. Mol. Immunol. *43*, 31-44.

Lachmann, P.J. (1991). The control of homologous lysis. Immunol. Today *12*, 312-315.

Lachmann, P.J. (1992): Complement. In Oxford Textbook of Pathology, J.O.D. McGee, P.G. Isaacson, and N.A. Wright, eds. (Oxford: Oxford University Press), pp 259-266.

Lee, L.Y., Hook, M., Haviland, D., Wetsel, R.A., Yonter, E.O., Syribeys, P., Vernachio, J., and Brown, E.L. (2004). Inhibition of complement activation by a secreted *Staphylococcus aureus* protein. J. Infect. Dis. *190*, 571-519.

Lucas, A., and McFadden, G. (2004). Secreted immunomodulatory viral proteins as novel biotherapeutics. J. Immunol. *173*, 4765-4774.

Maurer, K., Krey, T., Moennig, V., Thiel, H.J., and Rumenapf, T. (2004). CD46 is a cellular receptor for bovine viral diarrhea virus. J. Virol. *78*, 1792-1799.

McDowell, J.V., Tran, E., Hamilton, D., Wolfgang, J., Miller, K., Marconi, R.T. (2003). Analysis of the ability of spirochete species associated with relapsing fever, avian borreliosis, and epizootic bovine abortion to bind factor H and cleave C3b. J. Clin. Microbiol. *41*, 3905-3910.

Meri, T., Blom, A.M., Hartmann, A., Lenk, D., Meri, S., and Zipfel, P.F.(2004). The hyphal and yeast forms of *Candida albicans* bind the complement regulator C4b-binding protein. Infect. Immun. *72*, 6633-6641.

Mold, C., Bradt, B.M., Nemerow, G.R., and Cooper, N.R. (1988). Epstein-Barr virus regulates activation and processing of the third component of complement. J. Exp. Med. *168*, 949-969.

Morgan, B.P., and Harris, C.L. (1998). Complement Regulatory Proteins (London: Academic Press).

Mullick, J., Singh, A.K., Panse, Y., Yadav, V., Bernet, J., Sahu, A. (2005). Identification of functional domains in kaposica, the complement control protein homolog of Kaposi's sarcoma-associated herpesvirus (human herpesvirus 8). J. Virol. *79*, 5850-5856.

Nemerow, G.R., Yamamoto, K.I., and Lint, T.F. (1979). Restriction of complement-mediated membrane damage by the eighth component of complement: a dual role for C8 in the complement attack sequence. J. Immunol. *123*, 1245-1252.

Parizade, M., Arnon, R., Lachmann, P.J., and Fishelson, Z. (1994). Functional and antigenic similaries between a 94-kDa protein of schistosoma mansoni (SCIP-1) and human CD59. J. Exp. Med. *179*, 1625-1636.

Pausa, M., Pellis, V., Cinco, M., Giulianini, P.G., Presani, G., Perticarari, S., Murgia, R., and Tedesco, F. (2003). Serum-resistant strains of *Borrelia burgdorferi* evade complement-mediated killing by expressing a CD59-like complement inhibitory molecule. J. Immunol. *170*, 3214-3222.

Perez-Caballero, D., Garcia-Laorden, I., Cortes, G., Wessels, M.R., de Cordoba, S.R., and Alberti, S. (2004). Interaction between complement regulators and *Streptococcus pyogenes*: binding of C4b-binding protein and factor H/factor H-like protein 1 to M18 strains involves two different cell surface molecules. J. Immunol. *173*, 6899-6904.

Price, J.D., Schaumburg, J., Sandin, C., Atkinson, J.P., Lindahl, G., and Kemper, C. (2005). Induction of a regulatory phenotype in human CD4+ T cells by streptococcal M protein. J. Immunol. *175*, 677-678.

Prodinger, W.M., Würzner, R., Stoiber, H., and Dierich, M.P. (2003). Complement. In Fundamental Immunology, W.E, Paul, ed. (Philadelphia: Lippincott-Raven), pp 1077-1103.

Rautemaa, R., Jarvis, G.A., Marnila, P., and Meri, S. (1998). Acquired resistance of Escherichia coli to complement lysis by binding of glycophosphoinositol-anchored protectin (CD59). Infect. Immun. *66*, 1928-1933.

Rautemaa, R., Rautelin, H., Puolakkainen, P., Kokkola, A., Karkkainen, P., and Meri, S. (2001). Survival of Helicobacter pylori from complement lysis by binding of GPI-anchored protectin (CD59). Gastroenterology *120*, 470-479.

Reed, S.L., Keene, W.E., McKerrow, J.H., and Gigli, I. (1989). Cleavage of C3 by a neutral cysteine proteinase of entamoeba histolytica. J. Immunol. *143*, 189-195.

Rooijakkers, S.H., van Wamel, W.J., van Steeg, T., Ruyken, M., Presanis, J.S., Heezius, H.C.J.M., Sim, R.B., van Kessel, K.P.M., and van Strijp, J.A.G. (2004). Lectin pathway inhibitor (LPI) of *Staphylococcuss aureus*: a unique molecule that blocks phagocytosis. Mol. Immunol. *41*, 301.

Rooijakkers, S.H., van Wamel, W.J., Ruyken, M., van Kessel, K.P., and van Strijp, J.A. (2005). Anti-opsonic properties of staphylokinase. Microbes Infect. *7*, 476-484.

Rother, K., Till, G.O., and Hänsch, G.M. (1998). The complement system (Berlin, Heidelberg, New York: Springer).

Rother, R.P., Rollins, S.A., Fodor, W.L., Albrecht, J.C., Setter, E., Fleckenstein, B., and Squinto, S.P. (1994). Inhibition of complement-mediated cytolysis by the terminal complement inhibitor of herpesvirus saimiri. J. Virol. *68*, 730-737.

Schwimmbeck, P.L., Yu, D.T., and Oldstone, M.B. (1987). Autoantibodies to HLA B27 in the sera of HLA B27 patients with ankylosing spondylitis and Reiter`s syndrome. Molecular mimicry with *Klebsiella pneumoniae* as potential mechanism of autoimmune disease. J. Exp. Med. *166*, 173-181.

Segerman, A., Atkinson, J.P., Marttila, M., Dennerquist, V., Wadell, G., and Arnberg, N. (2003). Adenovirus type 11 uses CD46 as a cellular receptor. J. Virol. *77*, 9183-9191.

Sirena, D., Lilienfeld, B., Eisenhut, M., Kalin, S., Boucke, K., Beerli, R.R., Vogt, L., Ruedl, C., Bachmann, M.F., Greber, U.F., and Hemmi, S. (2004). The human membrane cofactor CD46 is a receptor for species B adenovirus serotype 3. J. Virol. *78*, 4454-4462.

Spiller, O.B., Morgan, B.P., Tufaro, F., and Devine, D.V. (1996). Altered expression of host-encoded complement regulators on human cytomegalovirus-infected cells. Eur. J. Immunol. *26*, 1532-1538.

Spiller, O.B., Blackbourn, D.J., Mark, L., Proctor, D.G., and Blom, A.M. (2003). Functional activity of the complement regulator encoded by Kaposi's sarcoma-associated herpesvirus. J. Biol. Chem. *278*, 9283-9289.

Thompson, SA. (2002). Campylobacter surface-layers (S-layers) and immune evasion. Ann. Periodontol. 7, 43-53.

Van Den Berg, C.W., De Andrade, R.M., Magnoli, F.C., Marchbank, K.J., and Tambourgi, D.V. (2002). Loxosceles spider venom induces metalloproteinase mediated cleavage of MCP/CD46 and MHCI and induces protection against C-mediated lysis. Immunology *107*, 102-110.

Vogl, G., Jensen, D.B., Speth, C., Lass-Flörl, C., Eck, R., Zipfel, P.F., Dierich, M.P., and Würzner R. (2006). The mould Aspergillus binds complement regulator factor H. Mol. Immunol. *43*, 177.

Volanakis, J.E., and Frank, M.M. (1998). The human complement system in health and disease (New York: Marcel Dekker).

Walport, M. (2001). Complement. New Engl. J. Med. *344*, 1058-1066.

Wei, L., Pandiripally, V., Gregory, E., Clymer, M., and Cue, D. (2005). Impact of the SpeB protease on binding of the complement regulatory proteins factor H and factor H-like protein 1 by *Streptococcus pyogenes*. Infect. Immun. *73*, 2040-2050.

Würzner, R. (1999). Evasion of pathogens by avoiding recognition or eradication by complement, in part via molecular mimicry. Mol. Immunol. *36*, 249-260.

Würzner, R. (2000). Modulation of complement membrane attack by local C7 synthesis. Clin. Exp. Immunol. *121*, 8-10.

Würzner, R. (2003), Complement in infectious diseases. In Host Response Mechanisms in Infectious Diseases, H. Herwald, ed. (Basel: Karger), pp 1-17.

Würzner, R., and Zipfel, P.F. (2004). Microbial evasion mechanism against human complement. In Complement, J. Szebeni, ed. (Dordrecht: Kluwer), pp. 231-248.

Zaragoza, O., Taborda, C.P., and Casadevall, A. (2003). The efficacy of complement-mediated phagocytosis of Cryptococcus neoformans is dependent on the location of C3 in the polysaccharide capsule and involves both direct and indirect C3-mediated interactions. Eur. J. Immunol. *33*, 1957-1967.

Zhao, Z.S., Granucci, F., Yeh, L., Schaffer, P.A., and Cantor, H. (1998). Molecular mimicry by herpes simplex virus-type 1: autoimmune disease after viral infection. Science *279*, 1344-1347.

2

Microbial Subversion of NK Cell Function

Ulrich H. Koszinowski and Stipan Jonjic

Abstract

Natural killer (NK) cells play a crucial role in the resistance to several viruses and other intracellular pathogens. They are among the first cells to sense the release of proinflammatory cytokines, as well as the perturbations in the expression of MHC class I molecules and other surface molecules induced by microbial invasion of cells. One of the best illustrations of the importance of NK cells is represented by the various microbial functions evolved to counter NK cell response. Here we summarize the major mechanisms and receptors that NK cells can employ to positively recognize and kill virus-infected cells and the range of microbial strategies that reduce the efficacy of NK cell activation and their effector functions. The elements for viral NK cell- subversions are mainly based on the assumption deriving from numerous findings in readily controlled *in vitro* systems whereas the evidence for their relevance *in vivo* are still only fragmentary. The murine model of cytomegalovirus infection highlights the biological importance of these viral immunoevasins. The analysis of the viral proteins will continue to provide important details on the diversity of mechanisms by which NK cells can be activated and modulated.

1. Introduction

Innate immunity encompasses several immediate defense mechanisms devoted to sensing and controlling microbial replication during the time preceding the induction of the adaptive immune response. Unlike T and B lymphocytes, which undergo somatic recombination and clonal specification to generate antigen-specific receptors, the cells of the innate immune system, such as macrophages and dendritic cells (DC), use receptors that recognize structurally conserved molecular patterns on a broad spectrum of microorganisms and infected cells. Key molecules by which cells of the innate immune system sense the invasion of microorganisms are Toll-like receptors (TLRs) (Akira and Takeda, 2004). Natural killer (NK) cells

*For correspondence email koszinowski@mvp.uni-muenchen.de, jstipan@medri.hr

are a component of innate immunity but differ from other cells of the innate immune system by the expression of distinct receptors and effector functions, which make them an essential link between innate and adaptive immune responses. NK cells have a role as effector cells and act through cytolysis of target cells and cytokine release. In addition, NK cells play a role in the regulation of the immune response. The evolutionary struggles between microbes and NK cells are conserved in their genomes, which give us hints on how important NK cells are in the control of viral and other infections (Orange et al., 2002).

NK cells comprise up to 15% of peripheral blood lymphocytes, which are also found in peripheral tissues, including the spleen, liver, peritoneal cavity and placenta (Trinchieri, 1989). A hallmark of NK cells is their ability to monitor the surface expression of MHC class I molecules. Physiological MHC class I expression prevents NK cells to attack normal cells. NK cells do not only recognize infected or tumor cells that express low levels of self MHC class I molecules. They also recognize cells that express other NK cell activating ligands induced upon infection, transformation or cellular stress (Cerwenka and Lanier, 2001). Some NK cells may even express receptors which specifically recognize microbial proteins (Scalzo, 2002).

Human NK cells are defined as cells that express CD56 but not the T-cell antigen receptor complex. Based on the cell surface density of the CD56 and CD16 (FcγRIII) molecules, human NK cells are divided into two subsets. The major subset is characterized by a low level of surface CD56 ($CD56^{dim}$) and a high level of CD16 ($CD16^{bright}$). The minor subset expresses CD56 at a high density ($CD56^{bright}$) and a low density of CD16 ($CD16^{dim}$), or lacks CD16 altogether (Cooper et al., 2001). The relative biological role of these two subsets *in vivo* is not yet clear. Mouse NK cells apparently do not contain subsets analogous to those in humans and do not express CD56. Mouse NK cells express the NK1.1, DX5 and asialoganglio-N-tetraosylceramide (AGM1) markers. They are defined as $NK1.1^{+}CD3^{-}$ or $DX5^{+}CD3^{-}$ cells, although none of these markers is exclusively specific for NK cells, which are found on a small percentage of T cells as well (Raulet et al., 2001). AGM1 is expressed at high levels on both, human and mouse NK cells (Trinchieri, 1989). Like cytotoxic T cells, NK cells control infection by cytolysis and by secreting inflammatory cytokines. The release of cytotoxic granules containing perforin and granzymes is the dominant cytolytic effector function (Trapani and Smyth, 2002). In addition, they deploy other cytolytic functions through the Fas-Fas ligand pathway, the tumor necrosis factor (TNF)-related apoptosis-inducing ligand (TRAIL) pathway, the membrane TNF-α pathway, and the antibody-dependent cellular cytotoxicity (ADCC) through CD16 (Raulet, 2003a). Non-cytolytic effector functions of NK cells include the release of cytokines such as interferon-gamma (IFN-γ), TNF-α, the granulocyte/macrophage colony-stimulating factor (GM-CSF) and chemokines including MIP-1α/MIP-1β (macrophage inflammatory protein 1α and 1β), RANTES (factor regulated on activation-normal T cell expressed and secreted), and lymphotactin also known as ATAC (activation induced, T cell-derived and chemokine-related cytokine) (Biron et al., 1999; Robertson, 2002) (Figure 1). IFN-γ and TNF-α are the key effector

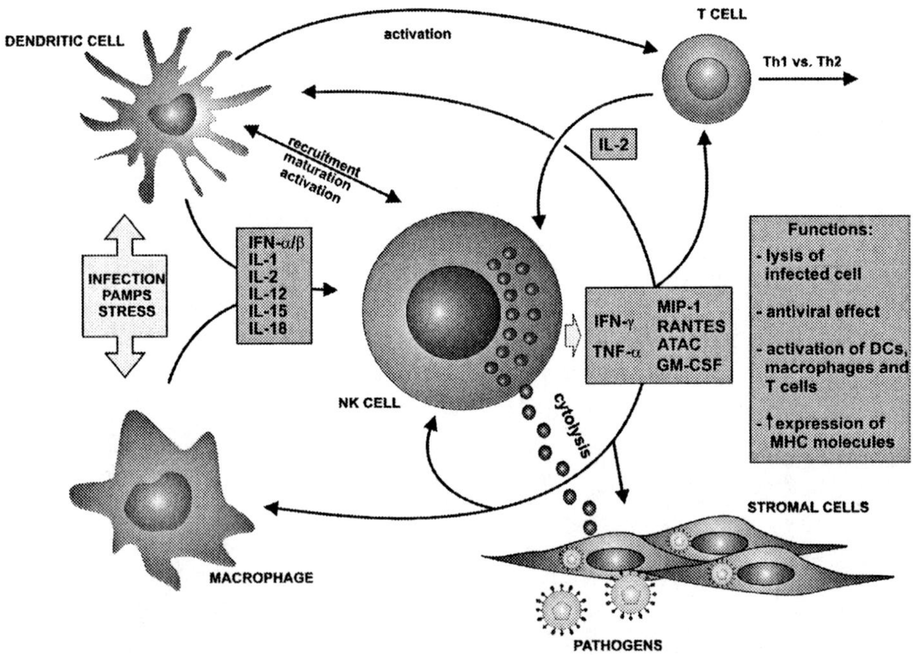

Figure 1. Functional interactions of NK cells. After dendritic cells and macrophages encounter microbial agents and other stimuli they secrete numerous cytokines and chemokines, which play a role in recruitment, maturation and activation of NK cells. In turn, in order to exert their non-cytolitic effector functions, NK cells secrete different cytokines and chemokines including those with antiviral properties, induce MHC expression by IFN-γ and activate dendritic cells, macrophages and T cells. On the other hand, activated NK cells can lyse target cells by secreting perforin and granzymes. While IFN-α/β is important for NK cell cytotoxicity, some other cytokines like IL-12, IL-15 and IL-18 are important for IFN-γ production. Interaction between dendritic and NK cells appears to be mutually important for their function and shaping of the adaptive immune response.

cytokines possessing direct antimicrobial activity and are also able to engage other effector functions including the activation of macrophages (Samuel, 2001).

The mutual interaction of NK and DC appears to be important to trigger their function and to shape the adaptive immune response (Cooper et al., 2004; Ferlazzo and Munz, 2004). DC promote migration of NK cells into the lymph node. NK cell recruitment and activation also contributes to efficient T cell priming through the interaction between the 2B4 and the CD48 molecules. Cell-to-cell contact is required for optimal interaction between DC and NK cells (Piccioli et al., 2002). DC promote NK cell proliferation and activation also through secretion of cytokines including IFN-α/β, IL-12, IL-15 and IL-18 (Andrews et al., 2003; Biron, 1999) (Figure 1). While IL-12, IL-15 and IL-18 are predominantly involved in induction of IFN-γ secretion, type I IFNs are required for NK cell cytotoxicity (Biron et al., 1999). On the other hand, NK cells control DC functions by destruction of immature DC (iDC) while mature DC (mDC) are spared, perhaps due to up-regulation of MHC class I molecules during DC maturation (Ferlazzo et al., 2001).

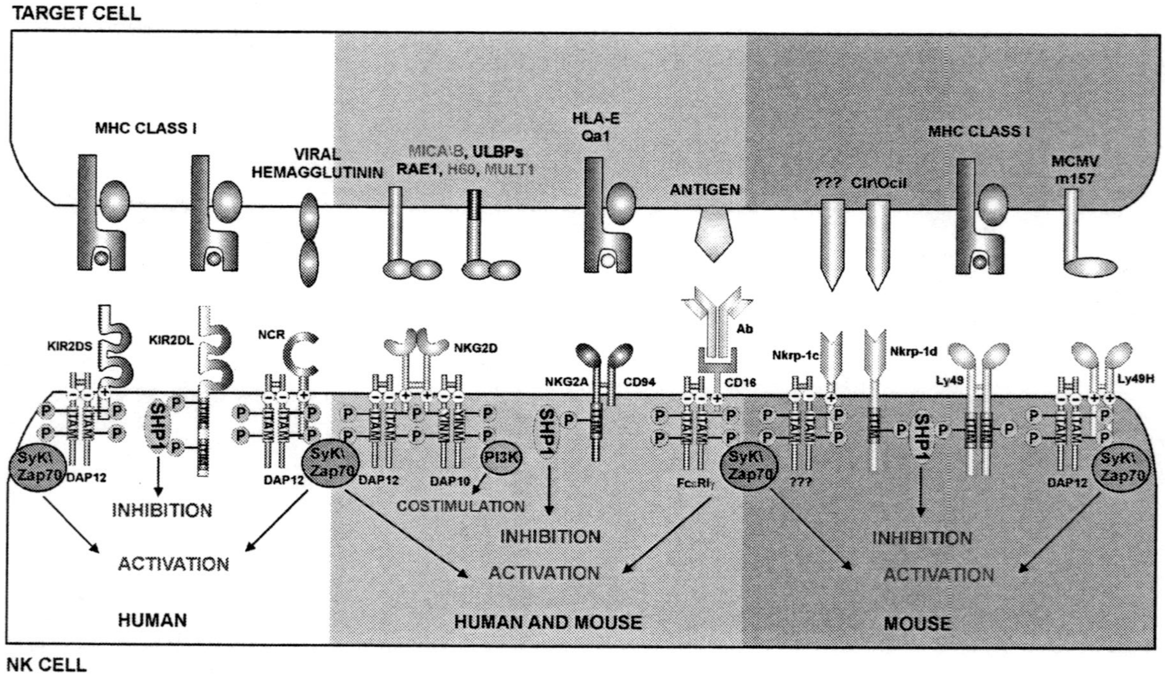

Figure 2. Major NK cell receptors and their ligands in man and mouse. NK cell receptors (NKRs) belong to two major families: (i) NKRs with the Ig-like domains which include KIRs, NCRs and LIRs (not shown) all expressed on human but not on mouse NK cells, (ii) NKRs with C-type lectine-like domains including Ly49 homodimers expressed only in mouse and CD94/NKG2 family, including NKG2D homodimer, expressed both in mouse and humans. Additional members of this family are Nkrp receptors, which are expressed in mice and until now only one Nkrp receptor is found in humans. CD16 is an activating NK cell receptor, which binds the Fc portion of IgG and mediates antibody-dependent cytotoxicity. In addition to the receptors depicted here, there are several other important NK cell receptors and co-receptors such as CD2, 2B4, CD11a-CD18, CD56 and CD69. MHC class I molecules serve as ligands for KIRs and Ly49 receptors, with the exception of Ly49H, which binds to MCMV m157 protein. MHC class I-related molecules (stress proteins) serve as a ligand for the NKG2D receptor. Cellular ligands for NCRs are not yet known although they may bind viral hemagglutinin. The ligands for Nkrp are also C-type lectin-related molecules encoded by natural killer complex. All inhibitory NK cell receptors possess immunoreceptor tyrosine-based inhibitory motif (ITIM) and associate with phosphatases such as SHP-1 resulting in the inhibition of NK cells. Activating receptors accomplish signaling by interaction with adaptor proteins that possess immunoreceptor tyrosine-based activation motif (ITAM) or YINM motif, which signal through Syk/ZAP70 and PI3 kinase, respectively.

1.1. NK cell receptors

NK cells do not use V(D)J recombination to generate antigen-specific receptors. Nevertheless, they express several inhibitory and activating receptors which tightly regulate their effector functions (Figure 2). NK cell receptors (NKR) belong to two major families: (i) receptors containing immunoglobulin (Ig)-like ectodomains and (ii) receptors containing C-type lectin-like domains. NKR possessing the Ig-like domains are type I transmembrane proteins which include killer cell Ig-like receptors (KIR), leukocyte Ig-like receptors (LIR) and natural cytotoxicity receptors (NCR) (Raulet, 2003a; Raulet et al., 2001). These receptors are found on human but not on rodent NK cells with the exception of the NCR member NKp46, which possess homologue in rodents (Biassoni et al., 1999). The second NKR family are type II transmembrane proteins with C-type lectin-like domains that include Ly49 homodimers (mice) and as heterodimers of CD94 covalently associated with an NKG2 isoform (mice and humans) (Raulet, 2003a; Raulet et al., 2001). Each of these families comprises both inhibitory and activating receptors.

Inhibitory receptors contain in their cytoplasmic domains an immunoreceptor tyrosine-based inhibitory motif (ITIM), with the consensus sequence V/IxYxxL/v (Long et al., 1997; Vivier and Daeron, 1997). Binding of these receptors to their ligands results in the phosphorylation of tyrosine residues in ITIM, leading to the recruitment and activation of protein tyrosine phosphatase SHP-1 that inhibits NK cell activation. Activating NK receptors lack an ITIM motif in their cytoplasmic domain but contain a charged amino acid residue within their transmembrane domain required for association with bearing signaling adaptor molecules (Moretta et al., 2001). The majority of NK cell receptors associate with the signaling adaptor molecule DAP12, and some interact with FcεRIγ, CD3ζ or DAP10 (Moretta et al., 2001). The cytoplasmic tail of these transmembrane proteins contains an immunoreceptor tyrosine-based activation motifs (ITAMs) or an YxxM motif. Binding of the activating receptors to their cognitive ligands leads to tyrosine phosphorylation of coupled adaptor proteins and to recruitment and activation of protein tyrosine kinases such as Syk/ZAP70 or PI3 (Lanier, 2003).

KIR are encoded by a family of genes that reside in the leukocyte receptor cluster (LRC) located on human chromosome 19. The receptor has two or three immunoglobulin domains (2D for 2 and 3D for 3 domains, respectively) and a long (L) or short (S) cytoplasmic tail (Vilches and Parham, 2002). KIR with long cytoplasmic tails containing pairs of ITIMs have inhibitory properties, whereas some KIR that serve as activating receptors have short cytoplasmic tails which associate with the signaling adaptor molecule DAP12 (Trowsdale et al., 2001; Vilches and Parham, 2002). Mouse NK cells do not possess KIR. They express a family of Ly49 receptors, which are members of the C-type lectin-like superfamily. They are functionally similar to KIR since both use MHC class I molecules as their ligands (Raulet, 2003a). The Ly49 family including a variety of receptors designated A through W maps within the natural killer complex (NKC) on mouse chromosome 6 (Karlhofer et al., 1992; Yokoyama et al., 1991). The majority of Ly49 receptors have inhibitory properties and few, including Ly49D and Ly49H are stimulatory.

The CD94/NKG2 heterodimer family of NK receptors expressed by human and mouse NK cells recognize MHC class Ib molecules (Houchins et al., 1991; Vance et al., 1999; Vance et al., 1998). The NKG2 chain endows these receptors with either inhibitory (CD94/NKG2A) or activating function (CD94/NKG2C and CD94/NKG2E). The activating NK cell receptor NKG2D is classified as a member of the NKG2 family although it is only distantly related and does not pair with CD94 (Houchins et al., 1991; Raulet, 2003b). Mouse NKG2D exists in two alternatively spliced isoforms, which differ from each other by 13 amino acids extending the cytoplasmic domain (Diefenbach et al., 2002; Gilfillan et al., 2002). The shorter isotype, NKG2D-S couples not only to DAP10, but also to DAP12, whereas NKG2D-L binds exclusively to DAP10. Mouse NKG2D-S is expressed on activated NK cells and activated CD8$^+$ T cells. Low amounts of NKG2D-S are found in resting NK cells and naive CD8$^+$ T cells. The NKG2D-L isoform is present in resting NK cells. CD8$^+$ T cells do not express DAP12; therefore, in these cells NKG2D pairs only with DAP10. In contrast to NK cells, NKG2D functions in T lymphocytes only as a costimulatory receptor. The relative contribution of DAP12 *versus* DAP10 to NKG2D signaling on the function of NKG2D *in vivo* has yet to be determined.

The MHC class I specific receptors listed above do not account for all NK cell specificities. The NKC encodes a number of lectin-like receptors described as the natural killer receptor protein-1 (Nkrp-1) family (Giorda et al., 1990; Lanier et al., 1994; Plougastel et al., 2001), which are conserved between rodents and humans and include mouse Nkrp-1c (NK1.1) (Karlhofer and Yokoyama, 1991), an activating receptor also known as a marker for NK cells in some mouse strains including C57BL/6 strain (Figure 2). Nkrp-1d is an inhibitory receptor, which contains ITIM, whereas Nkrp-1f is a putative activation receptor and lacks ITIM, but contains charged transmembrane residues. The inhibitory Nkrp-1d receptor mediates inhibition *via* recognition of specific C-type lectin-related (Clr) molecules also known as osteoclast inhibitory lectin (Ocil) molecules on dendritic cells and macrophages. Interestingly, *Nkrp1* and *Clr/Ocil* genes are collocated in the same region of the NKC (Iizuka et al., 2003). It has been speculated that the Nkrp1/Clr receptor-ligand recognition system may represent an alternative and evolutionary conserved mechanism for the discrimination between normal and damaged cells similar to the monitoring of MHC class I by KIR/Ly49 receptors.

Non-MHC-dependent NK cell cytotoxicity is also mediated by NCRs (Moretta et al., 2001). NCR are NKp46, NKp44 and NKp30 all of which are members of the immunoglobulin superfamily. Although their cellular ligands are still unknown, they may function by binding of viral hemagglutinin (Mandelboim et al., 2001). NKp30 is the activation receptor involved in the interaction between DCs and NK cells (Ferlazzo et al., 2002).

An important activating NK receptor, which connects acquired and innate immunity, is the receptor for the Fc portion of immunoglobulins (Merrill et al., 1981). Recognition of microbial antigens on the surface of infected cells by specific antibodies triggers killing of these cells through the engagement of the Fc receptor of the type III (FcγRIII) in a process named antibody-dependent cellular cytotoxicity (ADCC). FcγRIII or CD16, the first activating NK cell receptor identified, contains two extracellular Ig domains and is expressed either as a transmembrane or a GPI-linked protein (Gessner et al., 1998). It lacks intrinsic signaling properties and it associates with the adaptor protein FcεRIγ or CD3ζ (Budt et al., 2004). These adaptors contain an ITAM motif to induce the activating signaling cascade (Ravetch and Bolland, 2001). The binding of the Fc fragment of immunoglobulins on opsonized cells to Fc receptor on NK cells leads to the release of perforin and granzymes from cytolytic granula.

In addition to the above mentioned NK cell receptors other membrane proteins such as CD2, 2B4, CD11a-CD18, CD56 and CD69 can induce or modulate NK cell cytotoxicity and cytokine production (Colucci et al., 2002). For instance, signals induced upon binding of adhesion molecule LFA1 to ICAM1 on target cells can lead to cytoskeleton rearrangements and amplification of signals from other activating receptors. Thus, LFA1 can also function as an activation receptor on NK cells (Riteau et al., 2003). Although the engagement of inhibitory NK cell receptors seems to dominate over activation, the inhibition can be overridden *via* signals derived from some activating receptors. Thus, activation of NK cells can result from up-regulation of ligands for activating NK cell receptors and/or down-regulation of ligands for inhibitory NK cell receptors.

1.2. NK cell receptors ligands

MHC class I proteins are ligands for three major types of inhibitory NKRs: KIRs, Ly49 and CD94/NKG2 receptors (Figure 2). NK cells surveillance of classical MHC class Ia expression can be either direct or indirect. KIR and Ly49 receptors directly recognize MHC class Ia molecules on the cell surface, whereas CD94/NKG2 receptors indirectly monitor the expression of these molecules. The indirect CD94/NKG2A receptor control occurs *via* the conserved nine-amino acid peptide sequence contained in the signal peptide domain of MHC class Ia molecules which is presented by non-classical MHC class Ib molecules (HLA-E in humans and Qa-1 in mice).

NK cell activation is mediated *via* number of distinct activating NK cell receptors specific for constitutively expressed or inducible cellular ligands. In addition to MHC class I molecules, activating NK receptors recognize MHC class I-related molecules induced in transformed, stressed or infected cells (Figure 2). The existence of activating NK receptors specific for conventional and constitutively expressed MHC class I molecules is not well understood. They may either screen the down-modulation of MHC class I molecules or they may bind to MHC class I molecules that express microbial peptides. Major ligands for NKG2D are MHC class I-related molecules. In humans, NKG2D recognizes two polymorphic families of MHC class I-related molecules (Cerwenka and Lanier, 2003; Raulet, 2003b; Watzl, 2003), the MHC class I-related chain (MIC) family including MICA, MICB, and the UL-16 binding protein (ULBP1-4) family. The MICA and MICB proteins consist of the α1, α2 and α3 domains with MHC class I folds, and which, unlike classical MHC molecules, do not bind β2-microglobulin and peptide (Cerwenka and Lanier, 2003). The ULBPs are GPI-anchored proteins and possess only α1 and α2 domains (Cosman et al., 2001). Mouse NKG2D ligands are encoded by retinoic acid inducible genes 1 (RAE-1α, β, γ, δ, ε), the minor histocompatibility antigen

H60 gene and the recently identified murine UL-16 binding protein like transcript 1 (MULT-1) (Cerwenka and Lanier, 2003; Raulet, 2003b; Watzl, 2003). H60 was originally described as a minor histocompatibility antigen recognized by T cells from C57BL/6 mice in reaction to BALB.B splenocytes (Malarkannan et al., 1998) and RAE-1 has a role during embryonic development (Zou et al., 1996).

An open question is why multiple activating ligands interact with a single activating NK cell receptor. Perhaps the ligands for NKG2D are not redundant, but very specific based on expression patterns, inducibility and binding affinities. One would expect a tightly regulated expression of NKG2D ligands to prevent an autoimmune response caused by NK and $CD8^+$ T cell activation. The expression of MICA is under control of a heat shock promoter, which is induced by cellular stress, by bacterial and viral infection, and by transformation (Groh et al., 1996). Furthermore, ligands, which are differentially expressed during the development of cells and tissues may have different biological functions. For instance, the mouse RAE-1 proteins are expressed during embryonic development and are down-modulated in normal cells unless they are infected or transformed (Cerwenka et al., 2000; Lodoen et al., 2003). Another possibility is that these ligands are selectively regulated in different cells. Signaling through Toll-like receptors (TLRs) in macrophages induces the transcription of RAE-1 but not of H60 or MULT-1 genes (Hamerman et al., 2004). Given the importance of NKG2D-ligands for NK- and T cell function NKG2D ligands may represent a major target for the microbial escape from NK cell mediated control. NKG2D-ligand specific evasion functions would promote pathogen persistence and represent a virulence factor, particularly in tissues with limited capacity of self-renewal.

1.3. NK cells activation after microbial infections

NK cells are activated through soluble mediators, such as interferons and cytokines or by direct cell–to-cell contact (Figure 1). NK cells sense the loss of MHC class I expression from the surface of infected cells and can also sense cellular stress (Cerwenka and Lanier, 2001). NK cells can also be activated *via* TLRs which recognize specific molecular patterns expressed on pathogens (Becker et al., 2003; Tabeta et al., 2004). The expression of ligands for the activating NK cell receptors NKG2D and Ly49H overrides the inhibition mediated by MHC class I–specific receptors (Arase and Lanier, 2002; Carayannopoulos and Yokoyama, 2004). During cellular stress the peptide loaded on MHC class Ib peptide is replaced by a peptide derived from a heat shock protein, which represents a mechanism for recognition of stressed cells by NK cells (Michaelsson et al., 2002).

NK cells control has been reported for several viral infections, including herpesviruses, orthomyxoviruses, poxviruses, coronaviruses and picornaviruses (Biron et al., 1999). NK cells also play an important role in control of Mycobacteria, Listeria monocytogenes, Leishmania major and Toxoplasma (Hornef et al., 2002). Detailed evidence for the importance of NK cells in antiviral immunity has been provided for herpesvirus infections. NK cell-deficient humans suffer from recurrent infections with cytomegalovirus (CMV), varicella and herpes simplex virus (HSV) (Biron et al., 1989). In mouse strains resistant to murine CMV (MCMV), the depletion of NK cells abolishes the resistance (Bukowski et al., 1984; Welsh et al., 1991). Additionally, newborn mice are very sensitive to MCMV due to the immaturity of their NK cells. Acquisition of resistance at about third week of age correlates with the expression of Ly49 molecules on NK cells (Booss and Wheelock, 1975; Williams et al., 2000). Genetic resistance of mice to MCMV maps to the *Cmv1^r* locus on chromosome 6 that encodes the activating receptor Ly49H (Scalzo, 2002). Ly49H$^+$ resistant mice treated with anti-Ly49H antibodies become susceptible to MCMV infection (Daniels et al., 2001). The MHC-class I-like protein encoded by the MCMV *m157* gene is a ligand for Ly49H (Arase et al., 2002; Smith et al., 2002b). However, in certain MCMV-susceptible inbred mouse strains, m157 binds the Ly49I inhibitory receptor (Arase et al., 2002) suggesting that this protein has originally evolved to serve as the viral NK decoy in order to promote viral propagation and persistence. Resistance to ectromelia and HSV has also been linked to the NKC (Scalzo, 2002), but it remains to be determined whether Ly49H contributes to immunity to these viruses as well.

Some activating NK receptors also interact with viral proteins expressed on virions and virus-infected cells (Arnon et al., 2001; Mandelboim et al., 2001). NKp46 and NKp44 interact with sialic acid-binding influenza, parainfluenza and Sendai viral receptors, respectively. Multimerization of NKp46 by these viral proteins expressed on the surface of infected cells may activate and target NK cell effector functions. However, sialic acid on the viral lipid envelope of virions may cross-link NKp46 and thus trigger a futile activation of NK cells. Resistance to Epstein–Barr virus (EBV) infection and perhaps to EBV-induced oncogenesis may be related with NK cell activation through 2B4 (CD244), a cell surface receptor of the CD2 family (Parolini et al., 2000). CD48, the natural ligand for 2B4, is a broadly expressed GPI-linked protein, which is upregulated in EBV-transformed B cells. 2B4 provides a co-receptor activity to NK cells after binding to other activating NKR (Sivori et al., 2000). However, recent data indicate that 2B4 can also serve as an inhibitory NK cell receptor for non-MHC ligands (Lee et al., 2004).

2. Microbial subversion of NK cells by interfering with MHC class I expression

CD8$^+$ T cells recognize antigenic peptides in the context of MHC class I molecules. Several viruses, particularly herpesviruses, compromise the MHC class I presentation pathway (Alcami and Koszinowski, 2000). Different virus proteins affect different steps of the antigen presentation pathway by blocking MHC class I gene transcription, by degradation of newly formed class I molecules, by the interference with Tap-mediated peptide transport, by the inhibition of MHC transport to the cell surface, by the distortion of the MHC molecule shape, and by rapid down-regulation of these molecules from the cell surface (Furman and Ploegh, 2002; Hengel and Koszinowski, 1997; Petersen et al., 2003). Although the down-modulation of MHC class I molecules might protect against CD8$^+$ T cell surveillance, the same effect would activate NK cells (Ljunggren and Karre, 1990). Most cells express high levels of MHC class I molecules, which provide a strong inhibitory signal to NK cells through inhibitory NK receptors. Namely, three major families of inhibitory NK receptors, KIR, Ly49 and CD94/NKG2 use MHC class I molecules as their inhibitory ligands (Lanier, 2003). Viruses are under pressure to

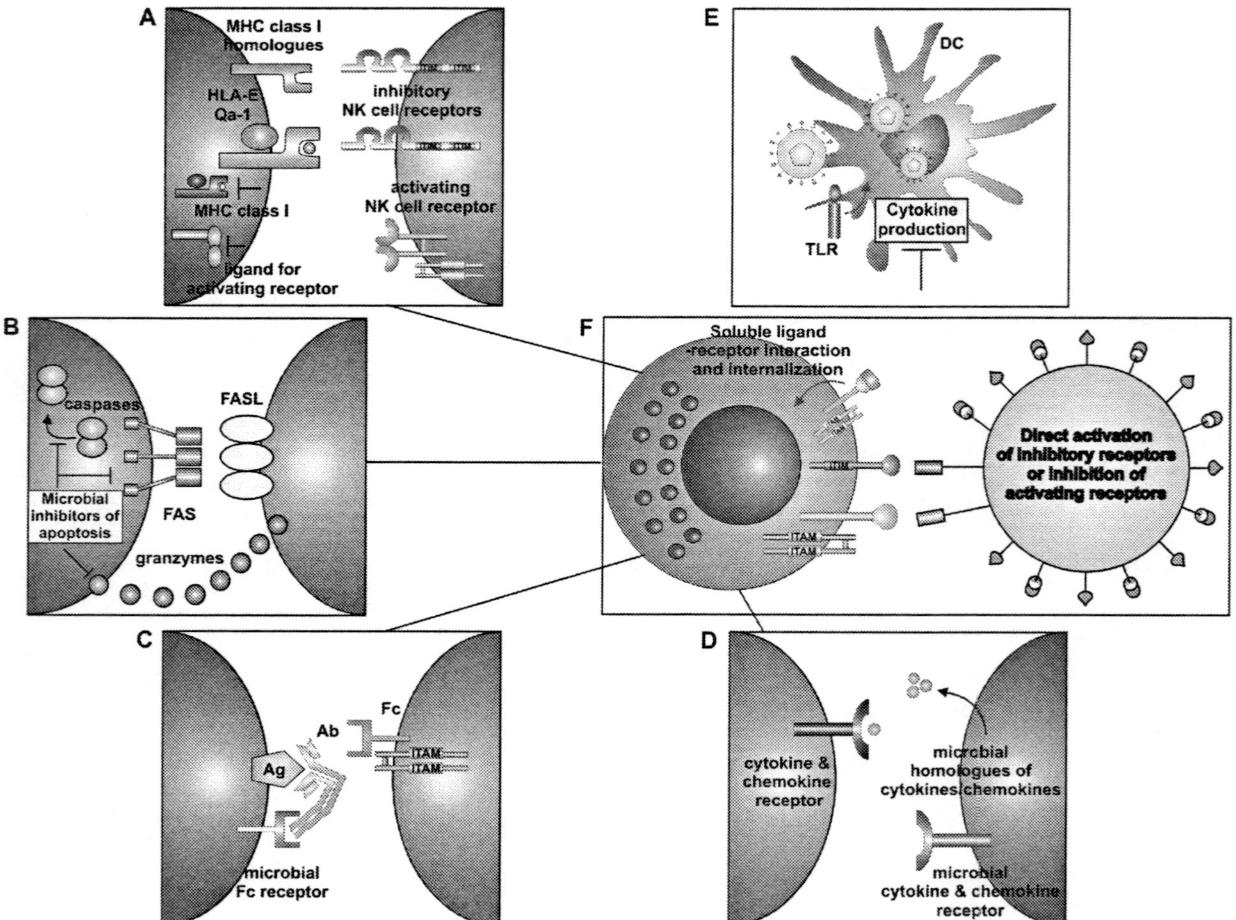
A
MHC class I homologues
HLA-E Qa-1
MHC class I
ligand for activating receptor
inhibitory NK cell receptors
activating NK cell receptor
B
caspases
Microbial inhibitors of apoptosis
FAS
FASL
granzymes
C
Ag
Ab
Fc
ITAM
ITAM
microbial Fc receptor
D
cytokine & chemokine receptor
microbial homologues of cytokines/chemokines
microbial cytokine & chemokine receptor
E
DC
TLR
Cytokine production
F
Soluble ligand -receptor interaction and internalization
ITAM
ITAM
Direct activation of inhibitory receptors or inhibition of activating receptors

Figure 3. Diversity of microbial mechanisms to subvert NK cells. Most of the microbial mechanisms of NK evasion act on the level of target cells and only few have a direct effect on NK cells. **A)** Subversion of NK cells by interfering with ligands for inhibitory and activating NK receptors. Viruses can encode MHC class I homologues, which serve as a ligand for inhibitory NK receptor, or they can differentially modulate MHC class I molecules by down-modulating those that present viral peptides to $CD8^+$ T lymphocytes, while preserving MHC class Ib molecules that bind to inhibitory CD94/NKG2 receptors. The peptide sequence of some viral proteins correspond to the sequence of lider peptide of MHC class I molecules, which normally associate with HLA-E or Qa-1. By doing so viruses can rescue these ligands for inhibitory NK receptors even if MHC class Ia are missing. Several cytomegalovirus genes encode proteins that are responsible for down-modulation of cellular ligands for activating NK cell receptors (see also figure 4 and 5). **B)** Viruses can escape NK cell-mediated apoptosis by encoding inhibitors that interfere with components of the apoptotic pathway and encode inhibitors of intracellular serin protease. **C)** To avoid antibody-dependent cellular cytotoxicity microbial Fc receptors can bind Fc portion of immunoglobulins specifically bound to microbial antigen on the cell surface and therefore prevent activation of NK cells *via* activating receptor CD16. **D)** Several pathogens explore the cytokines, chemokines and their receptor to modulate NK cell response. This includes microbial homologues of cytokines and chemokines and their receptors, and other microbial proteins, which affect cytokine/chemokine function either on the level of target cells or NK cells. **E)** Interaction with dendritic cells is crucial for NK cell effector and regulatory functions. By targeting dendritic cells for infection, microbes can affect their function in immune response and by modulating their receptors, like Toll like receptors, microbes could affect NK cells migration and activation. **F)** Some viruses can directly inhibit NK cells by ligation of NK cell inhibitory receptors or impair activating NK cell receptors. In addition, some viruses can infect NK cells and therefore compromise their functions.

adopt mechanisms, which enable infected cells to escape NK cells control despite of down-modulation of MHC class I molecules. To achieve this goal, viruses use at least two different mechanisms: (i) viral homologues for MHC class I molecules and (ii) modulation of MHC class I molecules (Figure 3).

2.1. Viral MHC class I homologues as ligands for inhibitory NK cell receptors

Viruses encode MHC class I homologues that are expressed on the infected cells and serve as surrogate ligands for NK cell inhibitory receptors (Figure 3A). Human CMV (HCMV) encodes an MHC class I homologue, UL18, that was originally described as a decoy ligand for inhibitory NKR (Reyburn et al., 1997). Namely, it has been reported that HCMV *UL18* expressed in the HLA-A,B and C deficient human cells can reduce their susceptibility to the NK cell lysis (Reyburn et al., 1997). This finding become controversial since the cellular receptor for UL18, LIR-1, is only marginally expressed on NK cells (Cosman et al., 1997) and since the UL18 protein does not contain a peptide sequence that would form a complex with HLA-E, the cellular ligand for inhibitory NK cell receptor, CD94/NKG2A (Tomasec et al., 2000). Also MCMV encodes an MHC class I homologue, m144, which can inhibit NK cell activation. A recombinant virus lacking the *m144* gene shows reduced pathogenesis that can be reversed by NK cell depletion (Farrell et al., 1997). In addition, the expression of *m144* on tumor cells has an inhibitory effect on NK cells (Cretney et al., 1999) and the transfection of *m144* into the cell line inhibits ADCC (Kubota et al., 1999). Yet, the mechanism of m144-mediated NK cell evasion is not resolved since its NK receptor remains elusive. The homologue of *m144* in the rat CMV has also been identified and the deletion mutant virus was found to be attenuated *in vivo* (Kloover et al., 2002). MCMV encodes the m157 protein that belongs to m145 protein family, several of which are involved in regulation of NK cells and show some structural homology to MHC class I molecules (see also *Specific Lessons from MCMV*). m157 is a ligand for Ly49H activating and Ly49I inhibitory NK receptors (Arase et al., 2002; Smith et al., 2002a). At least three additional members of the same gene family are involved in the down-regulation of cellular ligands for the activating NK cell receptor, NKG2D (see also *Specific Lessons from MCMV*). Molluscum contagiosum virus (MCV) encodes endoplasmic reticulum (ER) and Golgi-resident protein, MC080R, which is a MHC class I homologue with so far unknown effect on NK cell activity (Senkevich and Moss, 1998). In summary, although the expression of viral MHC class I homologues were initially proposed as an attractive immunoevasion strategy the evidence that these viral proteins directly inhibit NK cells by interacting with major inhibitory receptors is still lacking.

2.2. Differential modulation of MHC class I expression

To evade CD8 and NK cell control the regulation of MHC class I molecules might suffice: the down-modulation of MHC class Ia, being important for CD8 recognition while preserving the expression of MHC class Ib molecules (HLA-E in humans and Qa-1 in mice), the ligands for the inhibitory NK cell receptor CD94/NKG2A (Figure 3A). The latter MHC molecules are loaded with a peptide derived from a conserved signal sequence of MHC class Ia molecules (Kraft et al., 2000; Llano et al., 1998). Only cells that synthesize MHC class Ia molecules will generate a functional ligand to engage the CD94/NKG2 receptor. Since the presentation of MHC class Ia molecules is TAP-dependent (Hengel and Koszinowski, 1997) cells that either lack TAP or express viral function that compromise TAP functions should fail to generate the cellular ligand for CD94/NKG2. Therefore, the HLA-E-CD94/NKG2 ligand-receptor interaction is an indirect monitoring system for MHC class Ia expression in contrast to the direct recognition by KIR and Ly49 NK receptors. Thus, viral functions, which save the expression of HLA-E (Qa-1) molecules to serve as a ligand for inhibitory NK receptor despite of down-modulation of MHC class Ia should be an optimal concept to evade lysis by both NK cells and CD8$^+$ T cells. In principle, this concept is realized by the HCMV genome. A peptide derived from the signal sequence of the HCMV glycoprotein UL40 binds to HLA-E and upregulates its expression (Tomasec et al., 2000). The *UL40*-derived peptide is loaded onto HLA-E independently of the peptide transporter TAP and is recognized by CD94/NKG2A, which may prevent NK cell activation even in cells with a nonfunctional TAP (Wang et al., 2002). However, the concept has its problems. A peptide derived from the signal sequence of the stress induced heat shock protein 60 (hsp60) competes efficiently with an MHC class I–derived peptide and thus prevents binding of HLA-E by CD94/NKG2A (Michaelsson et al., 2002). Therefore, the stress induced hsp60-derived peptides can replace MHC class Ia-derived or *UL40*-derived peptide in HLA-E and prevent the inhibition of NK cells *via* binding of HLA-E and CD94/NKG2A. In addition, peptide replacement in Qa-1 may provide a mechanism for the activation of mouse NK cells (Kraft et al., 2000) to oppose viral evasion. Filling of HLA-E by the stress peptide would not prevent the inhibition of NK cells through KIR but should result in the activation of CD94/NKG2A-expressing NK cells due to the lack of inhibition. Although the concept of NK cell evasion by the UL40-derived peptide is attractive, there are data indicating that the enhancement of HLA-E expression by gpUL40 leader sequence peptides is insufficient to protect target cell to NK cells (Falk et al., 2002).

Viruses modulate MHC class I molecules to eliminate ligands for TCR on CD8$^+$ T cells while retaining ligands for NK inhibitory receptors. For example, the HCMV US2 and US11 proteins target certain MHC class I alleles for degradation, but spare the inhibitory receptor ligand, HLA-C and HLA-E (Llano et al., 2003; Schust et al., 1998). However, it is questionable whether this discrimination in down-regulating HLA alleles is beneficial for the virus. The selective loss of MHC class Ia alleles in HCMV-infected cells may be undetectable for CD94/NKG2A NK cell receptor, whereas NK cells regulated by a KIR with specificity for the down-regulated allele may become activated. Other viral proteins of HCMV, US3 and US6, are less selective and inhibit the cell surface expression of HLA-C and HLA-G (Ahn et al., 1997; Jun et al., 2000). The down-modulating function of US3 and US6 could make infected cells generally susceptible to NK cell lysis.

By selective down-regulation of HLA-A and HLA-B while preserving expression of HLA-C HIV Nef is another example of viral immunoevasion protein that affects both CD8$^+$ CTL and NK cells. The Nef protein is expressed early after infection and causes rapid internalization and lysosomal degradation of HLA-A and HLA-B, but has little or no effect on HLA-C or HLA-E (Cohen et al., 1999; Le Gall et al., 1998). Through the effect upon HLA-A and HLA-B the virus may avoid recognition by CD8⁺ T cells, while at the same time the normal expression of HLA-C and HL-A E would prevent NK cell activation. The resistance of HLA-C to Nef-induced down-regulation was explained by specific tyrosine-to-cysteine and aspartic-acid-to-asparagine substitutions in the cytoplasmic tail (Le Gall et al., 2000). Kaposi's sarcoma herpesvirus (KSHV)-encoded K3 and K5 proteins are E2 ubiquitin ligase responsible for rapid endocytosis of surface class I proteins and subsequent endosomal degradation (Coscoy et al., 2001; Ishido et al., 2000b). However, while K3 down-modulates all HLA class I molecules, K5 is selective and down-modulates only HLA-A and HLA-B (Ishido et al., 2000a). Three MCMV genes *m04*, *m06*, *m152* also selectively modulate class I expression (Wagner et al., 2002). The product of *m04*, gp34, might interfere also with NK cell activation. The gp34 forms complexes with MHC class I proteins which are expressed on the cell surface in a form unrecognizable by CTL (Kavanagh et al., 2001) while they still might inhibit NK cells.

3. Modulation of ligands for activating NK cell receptors

There is evidence that sensitivity to NK cells is not only, and perhaps not even primarily, determined by the level of MHC class I molecule expression. A prominent example is the activating NK receptor NKG2D, which upon binding to its ligand on target cells triggers NK cell activation, even if cells express normal levels of MHC class I (Cerwenka and Lanier, 2003; Raulet, 2003b; Watzl, 2003). NKG2D ligands are usually not constitutively expressed on normal cells but are induced by stress, infection, and transformation (Groh et al., 2001; Tieng et al., 2002).

NKG2D is expressed on all NK cells and the inducible expression of its ligands provides efficient activation of NK cells (Cerwenka and Lanier, 2003; Raulet, 2003b; Watzl, 2003). Activation of NK cells *via* NKG2D should represent a dominant mechanism to cope with viruses and other intracellular pathogens. In analogy to MHC class I modulation to compromise CD8⁺ T cell function, also the cellular ligands for the NKG2D receptor should be targets for the viral escape from NK cells control. Indeed, NKG2D ligands are targets for regulation by HCMV and MCMV immunoevasins (Cosman et al., 2001; Krmpotic et al., 2002). The ULBP, one of the member of NKG2D-ligand family, was originally identified by its ability to bind HCMV protein UL16 (Cosman et al., 2001). ULBP-1 and -2, as well as MICB were shown to bind to UL16. Furthermore, soluble UL16 is capable of binding ULBPs, and blocks the interaction between the NKG2D and its cognate activating ligand. However, the soluble protein failed to inhibit target killing *via* NKG2D-ligand interaction (Cosman et al., 2001; Kubin et al., 2001). More recent results indicate that the binding of UL16 to ULBP-1 takes place already inside the cells in the ER and cis-Golgi, which results in retention and sequestration of ULBP-1 (Dunn et al., 2003; Wu et al., 2003). HCMV also down-modulates LFA-3, a ligand for the activating NK cell receptor CD2, which increases the resistance of infected cells to NK cell-mediated control (Fletcher et al., 1998). The K5 protein of KSHV down-regulates ICAM-1, which serves as a ligand for the activating

NK cell receptors LFA-1, (Coscoy and Ganem, 2001; Ishido et al., 2000a). As already mentioned K5 of KSHV is an E2 ubiquitin ligase and probably directs the ubiquitination and degradation of ligands for LFA-1 (Coscoy et al., 2001).

3.1. Specific lessons from MCMV – down-regulation of NKG2D ligands and selection of NK cell escape variants

The MCMV infection shares many features with HCMV and serves as a prototypic model to study acute, latent, or recurrent virus infections in the mouse. With regard to susceptibility to MCMV, laboratory mouse strains fall into two categories: strains resistant to the virus, such as C57BL/6 strain, defined by the *Cmv1*^{r} gene locus associated with MCMV resistance and susceptible strains like BALB/c mice defined by the *Cmv1*^{s} gene locus, which is associated with MCMV susceptibility (Scalzo, 2002). Scalzo and colleagues linked resistance/susceptibility to MCMV infection with the NK cell response (Scalzo et al., 1990; 1992). MCMV-sensitive strains do not generate an efficient NK cell response to MCMV, but are capable of generating an NK cell response to other pathogens (Peng et al., 2001). MCMV encodes three glycoproteins that down-modulate MHC class I molecules (Wagner et al., 2002), and the loss of MHC class I expression should increase the susceptibility to NK cells as suggested by "missing self" hypothesis (Ljunggren and Karre, 1990). The results, however, contradicted this prediction. Namely, we and others have shown that *Cmv1*^{s} mice are mounting a very weak NK cell response against MCMV (Daniels et al., 2001; Krmpotic et al., 2002; Scalzo, 2002).

We presumed that the lack of NK cell response is due to viral evasion mechanisms resulting in the prevention of NK cell activation. To test this hypothesis we used genetic and immunological approaches. MCMV deletion mutants were screened for their ability to induce NK cell response in *Cmv1*^{s} mice. The deletion mutant which lacked the *m152* gene, whose product gp40 mediates the retention of MHC I molecules in the ERCIG/cis-Golgi compartment (Ziegler et al., 1997), also lost its resistance to NK cells and became NK sensitive (Figure 4). Subsequently, it was found that gp40 regulates the expression of RAE-1 (Lodoen et al., 2003). Therefore, gp40 has a dual function – by arresting MHC class I it prevents the activation of CD8$^+$ T cells and by down-regulating expression of RAE-1 it blocks the NK cells activation *via* NKG2D (Krmpotic et al., 2002; Krmpotic et al., 1999; Lodoen et al., 2003). Mice have at least two other ligands for the NKG2D receptor, H60 and MULT-1. Subsequent studies showed that there are at least two other viral genes involved in the down-regulation of NKG2D ligands, which are, similar to *m152,* members of the *m145* gene family (Hasan et al., 2005). The MCMV *m155* gene encodes the protein which down-regulates the expression of the H60 while the *m145* gene product regulates the expression of the MULT-1 protein (Krmpotic et al., 2005). The deletion of each individual gene, *m152*, *m155* and *m145* has an effect upon virus susceptibility to NK cells *in vivo*. The fact that the virus engages a specific gene for the regulation of each NKG2D ligand points to the importance of these mechanisms in the immunobiology of the virus. Notably, the NKG2D receptor is expressed as a co-receptor on T lymphocytes as well. Therefore, the biological importance of MCMV genes involved in the down-modulation of NKG2D ligands is likely to affect also other immune functions.

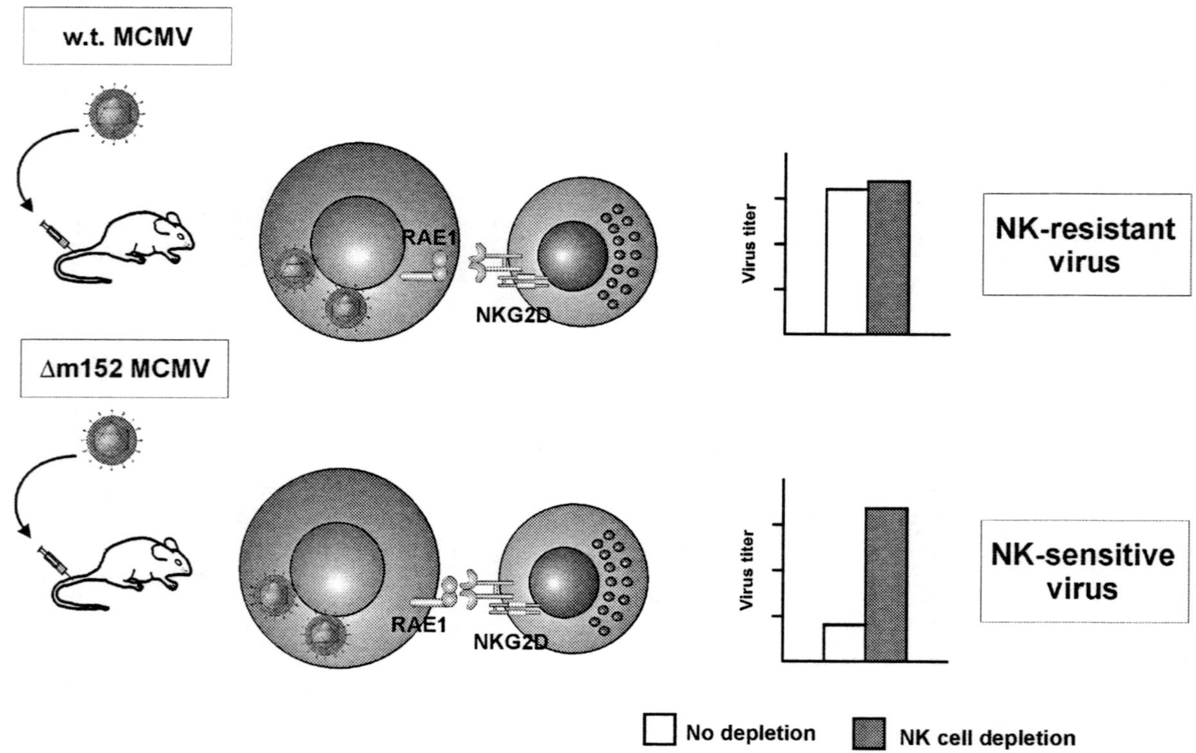

Figure 4. MCMV evasion from NK cells by down-modulation of NKG2D ligands. In *Cmv1*^{s} mice NK cell response to MCMV is compromised by the viral proteins responsible for down-modulation of inducible cellular ligands for the activating NK receptor, NKG2D. Therefore, depletion of NK cells does not increase the virus titer. Deletion of viral genes encoding proteins that interfere with expression of NKG2D ligands (*e.g. m152* which down-modulates RAE-1) convert NK-resistant virus to NK-sensitive one. This is illustrated by virus attenuation, which can be reversed by depletion of NK cells.

The situation is different in the MCMV-resistant *Cmv1^r^* mice. It has recently been shown that the *Cmv1^r^* resistance locus encodes Ly49H, the activating NK receptor, which is expressed on 50% of NK cells in *Cmv1^r^* mice (Brown et al., 2001; Daniels et al., 2001; Lee et al., 2001a). Unlike other receptors from the Ly49 family, the ligand for Ly49H is not a MHC class I molecule but a product of the MCMV *m157* gene, again a member of the *m145* gene family (Arase et al., 2002; Smith et al., 2002b). The *m157*/Ly49H interaction triggers the cytolitic response of NK cells as well as production of IFN-γ (Smith et al., 2002b).While the existence of genes whose products negatively regulate immune effector functions was expected, a ligand for the activating NK cell receptor came as a surprise. A possible explanation is the fact that the m157 protein has different functions in different mouse strains. In certain mouse strains the m157 protein can serve also as a ligand for Ly49I, the inhibitory NK cell receptor (Arase et al., 2002). Therefore, it can be assumed that the paradoxical situation that the virus encodes a ligand that strongly activates the NK response of the host is related to the origin of the MCMV strain generally used. Perhaps it was isolated from a mouse in which the *Ly49* gene encodes an inhibitory function.

Regardless of possible evolutionary explanations, a key question was whether the virus would maintain the gene function when passaged in a mouse strain in which the gene function provides a constant negative selection pressure through Ly49H$^+$ NK cells *in vivo* (Figure 4). Two groups have recently shown that under such conditions the *m157* gene undergoes frequent mutations and deletions and looses its capability to activate NK cells (French et al., 2004; Voigt et al., 2003). This is in line with our findings, which show that the targeted deletion of the *m157* gene results in the gain-of-virulence in Ly49H$^+$ mice (Bubic et al., 2004). The genetic escape mutations occurred rapidly when the virus is exposed to control by Ly49H$^+$ cells, but do not occur in Ly49H$^-$ mice (Figure 5).

4. Regulation of apoptosis as a mechanism to escape from NK cells

Apoptosis, the programmed cell death, controls the development and health of multicellular organisms and is also involved in the control of infection. Microbial interference with apoptosis may also represent efficient mean to escape NK cell control (Figure 3B). NK cells use two effector apoptotic pathways, the granule exocytosis pathway and death receptor pathway (Trapani and Smyth, 2002). Exocytosis of cytoplasmic granules that contain perforin and a family of serine proteases (granzymes) with various substrate specificities leads to induction of apoptosis of target cells. This pathway can be caspase dependent or independent (Sarin et al., 1997). The second apoptotic pathway is mediated through engagement of target-cell death receptors, such as Fas (CD95), by its cognate Fas ligand (Fasl), on the CTL or NK cell membrane, which results in caspase-dependent apoptosis (Nagata and Golstein, 1995).

Viruses can escape NK cell-mediated apoptosis by encoding inhibitors that interfere with specific components of the apoptotic pathway and endogenous intracellular serine protease inhibitors (serpins) (Guerin et al., 2002; Roulston et al., 1999). Examples of virus serpins are the *crmA* gene from cowpox virus and *P35*

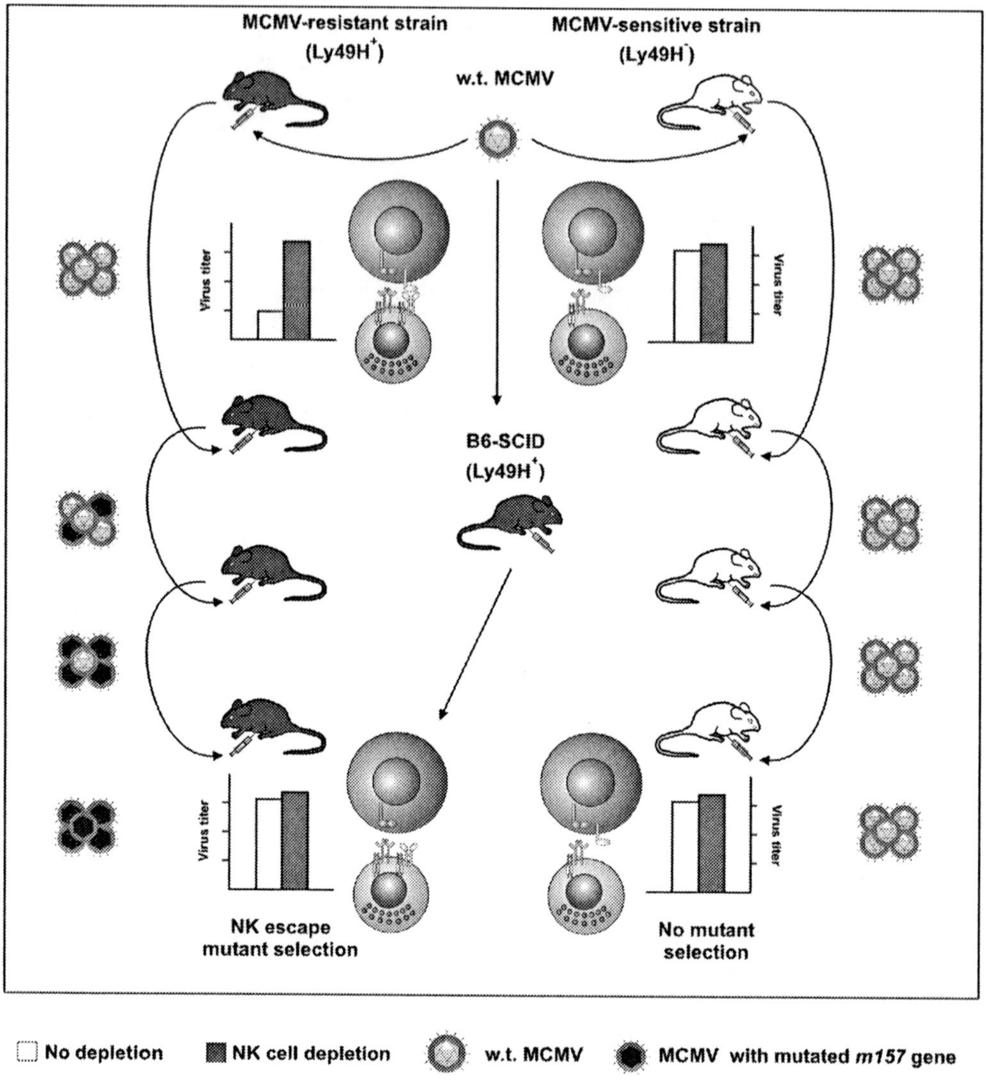

Figure 5. Escape from NK cell control by gene mutation. Ly49H$^+$ mouse strains are resistant to MCMV infection because NK cells are activated *via* interaction between the MCMV *m157* gene and Ly49H. Under the selective pressure by Ly49H$^+$ NK cells, for instance in C57BL/6 SCID mice or by serial passage through C57BL/6 mice, MCMV induces mutations and deletion of *m157* gene leading to the escape from NK cell-mediated control. Therefore, in contrast to w.t. virus, which is NK-sensitive, a point illustrated by the increased virus titer after NK cell depletion, MCMV with mutated *m157* gene gain the virulence and is resistant to NK cell control (left and central panel). To the contrary virus passaged through Ly49H$^-$ mice did not cause any mutation of *m157* gene and the virus remained NK-resistant in these mouse strains, as shown on figure 4 (right panel).

from baculoviruses, which block caspase activity and render infected cell resistant to Fas-mediated killing (Bump et al., 1995; Tewari and Dixit, 1995). Furthermore, myxoma virus possess a gene encoding for an ER resident protein called myxoma virus leukemia-associated protein (MV-LAP), which down-regulates surface Fas molecule (Guerin et al., 2002). Adenovirus L4-100 kDa protein is a granzyme-B substrate that prevents perforine/granzyme granule-induced apoptosis by potently inhibiting granzyme B (Andrade et al., 2001). Adenoviral complex E3-10.4K/14.5K selectively induces the loss of Fas surface expression and blocks Fas-induced apoptosis of infected cells (Elsing and Burgert, 1998; Shisler et al., 1997). Another adenovirus protein that influences apoptosis is the receptor internalization and degradation (RID) protein complex that mediates the internalization of cell-surface Fas for destruction inside the lysosomes (Tollefson et al., 1998). Herpes simplex virus 1 (HSV-1) *US5* gene encoding glycoprotein J107 inhibits perforin/granzyme-mediated apoptosis by an unknown mechanism (Jerome et al., 2001; 1999). Normally, a Syk/ZAP70–PI3K-dependent signal cascade is triggered in NK cells by target-cell recognition, which mobilizes the lytic granules towards the contacted target cell. Potentially, the inhibition of PI3K in NK cells by microbes would be an efficient mechanism to block the perforin and granzyme B effects. Viruses can also modulate mitochondrial components of apoptosis (Brune et al., 2003; Menard et al., 2003). For instance, vMIA, a mitochondria-localized inhibitor of apoptosis encoded by HCMV *UL37x1*, is capable of suppressing apoptosis induced by diverse stimuli, including Fas-mediated apoptosis (Goldmacher et al., 1999). It is unknown, however, whether these viral proteins affect NK cells-induced apoptotic pathways.

5. Microbial-encoded Fc receptors

Cellular Fc receptors play a key role for the antimicrobial activity of antigen-bound IgG antibodies. Pathogens use the principle of binding the IgG Fc fragment in order to oppose Fc-controlled effector functions (Figure 3C). Fc binding proteins are found in gram-positive bacteria, in protozoa like schistosomes and trypanosomes, in coronaviruses, hepatitis C virus (HCV), and in herpesviruses (Budt et al., 2004). In HSV-1 human IgG is bound by a complex formed by the glycoprotein gE and gI (Johnson et al., 1988) encoded by the *US7/US8* genes. Homologous molecules to gE and gI are also found in HSV-2 (Atalay et al., 2002), varicella zoster virus (VZV) (Litwin et al., 1992), pseudorabies virus (PRV) (Petrovskis et al., 1986) and Marek's disease virus (Jang et al., 1998). The gE-gI complex forms a high affinity receptor, which binds monomeric IgG thereby preventing HSV-infected cells from ADCC (Dubin et al., 1991). The effect is explained by the "bipolar bridging" which reflects IgG binding to an antigenic epitope *via* the Fab domain, and binding of the Fc domain to the microbial Fc receptor (Figure 4C) (Frank and Friedman, 1989). Thus, the engagement of the IgG Fc domain prevents it from binding to the Fc receptor on NK cells.

Three viral genes encoding two different Fc binding glycoproteins, gp68 and gp34, were identified within the HCMV genome (Atalay et al., 2002; Lilley et al., 2001). Gp68 is encoded by *UL119-118* genes and gp34 is encoded by two identical gene copies, *TRL11* and *IRL11*. In MCMV a single gene, *m138* encodes for a viral

FcγR (Thäle et al., 1994). Interestingly, the deletion mutant for MCMV *m138* is clearly attenuated *in vivo* but this attenuation is independent of antibodies (Crnkovic-Mertens et al., 1998). Further studies are required to assess whether MCMV *m138* might have another function, which explains the attenuation phenotype.

6. NK cell control through modulation of cytokines and chemokines

Cytokines and chemokines play a central role in the development, maturation and activation of NK cells. Since NK cells-derived cytokines and chemokines are important non-cytolytic means of infection control (Figure 1), the microbial binding to cytokines and chemokines or the interference with mediator expression indirectly modulates NK cell functions (Figure 3D).

The key cytokines involved in activation of NK cells released by DC, macrophages and stromal cells are IL-1, IL-2, IL-12, IL-15 and IL-18 (Biron et al., 1999; Cooper et al., 2004) and also IFN-α/β and TNF-α (Moretta, 2002). IL-18 is a proinflammatory cytokine that induces IFN-γ production and plays a critical role in NK cell activation (Takeda et al., 1998). It is structurally related to IL-1β and its inactive precursor is cleaved by the cysteine protease IL1β-converting enzyme, or caspase 1 (Ghayur et al., 1997). The IL-18 receptor belongs to the IL-1 receptor family. Its signaling activates NF-κB resulting in the synthesis of cytokines including IL-8, IL-1β and TNF-α (Robinson et al., 1997). IL-12 which is mainly produced by DC is a heterodimeric pro-inflammatory cytokine that provides an important link between innate and adaptive immunity (Trinchieri, 2003). By inducing an efficient Th1 immune response IL-12 seems to be important for resistance to most of bacteria, intracellular protozoa and fungal pathogens. The role of IL-12 in inducing NK cell activation is similar to that of IL-18 and the deficiency of either of these cytokines is mutually compensated (Takeda et al., 1998). Together with IL-18, IL-12 is important for the production of IFN-γ by NK cells. Both IFN-α and IFN–β are important in NK cell activation and cytotoxicity. These cytokines are induced during viral infection and by stimuli like double stranded RNA that arise in the life cycle of many viruses and engages TLRs (Alexopoulou et al., 2001; Biron et al., 1999). For instance, IFN-α/β are induced during early MCMV infection, which correlates with cytotoxic activity of NK cells (Orange and Biron, 1996).

There are several mechanisms by which viruses and other pathogens modulate cytokines and their receptors. IL-10 homologues encoded by HCMV and EBV (Kotenko et al., 2000; Moore et al., 1990) may affect NK cell activation by inhibiting production of Th1 type cytokines (D'Andrea et al., 1993). IL-18 is a frequent target for viral manipulation with direct impact on NK cell function. The p13, encoded by ectromelia virus, is a homologue of the mammalian IL-18 binding protein, which blocks the activity of IL-18 receptor by binding to IL-18 (Born et al., 2000). Deletion of the *p13* gene causes attenuation of virus *in vivo* due to enhanced sensitivity to NK cells. MCV also encodes a functional homologue of IL-18 binding protein with a potential ability to modulate NK cells (Xiang and Moss, 1999). Human papilloma virus 16 (HPV16) proteins E6 and E7 inhibit IL-18-induced IFN-γ production by specific and competitive binding to IL-18Rα

(Lee et al., 2001b). It is currently unknown whether this activity has a role in NK cell control *in vivo*. Some bacterial pathogens have also evolved mechanisms to modulate cytokine production and the subsequent immune response (Hornef et al., 2002). By up-regulating IL-6 and by inducing potent immunosuppressive cytokines such as IL-10 and TGF-β, Mycobacterium can inhibit cellular immune response (Moore et al., 1993). Bordetella pertussis stimulates the secretion of IL-10 by DC and inhibits IFN-γ production, *Y. enterocolitica* suppresses TNF-α production (Hornef et al., 2002). Altogether, the bacterial modulation of cytokines may have a strong indirect influence on the NK cell response.

Chemokines are important mediators of the early inflammatory response and play a role in NK cell biology and their response to infection (Robertson, 2002). For instance, MIP-1α is crucial factor for NK cell migration and NK cell-dependent antiviral protection in the liver during MCMV infection (Salazar-Mather et al., 1998). Similarly, susceptibility of BALB/c mice to Leishmania major could be explained by the defective production of chemokines such as IP-10 (CXCL10) causing a poor NK cell response (Vester et al., 1999). Functional homologues of cellular chemokines have been identified in a number of viruses, which suggests that the subversion of the host chemokine response contributes to the pathogenesis of these viruses in general (Inngjerdingen et al., 2001). MCK-2 (*m131-m129*) encoded by MCMV is a proinflammatory chemokine that recruits mononuclear cells to the site of infection and is important for virus dissemination (Fleming et al., 1999; Saederup et al., 2001). The virus lacking *m131-m129* is attenuated *in vivo* in a NK cell-dependent manner, although it is currently unknown by which mode MCK-2 affects NK cell function. KSHV encodes two antagonists: vMIP-I, a CCR8 antagonist that binds to chemokine receptors on NK cells and a broad-spectrum chemokine antagonist vMIP-II that probably blocks chemotactic responses to RANTES, MIP-1α and MIP-1β and thus has a putative significance for NK cell function (Kledal et al., 1997; Means et al., 2002). MC148 of MCV is a specific antagonist of MCP-1 (Luttichau et al., 2000). Also viral proteins without sequence homology to known chemokines block various chemokines important for NK cell function. For instance, murine γ-herpesvirus 68 (MHV68) encodes a broad-spectrum soluble chemokine-binding protein (Parry et al., 2000), and vaccinia virus (VV) encodes a soluble CC-chemokine-specific binding protein (Alcami et al., 1998).

7. TLRs as potential targets for microbial immunoevasion

In the past few years, data have been accumulated showing the involvement of receptors commonly referred as Toll-like receptors (TLRs) in the defense against pathogens (Akira and Takeda, 2004). TLRs are transmembrane proteins with a cystein and leucine rich extracellular domain and the cytoplasmatic TIR region which shows structural similarity with the IL-1 receptor (Medzhitov et al., 1997). Upon ligand binding they activate signaling cascade leading to the synthesis of proinflammatory cytokines. Thirteen members of the TLR family have been identified in mammals so far (Ulevitch, 2004). TLRs recognize pathogen associated molecular patterns (PAMPs) including bacterial lipopolysaccharide, bacterial lipoproteins and glycolipids, peptidoglycan, bacterial flagellin, bacterial

DNA, viral DNA and RNA and endogenous ligands such as the hsp60 (Akira and Takeda, 2004). When TLRs are engaged, usually the MyD88-IRAK-TRAF6 signaling pathway is activated (Medzhitov et al., 1998) resulting in the production of cytokines and co-stimulatory molecules CD80/CD86 that influence Th1 and Th2 differentiation (Banchereau and Steinman, 1998).

Recently a role of TLRs has been established for viral infections as well (Bowie et al., 2000; Compton et al., 2003; Tabeta et al., 2004). TLR9 recognizes unmethylated CpG motifs in double-stranded DNA (Hemmi et al., 2000), and TLR3 recognizes double-stranded RNA (Alexopoulou et al., 2001). Mice devoid of CpG–mediated signaling are more sensitive to MCMV (Tabeta et al., 2004), suggesting that insufficient NK cell activation might be the reason for the susceptibility. Insufficient signaling through TLR should lead to insufficient type I IFN production by dendritic cells and impaired NK cell activation. There is evidence that NK cells express TLRs, which trigger NK cell activation. TLR1, TLR6 and TLR9 are expressed on human NK cells (Hornung et al., 2002) and TLR3 is expressed on mouse NK cells (Hart and Gardiner, 2004). The inflammatory cytokines induced as results of TLR signaling are associated with serious systemic disorders such as endotoxic shock. In order to prevent such disorder the organisms have involved mechanisms for a negative regulation of TLR signaling (Akira and Takeda, 2004). In line with these observations, TLRs and/or their signaling adaptor molecules are expected to represent targets for negative regulation by viruses and other intracellular pathogens in order to prevent NK cell activation (Figure 3E). This has been already confirmed for VV, which interferes with the MyD88-IRAK-TRAF6 signaling pathway with broad effects on host defenses including the NK cell response (Bowie et al., 2000). In addition, microorganism may evade NK cell-mediated control *via* TLRs by methylating CpG motifs or by mutating the ligands in a way to preserve their functional characteristics but prevent binding to TLRs. TLRs9 is also involved in bacterial control and it will be interesting to study whether bacterial infection affects TLRs expression on the cell surface. Bacteria reprogram their ligands to evade recognition by TLRs but the effect on TLRs expression on the cell surface has not yet been studied (Guo et al., 1997).

8. Inhibition of the NK cell response by effects on dendritic cells

The interaction between the DC and NK cells is critical for control of several pathogens (Andrews et al., 2003). Therefore, antigen presenting cells are an optimal target for pathogens to subvert the host immune response (Figure 3E). DC are initiators of the adaptive immune response and important for the stimulation of NK cells (Figure 1). Targeting of DC by pathogens may cause insufficient cytokine production, including IL-18 and IL-12, with consequences for NK cell activation (Andrews et al., 2003). On the other hand, the NK cell response is needed to protect DC. Furthermore, DC can promote migration of NK cells into lymph nodes and therefore contribute to efficient T cell priming (MartIn-Fontecha et al., 2003). Moreover, DC colocalise with NK cells in lymph nodes and through IL-12 and IL-15 promote their proliferation and IFN-γ production, which is important for shaping the adaptive immune response (Ferlazzo and Munz, 2004). CD40/CD40 ligand interaction is essential for IL-12 production by DC (Fujii et al., 2004). Having in mind the important role of DC for NK cell activation one may predict that microbial regulation of CD40 molecule may compromise not only T cell but also NK cell response.

9. Direct microbial effects on NK cells

Little is known on the regulation of NKR on NK cells *in vivo*. Studies on microbial evasion mechanisms have mainly concentrated on the regulation of cellular ligands for NK cell receptors and on indirect mechanisms involving cytokines and chemokines. Yet, there are examples for direct NK cell evasion (Figure 3F). Some tumors secrete soluble forms of cellular ligands for activating NK cell receptors and thus compromise NK cell function (Groh et al., 2002). Tumors have been shown to secrete MICA and MICB as soluble ligands for the NKG2D receptor. Upon binding to NKG2D, these soluble ligands induce down-regulation of NKG2D receptors on NK and T cells.

HCV directly inhibits NK cells *via* ligation of an inhibitory NK receptor, CD81. CD81 is a broadly expressed tetraspanin molecule and a potential cellular receptor for HCV with a high affinity for the HCV E2 envelope glycoprotein (Wack et al., 2001). Ligation of CD81 on NK cells to immobilized E2 impairs NK cell cytotoxicity and cytokine release (Crotta et al., 2002; Tseng and Klimpel, 2002).

A comparable viral strategy represents the binding to LFA-1 on NK cells by HIV Tat which blocks NK cell activation and function (Poggi et al., 2002; Zocchi et al., 1998). HIV may evade NK cell also by direct infection of NK cells (Chehimi et al., 1991). HSV has also been reported to spread from infected fibroblasts to cultured NK cells, and thus inhibit NK cell cytotoxicity (York and Johnson, 1993).

10. Conclusions and perspectives

In recent years, a remarkable progress has been made in the understanding of NK cell biology and their role in the control of microbial infections. Numerous NK cell activation receptors and their ligands, induced by stress, infection or by microbial genes have been identified. Complex regulatory cytokine and chemokine pathways that shape the NK cell response to infection have been revealed. Insights were gained by studying microbial (mostly viral) strategies for NK cell escape. Results obtained with the MCMV model indicate that there is a constant pressure for NK cell evasion. The discovery of pattern recognition receptors such as TLRs in NK cells will broaden our concepts of NK cell biology.

There are questions that need answers: Is there any virus infection that is exempt from NK cell control? Is there a function for NK cells during chronic and latent infection and how do the microbial immunoevasins compromise this function? Do NK cells and microbial immunoevasins have a role during microbial spread and transmission? How is NK cell evasion linked with the adaptive immune response and prevention of potential autoimmunity? What is the rationale for the high level of functional redundancy of some NK cell receptors and for existence of several ligands for a single NK cell receptor?

The experimental methods and approaches currently in use for studying the role of microbial immunoevasins mainly include the transfection of isolated immunoevasion genes, and their expression through vectors. The construction of pathogen mutants and their testing in the natural host has been carried out only for very few examples. Therefore, the majority of our concepts stem from *in vitro* data and extrapolations to the situation *in vivo*. Only few of these hypotheses have been rigorously tested and challenged so far. Therefore, our understanding of microbial interference with NK cell biology and effector function is only indirect for the majority of examples. Major insights have to come from the analysis *in vivo*. The advent of gene targeted mice that are constitutively or conditionally deficient for defined NK receptors and their ligands or any other function essential for NK cells will represent the superior tools. Similarly, pathogens constructed to be deficient in defined genes that directly or indirectly affect NK cells have to complement the armamentum.

It is a truism that the dynamic system of innate immunity has many cellular players and includes a multitude of soluble factors that act either locally or at a distance. It is also a truism that innate immunity is connected with adaptive immunity. This invites the search for functional linkages. A linkage of one function with another, however, neither proves a direct interaction nor its functional importance. Given the complexity and redundancy of regulatory circuits, future efforts should concentrate on the description of functional hierarchies. Then, the collection of many facts can reveal the biological framework. This might lead to a more efficient prophylaxis and treatment of virus-induced pathology.

References

Ahn, K., Gruhler, A., Galocha, B., Jones, T. R., Wiertz, E. J., Ploegh, H. L., Peterson, P. A., Yang, Y., and Fruh, K. (1997). The ER-luminal domain of the HCMV glycoprotein US6 inhibits peptide translocation by TAP. Immunity *6*, 613-621.

Akira, S., and Takeda, K. (2004). Toll-like receptor signalling. Nat. Rev. Immunol. *4*, 499-511.

Alcami, A., and Koszinowski, U. H. (2000). Viral mechanisms of immune evasion. Immunol. Today *21*, 447-455.

Alcami, A., Symons, J. A., Collins, P. D., Williams, T. J., and Smith, G. L. (1998). Blockade of chemokine activity by a soluble chemokine binding protein from vaccinia virus. J. Immunol. *160*, 624-633.

Alexopoulou, L., Holt, A. C., Medzhitov, R., and Flavell, R. A. (2001). Recognition of double-stranded RNA and activation of NF-kappaB by Toll-like receptor 3. Nature *413*, 732-738.

Andrade, F., Bull, H. G., Thornberry, N. A., Ketner, G. W., Casciola-Rosen, L. A., and Rosen, A. (2001). Adenovirus L4-100K assembly protein is a granzyme B substrate that potently inhibits granzyme B-mediated cell death. Immunity *14*, 751-761.

Andrews, D. M., Scalzo, A. A., Yokoyama, W. M., Smyth, M. J., and Degli-Esposti, M. A. (2003). Functional interactions between dendritic cells and NK cells during viral infection. Nat. Immunol. *4*, 175-181.

Arase, H., and Lanier, L. L. (2002). Virus-driven evolution of natural killer cell receptors. Microbes. Infect. *4*, 1505-1512.

Arase, H., Mocarski, E. S., Campbell, A. E., Hill, A. B., and Lanier, L. L. (2002). Direct recognition of cytomegalovirus by activating and inhibitory NK cell receptors. Science *296*, 1323-1326.

Arnon, T. I., Lev, M., Katz, G., Chernobrov, Y., Porgador, A., and Mandelboim, O. (2001). Recognition of viral hemagglutinins by NKp44 but not by NKp30. Eur. J. Immunol. *31*, 2680-2689.

Atalay, R., Zimmermann, A., Wagner, M., Borst, E., Benz, C., Messerle, M., and Hengel, H. (2002). Identification and expression of human cytomegalovirus transcription units coding for two distinct Fcgamma receptor homologs. J. Virol. *76*, 8596-8608.

Banchereau, J., and Steinman, R. M. (1998). Dendritic cells and the control of immunity. Nature *392*, 245-252.

Becker, I., Salaiza, N., Aguirre, M., Delgado, J., Carrillo-Carrasco, N., Kobeh, L. G., Ruiz, A., Cervantes, R., Torres, A. P., Cabrera, N., et al. (2003). Leishmania lipophosphoglycan (LPG) activates NK cells through toll-like receptor-2. Mol. Biochem. Parasitol. *130*, 65-74.

Biassoni, R., Pessino, A., Bottino, C., Pende, D., Moretta, L., and Moretta, A. (1999). The murine homologue of the human NKp46, a triggering receptor involved in the induction of natural cytotoxicity. Eur. J. Immunol. *29*, 1014-1020.

Biron, C. A. (1999). Initial and innate responses to viral infections--pattern setting in immunity or disease. Curr. Opin. Microbiol. *2*, 374-381.

Biron, C. A., Byron, K. S., and Sullivan, J. L. (1989). Severe herpesvirus infections in an adolescent without natural killer cells. N. Engl. J. Med. *320*, 1731-1735.

Biron, C. A., Nguyen, K. B., Pien, G. C., Cousens, L. P., and Salazar-Mather, T. P. (1999). Natural killer cells in antiviral defense: function and regulation by innate cytokines. Annu. Rev. Immunol. *17*, 189-220.

Booss, J., and Wheelock, E. F. (1975). Correlation of survival from murine cytomegalovirus infection with spleen cell responsiveness to Concanavallin A. Proc. Soc. Exp. Biol. Med. *149*, 443-445.

Born, T. L., Morrison, L. A., Esteban, D. J., VandenBos, T., Thebeau, L. G., Chen, N., Spriggs, M. K., Sims, J. E., and Buller, R. M. (2000). A poxvirus protein that binds to and inactivates IL-18, and inhibits NK cell response. J. Immunol. *164*, 3246-3254.

Bowie, A., Kiss-Toth, E., Symons, J. A., Smith, G. L., Dower, S. K., and O'Neill, L. A. (2000). A46R and A52R from vaccinia virus are antagonists of host IL-1 and toll-like receptor signaling. Proc. Natl. Acad. Sci. USA. *97*, 10162-10167.

Brown, M. G., Dokun, A. O., Heusel, J. W., Smith, H. R., Beckman, D. L., Blattenberger, E. A., Dubbelde, C. E., Stone, L. R., Scalzo, A. A., and Yokoyama, W. M. (2001). Vital involvement of a natural killer cell activation receptor in resistance to viral infection. Science *292*, 934-937.

Brune, W., Nevels, M., and Shenk, T. (2003). Murine cytomegalovirus m41 open reading frame encodes a Golgi-localized antiapoptotic protein. J. Virol. *77*, 11633-11643.

Bubic, I. W., M., Krmpotic, A., Saulig, T., Kim, S., Yokoyama, W.M., Jonjic, S., Koszinowski, U.H. (2004). Gain of virulence caused by loss of a gene in murine cytomegalovirus. J. Virol. *78*, 7536-7544.

Budt, M., Reinhard, H., Bigl, A., and Hengel, H. (2004). Herpesviral Fcgamma receptors: culprits attenuating antiviral IgG? Int. Immunopharmacol. *4*, 1135-1148.

Bukowski, J. F., Woda, B. A., and Welsh, R. M. (1984). Pathogenesis of murine cytomegalovirus infection in natural killer cell-depleted mice. J. Virol. *52*, 119-128.

Bump, N. J., Hackett, M., Hugunin, M., Seshagiri, S., Brady, K., Chen, P., Ferenz, C., Franklin, S., Ghayur, T., Li, P., and et al. (1995). Inhibition of ICE family proteases by baculovirus antiapoptotic protein p35. Science *269*, 1885-1888.

Carayannopoulos, L. N., and Yokoyama, W. M. (2004). Recognition of infected cells by natural killer cells. Curr. Opin. Immunol. *16*, 26-33.

Cerwenka, A., Bakker, A. B., McClanahan, T., Wagner, J., Wu, J., Phillips, J. H., and Lanier, L. L. (2000). Retinoic acid early inducible genes define a ligand family for the activating NKG2D receptor in mice. Immunity *12*, 721-727.

Cerwenka, A., and Lanier, L. L. (2001). Natural killer cells, viruses and cancer. Nat. Rev. Immunol. *1*, 41-49.

Cerwenka, A., and Lanier, L. L. (2003). NKG2D ligands: unconventional MHC class I-like molecules exploited by viruses and cancer. Tissue Antigens *61*, 335-343.

Chehimi, J., Bandyopadhyay, S., Prakash, K., Perussia, B., Hassan, N. F., Kawashima, H., Campbell, D., Kornbluth, J., and Starr, S. E. (1991). *In vitro* infection of natural killer cells with different human immunodeficiency virus type 1 isolates. J. Virol. *65*, 1812-1822.

Cohen, G. B., Gandhi, R. T., Davis, D. M., Mandelboim, O., Chen, B. K., Strominger, J. L., and Baltimore, D. (1999). The selective downregulation of class I major histocompatibility complex proteins by HIV-1 protects HIV-infected cells from NK cells. Immunity *10*, 661-671.

Colucci, F., Di Santo, J. P., and Leibson, P. J. (2002). Natural killer cell activation in mice and men: different triggers for similar weapons? Nat. Immunol. *3*, 807-813.

Compton, T., Kurt-Jones, E. A., Boehme, K. W., Belko, J., Latz, E., Golenbock, D. T., and Finberg, R. W. (2003). Human cytomegalovirus activates inflammatory cytokine responses via CD14 and Toll-like receptor 2. J. Virol. *77*, 4588-4596.

Cooper, M. A., Fehniger, T. A., Fuchs, A., Colonna, M., and Caligiuri, M. A. (2004). NK cell and DC interactions. Trends. Immunol. *25*, 47-52.

Cooper, M. A., Fehniger, T. A., Turner, S. C., Chen, K. S., Ghaheri, B. A., Ghayur, T., Carson, W. E., and Caligiuri, M. A. (2001). Human natural killer cells: a unique innate immunoregulatory role for the CD56(bright) subset. Blood *97*, 3146-3151.

Coscoy, L., and Ganem, D. (2001). A viral protein that selectively downregulates ICAM-1 and B7-2 and modulates T cell costimulation. J. Clin. Invest. *107*, 1599-1606.

Coscoy, L., Sanchez, D. J., and Ganem, D. (2001). A novel class of herpesvirus-encoded membrane-bound E3 ubiquitin ligases regulates endocytosis of proteins involved in immune recognition. J. Cell. Biol. *155*, 1265-1273.

Cosman, D., Fanger, N., Borges, L., Kubin, M., Chin, W., Peterson, L., and Hsu, M. L. (1997). A novel immunoglobulin superfamily receptor for cellular and viral MHC class I molecules. Immunity *7*, 273-282.

Cosman, D., Mullberg, J., Sutherland, C. L., Chin, W., Armitage, R., Fanslow, W., Kubin, M., and Chalupny, N. J. (2001). ULBPs, novel MHC class I-related molecules, bind to CMV glycoprotein UL16 and stimulate NK cytotoxicity through the NKG2D receptor. Immunity *14*, 123-133.

Cretney, E., Degli-Esposti, M. A., Densley, E. H., Farrell, H. E., Davis-Poynter, N. J., and Smyth, M. J. (1999). m144, a murine cytomegalovirus (MCMV)-encoded major histocompatibility complex class I homologue, confers tumor resistance to natural killer cell-mediated rejection. J. Exp. Med. *190*, 435-444.

Crnkovic-Mertens, I., Messerle, M., Milotic, I., Szepan, U., Kucic, N., Krmpotic, A., Jonjic, S., and Koszinowski, U. H. (1998). Virus attenuation after deletion of the cytomegalovirus Fc receptor gene is not due to antibody control. J. Virol. *72*, 1377-1382.

Crotta, S., Stilla, A., Wack, A., D'Andrea, A., Nuti, S., D'Oro, U., Mosca, M., Filliponi, F., Brunetto, R. M., Bonino, F., et al. (2002). Inhibition of natural killer cells through engagement of CD81 by the major hepatitis C virus envelope protein. J. Exp. Med. *195*, 35-41.

D'Andrea, A., Aste-Amezaga, M., Valiante, N. M., Ma, X., Kubin, M., and Trinchieri, G. (1993). Interleukin 10 (IL-10) inhibits human lymphocyte interferon gamma-production by suppressing natural killer cell stimulatory factor/IL-12 synthesis in accessory cells. J. Exp. Med. *178*, 1041-1048.

Daniels, K. A., Devora, G., Lai, W. C., O'Donnell, C. L., Bennett, M., and Welsh, R. M. (2001). Murine cytomegalovirus is regulated by a discrete subset of natural killer cells reactive with monoclonal antibody to Ly49H. J. Exp. Med. *194*, 29-44.

Diefenbach, A., Tomasello, E., Lucas, M., Jamieson, A. M., Hsia, J. K., Vivier, E., and Raulet, D. H. (2002). Selective associations with signaling proteins determine stimulatory versus costimulatory activity of NKG2D. Nat. Immunol. *3*, 1142-1149.

Dubin, G., Socolof, E., Frank, I., and Friedman, H. M. (1991). Herpes simplex virus type 1 Fc receptor protects infected cells from antibody-dependent cellular cytotoxicity. J. Virol. *65*, 7046-7050.

Dunn, C., Chalupny, N. J., Sutherland, C. L., Dosch, S., Sivakumar, P. V., Johnson, D. C., and Cosman, D. (2003). Human cytomegalovirus glycoprotein UL16 causes intracellular sequestration of NKG2D ligands, protecting against natural killer cell cytotoxicity. J. Exp. Med. *197*, 1427-1439.

Elsing, A., and Burgert, H. G. (1998). The adenovirus E3/10.4K-14.5K proteins down-modulate the apoptosis receptor Fas/Apo-1 by inducing its internalization. Proc. Natl. Acad. Sci. USA. *95*, 10072-10077.

Falk, C. S., Mach, M., Schendel, D. J., Weiss, E. H., Hilgert, I., and Hahn, G. (2002). NK cell activity during human cytomegalovirus infection is dominated by US2-11-mediated HLA class I down-regulation. J. Immunol. *169*, 3257-3266.

Farrell, H. E., Vally, H., Lynch, D. M., Fleming, P., Shellam, G. R., Scalzo, A. A., and Davis-Poynter, N. J. (1997). Inhibition of natural killer cells by a cytomegalovirus MHC class I homologue *in vivo*. Nature *386*, 510-514.

Ferlazzo, G., and Munz, C. (2004). NK cell compartments and their activation by dendritic cells. J. Immunol. *172*, 1333-1339.

Ferlazzo, G., Semino, C., and Melioli, G. (2001). HLA class I molecule expression is up-regulated during maturation of dendritic cells, protecting them from natural killer cell-mediated lysis. Immunol. Lett. *76*, 37-41.

Ferlazzo, G., Tsang, M. L., Moretta, L., Melioli, G., Steinman, R. M., and Munz, C. (2002). Human dendritic cells activate resting natural killer (NK) cells and are recognized via the NKp30 receptor by activated NK cells. J. Exp. Med. *195*, 343-351.

Fleming, P., Davis-Poynter, N., Degli-Esposti, M., Densley, E., Papadimitriou, J., Shellam, G., and Farrell, H. (1999). The murine cytomegalovirus chemokine homolog, m131/129, is a determinant of viral pathogenicity. J. Virol. *73*, 6800-6809.

Fletcher, J. M., Prentice, H. G., and Grundy, J. E. (1998). Natural killer cell lysis of cytomegalovirus (CMV)-infected cells correlates with virally induced changes in cell surface lymphocyte function-associated antigen-3 (LFA-3) expression and not with the CMV-induced down-regulation of cell surface class I HLA. J. Immunol. *161*, 2365-2374.

Frank, I., and Friedman, H. M. (1989). A novel function of the herpes simplex virus type 1 Fc receptor: participation in bipolar bridging of antiviral immunoglobulin G. J. Virol. *63*, 4479-4488.

French, A. R., Pingel, J. T., Wagner, M., Bubic, I., Yang, L., Kim, S., Koszinowski, U., Jonjic, S., and Yokoyama, W. M. (2004). Escape of mutant double-stranded DNA virus from innate immune control. Immunity *20*, 747-756.

Fujii, S., Liu, K., Smith, C., Bonito, A. J., and Steinman, R. M. (2004). The linkage of innate to adaptive immunity via maturing dendritic cells *in vivo* requires CD40 ligation in addition to antigen presentation and cd80/86 costimulation. J. Exp. Med. *199*, 1607-1618.

Furman, M. H., and Ploegh, H. L. (2002). Lessons from viral manipulation of protein disposal pathways. J. Clin. Invest. *110*, 875-879.

Gessner, J. E., Heiken, H., Tamm, A., and Schmidt, R. E. (1998). The IgG Fc receptor family. Ann. Hematol. *76*, 231-248.

Ghayur, T., Banerjee, S., Hugunin, M., Butler, D., Herzog, L., Carter, A., Quintal, L., Sekut, L., Talanian, R., Paskind, M., et al. (1997). Caspase-1 processes IFN-gamma-inducing factor and regulates LPS-induced IFN-gamma production. Nature *386*, 619-623.

Gilfillan, S., Ho, E. L., Cella, M., Yokoyama, W. M., and Colonna, M. (2002). NKG2D recruits two distinct adapters to trigger NK cell activation and costimulation. Nat. Immunol. *3*, 1150-1155.

Giorda, R., Rudert, W. A., Vavassori, C., Chambers, W. H., Hiserodt, J. C., and Trucco, M. (1990). NKR-P1, a signal transduction molecule on natural killer cells. Science *249*, 1298-1300.

Goldmacher, V. S., Bartle, L. M., Skaletskaya, A., Dionne, C. A., Kedersha, N. L., Vater, C. A., Han, J. W., Lutz, R. J., Watanabe, S., Cahir McFarland, E. D., et al. (1999). A cytomegalovirus-encoded mitochondria-localized inhibitor of apoptosis structurally unrelated to Bcl-2. Proc. Natl. Acad. Sci. USA. *96*, 12536-12541.

Groh, V., Bahram, S., Bauer, S., Herman, A., Beauchamp, M., and Spies, T. (1996). Cell stress-regulated human major histocompatibility complex class I gene expressed in gastrointestinal epithelium. Proc. Natl. Acad. Sci. USA. *93*, 12445-12450.

Groh, V., Rhinehart, R., Randolph-Habecker, J., Topp, M. S., Riddell, S. R., and Spies, T. (2001). Costimulation of CD8alphabeta T cells by NKG2D via engagement by MIC induced on virus-infected cells. Nat. Immunol. *2*, 255-260.

Groh, V., Wu, J., Yee, C., and Spies, T. (2002). Tumour-derived soluble MIC ligands impair expression of NKG2D and T-cell activation. Nature *419*, 734-738.

Guerin, J. L., Gelfi, J., Boullier, S., Delverdier, M., Bellanger, F. A., Bertagnoli, S., Drexler, I., Sutter, G., and Messud-Petit, F. (2002). Myxoma virus leukemia-associated protein is responsible for major histocompatibility complex class I and Fas-CD95 down-regulation and defines scrapins, a new group of surface cellular receptor abductor proteins. J. Virol. *76*, 2912-2923.

Guo, L., Lim, K. B., Gunn, J. S., Bainbridge, B., Darveau, R. P., Hackett, M., and Miller, S. I. (1997). Regulation of lipid A modifications by *Salmonella typhimurium* virulence genes phoP-phoQ. Science *276*, 250-253.

Hamerman, J. A., Ogasawara, K., and Lanier, L. L. (2004). Cutting edge: toll-like receptor signaling in macrophages induces ligands for the NKG2D receptor. J. Immunol. *172*, 2001-2005.

Hart, O., and Gardiner, C. M. (2004). The role of TOLL-like receptors (TLRs) in natural killer cell function, Paper presented at: 8th Annual Meeting of the Society for Natural Immunity and 20th International Natural Killer Cell Workshop (Noordwijkerhout, The Netherlands: The Society for Natural Immunity).

Hasan, M., Krmpotic, A., Ruzsics, Z., Bubic, I., Lenac, T., Halenius, A., Loewendorf, A., Messerle, M., Hengel, H., Jonjic, S., and Koszinowski, U. H. (2005). Selective down-regulation of the NKG2D ligand H60 by mouse cytomegalovirus m155 glycoprotein. J. Virol. *79*, 2920-2930.

Hemmi, H., Takeuchi, O., Kawai, T., Kaisho, T., Sato, S., Sanjo, H., Matsumoto, M., Hoshino, K., Wagner, H., Takeda, K., and Akira, S. (2000). A toll-like receptor recognizes bacterial DNA. Nature *408*, 740-745.

Hengel, H., and Koszinowski, U. H. (1997). Interference with antigen processing by viruses. Curr. Opin. Immunol. *9*, 470-476.

Hornef, M. W., Wick, M. J., Rhen, M., and Normark, S. (2002). Bacterial strategies for overcoming host innate and adaptive immune responses. Nat. Immunol. *3*, 1033-1040.

Hornung, V., Rothenfusser, S., Britsch, S., Krug, A., Jahrsdorfer, B., Giese, T., Endres, S., and Hartmann, G. (2002). Quantitative expression of toll-like receptor 1-10 mRNA in cellular subsets of human peripheral blood mononuclear cells and sensitivity to CpG oligodeoxynucleotides. J. Immunol. *168*, 4531-4537.

Houchins, J. P., Yabe, T., McSherry, C., and Bach, F. H. (1991). DNA sequence analysis of NKG2, a family of related cDNA clones encoding type II integral membrane proteins on human natural killer cells. J. Exp. Med. *173*, 1017-1020.

Iizuka, K., Naidenko, O. V., Plougastel, B. F., Fremont, D. H., and Yokoyama, W. M. (2003). Genetically linked C-type lectin-related ligands for the NKRP1 family of natural killer cell receptors. Nat. Immunol. *4*, 801-807.

Inngjerdingen, M., Damaj, B., and Maghazachi, A. A. (2001). Expression and regulation of chemokine receptors in human natural killer cells. Blood *97*, 367-375.

Ishido, S., Choi, J. K., Lee, B. S., Wang, C., DeMaria, M., Johnson, R. P., Cohen, G. B., and Jung, J. U. (2000a). Inhibition of natural killer cell-mediated cytotoxicity by Kaposi's sarcoma-associated herpesvirus K5 protein. Immunity *13*, 365-374.

Ishido, S., Wang, C., Lee, B. S., Cohen, G. B., and Jung, J. U. (2000b). Down-regulation of major histocompatibility complex class I molecules by Kaposi's sarcoma-associated herpesvirus K3 and K5 proteins. J. Virol. *74*, 5300-5309.

Jang, H. K., Ono, M., Kim, T. J., Izumiya, Y., Damiani, A. M., Matsumura, T., Niikura, M., Kai, C., and Mikami, T. (1998). The genetic organization and transcriptional analysis of the short unique region in the genome of nononcogenic Marek's disease virus serotype 2. Virus Res. *58*, 137-147.

Jerome, K. R., Chen, Z., Lang, R., Torres, M. R., Hofmeister, J., Smith, S., Fox, R., Froelich, C. J., and Corey, L. (2001). HSV and glycoprotein J. inhibit caspase activation and apoptosis induced by granzyme B or Fas. J. Immunol. *167*, 3928-3935.

Jerome, K. R., Fox, R., Chen, Z., Sears, A. E., Lee, H., and Corey, L. (1999). Herpes simplex virus inhibits apoptosis through the action of two genes, Us5 and Us3. J. Virol. *73*, 8950-8957.

Johnson, D. C., Frame, M. C., Ligas, M. W., Cross, A. M., and Stow, N. D. (1988). Herpes simplex virus immunoglobulin G Fc receptor activity depends on a complex of two viral glycoproteins, gE and gI. J. Virol. *62*, 1347-1354.

Jun, Y., Kim, E., Jin, M., Sung, H. C., Han, H., Geraghty, D. E., and Ahn, K. (2000). Human cytomegalovirus gene products US3 and US6 down-regulate trophoblast class I MHC molecules. J. Immunol. *164*, 805-811.

Karlhofer, F. M., Ribaudo, R. K., and Yokoyama, W. M. (1992). MHC class I alloantigen specificity of Ly-49+ IL-2-activated natural killer cells. Nature *358*, 66-70.

Karlhofer, F. M., and Yokoyama, W. M. (1991). Stimulation of murine natural killer (NK) cells by a monoclonal antibody specific for the NK1.1 antigen. IL-2-activated NK cells possess additional specific stimulation pathways. J. Immunol. *146*, 3662-3673.

Kavanagh, D. G., Gold, M. C., Wagner, M., Koszinowski, U. H., and Hill, A. B. (2001). The multiple immune-evasion genes of murine cytomegalovirus are not redundant: m4 and m152 inhibit antigen presentation in a complementary and cooperative fashion. J. Exp. Med. *194*, 967-978.

Kledal, T. N., Rosenkilde, M. M., Coulin, F., Simmons, G., Johnsen, A. H., Alouani, S., Power, C. A., Luttichau, H. R., Gerstoft, J., Clapham, P. R., et al. (1997). A broad-spectrum chemokine antagonist encoded by Kaposi's sarcoma-associated herpesvirus. Science *277*, 1656-1659.

Kloover, J. S., Grauls, G. E., Blok, M. J., Vink, C., and Bruggeman, C. A. (2002). A rat cytomegalovirus strain with a disruption of the r144 MHC class I-like gene is attenuated in the acute phase of infection in neonatal rats. Arch. Virol. *147*, 813-824.

Kotenko, S. V., Saccani, S., Izotova, L. S., Mirochnitchenko, O. V., and Pestka, S. (2000). Human cytomegalovirus harbors its own unique IL-10 homolog (cmvIL-10). Proc. Natl. Acad. Sci. USA. *97*, 1695-1700.

Kraft, J. R., Vance, R. E., Pohl, J., Martin, A. M., Raulet, D. H., and Jensen, P. E. (2000). Analysis of Qa-1(b) peptide binding specificity and the capacity of CD94/NKG2A to discriminate between Qa-1-peptide complexes. J. Exp. Med. *192*, 613-624.

Krmpotic, A., Busch, D. H., Bubic, I., Gebhardt, F., Hengel, H., Hasan, M., Scalzo, A. A., Koszinowski, U. H., and Jonjic, S. (2002). MCMV glycoprotein gp40 confers virus resistance to CD8+ T cells and NK cells *in vivo*. Nat. Immunol. *3*, 529-535.

Krmpotic, A., Hasan, M., Loewendorf, A., Saulig, T., Halenius, A., Lenac, T., Polic, B., Bubic, I., Kriegeskorte, A., Pernjak-Pugel, E., et al. (2005). NK cell activation through the NKG2D ligand MULT-1 is selectively prevented by the glycoprotein encoded by mouse cytomegalovirus gene *m145*. J. Exp. Med. *201*, 211-220.

Krmpotic, A., Messerle, M., Crnkovic-Mertens, I., Polic, B., Jonjic, S., and Koszinowski, U. H. (1999). The immunoevasive function encoded by the mouse cytomegalovirus gene m152 protects the virus against T cell control *in vivo*. J. Exp. Med. *190*, 1285-1296.

Kubin, M., Cassiano, L., Chalupny, J., Chin, W., Cosman, D., Fanslow, W., Mullberg, J., Rousseau, A. M., Ulrich, D., and Armitage, R. (2001). ULBP1, 2, 3: novel MHC class I-related molecules that bind to human cytomegalovirus glycoprotein UL16, activate NK cells. Eur. J. Immunol. *31*, 1428-1437.

Kubota, A., Kubota, S., Farrell, H. E., Davis-Poynter, N., and Takei, F. (1999). Inhibition of NK cells by murine CMV-encoded class I MHC homologue m144. Cell. Immunol. *191*, 145-151.

Lanier, L. L. (2003). Natural killer cell receptor signaling. Curr. Opin. Immunol. *15*, 308-314.

Lanier, L. L., Chang, C., and Phillips, J. H. (1994). Human NKR-P1A. A disulfide-linked homodimer of the C-type lectin superfamily expressed by a subset of NK and T lymphocytes. J. Immunol. *153*, 2417-2428.

Le Gall, S., Buseyne, F., Trocha, A., Walker, B. D., Heard, J. M., and Schwartz, O. (2000). Distinct trafficking pathways mediate Nef-induced and clathrin-dependent major histocompatibility complex class I down-regulation. J. Virol. *74*, 9256-9266.

Le Gall, S., Erdtmann, L., Benichou, S., Berlioz-Torrent, C., Liu, L., Benarous, R., Heard, J. M., and Schwartz, O. (1998). Nef interacts with the mu subunit of clathrin adaptor complexes and reveals a cryptic sorting signal in MHC I molecules. Immunity *8*, 483-495.

Lee, K.-M., McNerney, M. E., Stepp, S. E., Mathew, P. A., Schatzle, J. D., Bennett, M., and Kumar, V. (2004). 2B4 acts as a non-major histocompatibility complex binding inhibitory receptor on mouse natural killer cells. J. Exp. Med. *199*, 1245-1254.

Lee, S. H., Girard, S., Macina, D., Busa, M., Zafer, A., Belouchi, A., Gros, P., and Vidal, S. M. (2001a). Susceptibility to mouse cytomegalovirus is associated with deletion of an activating natural killer cell receptor of the C-type lectin superfamily. Nat. Genet. *28*, 42-45.

Lee, S. J., Cho, Y. S., Cho, M. C., Shim, J. H., Lee, K. A., Ko, K. K., Choe, Y. K., Park, S. N., Hoshino, T., Kim, S., et al. (2001b). Both E6 and E7 oncoproteins of human papillomavirus 16 inhibit IL-18-induced IFN-gamma production in human peripheral blood mononuclear and NK cells. J. Immunol. *167*, 497-504.

Lilley, B. N., Ploegh, H. L., and Tirabassi, R. S. (2001). Human cytomegalovirus open reading frame TRL11/IRL11 encodes an immunoglobulin G Fc-binding protein. J. Virol. *75*, 11218-11221.

Litwin, V., Jackson, W., and Grose, C. (1992). Receptor properties of two varicella-zoster virus glycoproteins, gpI and gpIV, homologous to herpes simplex virus gE and gI. J. Virol. *66*, 3643-3651.

Llano, M., Guma, M., Ortega, M., Angulo, A., and Lopez-Botet, M. (2003). Differential effects of US2, US6 and US11 human cytomegalovirus proteins on HLA class Ia and HLA-E expression: impact on target susceptibility to NK cell subsets. Eur. J. Immunol. *33*, 2744-2754.

Llano, M., Lee, N., Navarro, F., Garcia, P., Albar, J. P., Geraghty, D. E., and Lopez-Botet, M. (1998). HLA-E-bound peptides influence recognition by inhibitory and triggering CD94/NKG2 receptors: preferential response to an HLA-G-derived nonamer. Eur. J. Immunol. *28*, 2854-2863.

Lodoen, M., Ogasawara, K., Hamerman, J. A., Arase, H., Houchins, J. P., Mocarski, E. S., and Lanier, L. L. (2003). NKG2D-mediated natural killer cell protection against cytomegalovirus is impaired by viral gp40 modulation of retinoic acid early inducible 1 gene molecules. J. Exp. Med. *197*, 1245-1253.

Long, E. O., Burshtyn, D. N., Clark, W. P., Peruzzi, M., Rajagopalan, S., Rojo, S., Wagtmann, N., and Winter, C. C. (1997). Killer cell inhibitory receptors: diversity, specificity, and function. Immunol. Rev. *155*, 135-144.

Luttichau, H. R., Stine, J., Boesen, T. P., Johnsen, A. H., Chantry, D., Gerstoft, J., and Schwartz, T. W. (2000). A highly selective CC chemokine receptor (CCR)8 antagonist encoded by the poxvirus molluscum contagiosum. J. Exp. Med. *191*, 171-180.

Ljunggren, H. G., and Karre, K. (1990). In search of the 'missing self': MHC molecules and NK cell recognition. Immunol. Today *11*, 237-244.

Malarkannan, S., Shih, P. P., Eden, P. A., Horng, T., Zuberi, A. R., Christianson, G., Roopenian, D., and Shastri, N. (1998). The molecular and functional characterization of a dominant minor H antigen, H60. J. Immunol. *161*, 3501-3509.

Mandelboim, O., Lieberman, N., Lev, M., Paul, L., Arnon, T. I., Bushkin, Y., Davis, D. M., Strominger, J. L., Yewdell, J. W., and Porgador, A. (2001). Recognition of haemagglutinins on virus-infected cells by NKp46 activates lysis by human NK cells. Nature *409*, 1055-1060.

MartIn-Fontecha, A., Sebastiani, S., Hopken, U. E., Uguccioni, M., Lipp, M., Lanzavecchia, A., and Sallusto, F. (2003). Regulation of dendritic cell migration to the draining lymph node: impact on T lymphocyte traffic and priming. J. Exp. Med. *198*, 615-621.

Means, R. E., Choi, J. K., Nakamura, H., Chung, Y. H., Ishido, S., and Jung, J. U. (2002). Immune evasion strategies of Kaposi's sarcoma-associated herpesvirus. Curr. Top. Microbiol. Immunol. *269*, 187-201.

Medzhitov, R., Preston-Hurlburt, P., and Janeway, C. A., Jr. (1997). A human homologue of the Drosophila Toll protein signals activation of adaptive immunity. Nature *388*, 394-397.

Medzhitov, R., Preston-Hurlburt, P., Kopp, E., Stadlen, A., Chen, C., Ghosh, S., and Janeway, C. A., Jr. (1998). MyD88 is an adaptor protein in the hToll/IL-1 receptor family signaling pathways. Mol. Cell. *2*, 253-258.

Menard, C., Wagner, M., Ruzsics, Z., Holak, K., Brune, W., Campbell, A. E., and Koszinowski, U. H. (2003). Role of murine cytomegalovirus US22 gene family members in replication in macrophages. J. Virol. *77*, 5557-5570.

Merrill, J. E., Ullberg, M., and Jondal, M. (1981). Influence of IgG and IgM receptor triggering on human natural killer cell cytotoxicity measured on the level of the single effector cell. Eur. J. Immunol. *11*, 536-541.

Michaelsson, J., Teixeira de Matos, C., Achour, A., Lanier, L. L., Karre, K., and Soderstrom, K. (2002). A signal peptide derived from hsp60 binds HLA-E and interferes with CD94/NKG2A recognition. J. Exp. Med. *196*, 1403-1414.

Moore, K. W., O'Garra, A., de Waal Malefyt, R., Vieira, P., and Mosmann, T. R. (1993). Interleukin-10. Annu. Rev. Immunol. *11*, 165-190.

Moore, K. W., Vieira, P., Fiorentino, D. F., Trounstine, M. L., Khan, T. A., and Mosmann, T. R. (1990). Homology of cytokine synthesis inhibitory factor (IL-10) to the Epstein-Barr virus gene BCRFI. Science *248*, 1230-1234.

Moretta, A. (2002). Natural killer cells and dendritic cells: rendezvous in abused tissues. Nat. Rev. Immunol. *2*, 957-964.

Moretta, A., Bottino, C., Vitale, M., Pende, D., Cantoni, C., Mingari, M. C., Biassoni, R., and Moretta, L. (2001). Activating receptors and coreceptors involved in human natural killer cell-mediated cytolysis. Annu. Rev. Immunol. *19*, 197-223.

Nagata, S., and Golstein, P. (1995). The Fas death factor. Science *267*, 1449-1456.

Orange, J. S., and Biron, C. A. (1996). Characterization of early IL-12, IFN-alphabeta, and TNF effects on antiviral state and NK cell responses during murine cytomegalovirus infection. J. Immunol. *156*, 4746-4756.

Orange, J. S., Fassett, M. S., Koopman, L. A., Boyson, J. E., and Strominger, J. L. (2002). Viral evasion of natural killer cells. Nat. Immunol. *3*, 1006-1012.

Parolini, S., Bottino, C., Falco, M., Augugliaro, R., Giliani, S., Franceschini, R., Ochs, H. D., Wolf, H., Bonnefoy, J. Y., Biassoni, R., et al. (2000). X-linked lymphoproliferative disease. 2B4 molecules displaying inhibitory rather than activating function are responsible for the inability of natural killer cells to kill Epstein-Barr virus-infected cells. J. Exp. Med. *192*, 337-346.

Parry, C. M., Simas, J. P., Smith, V. P., Stewart, C. A., Minson, A. C., Efstathiou, S., and Alcami, A. (2000). A broad spectrum secreted chemokine binding protein encoded by a herpesvirus. J. Exp. Med. *191*, 573-578.

Peng, Y., Falck-Pedersen, E., and Elkon, K. B. (2001). Variation in adenovirus transgene expression between BALB/c and C57BL/6 mice is associated with differences in interleukin-12 and gamma interferon production and NK cell activation. J. Virol. *75*, 4540-4550.

Petersen, J. L., Morris, C. R., and Solheim, J. C. (2003). Virus evasion of MHC class I molecule presentation. J. Immunol. *171*, 4473-4478.

Petrovskis, E. A., Timmins, J. G., and Post, L. E. (1986). Use of lambda gt11 to isolate genes for two pseudorabies virus glycoproteins with homology to herpes simplex virus and varicella-zoster virus glycoproteins. J. Virol. *60*, 185-193.

Piccioli, D., Sbrana, S., Melandri, E., and Valiante, N. M. (2002). Contact-dependent stimulation and inhibition of dendritic cells by natural killer cells. J. Exp. Med. *195*, 335-341.

Plougastel, B., Matsumoto, K., Dubbelde, C., and Yokoyama, W. M. (2001). Analysis of a 1-Mb BAC contig overlapping the mouse Nkrp1 cluster of genes: cloning of three new Nkrp1 members, Nkrp1d, Nkrp1e, and Nkrp1f. Immunogenetics *53*, 592-598.

Poggi, A., Carosio, R., Spaggiari, G. M., Fortis, C., Tambussi, G., Dell'Antonio, G., Dal Cin, E., Rubartelli, A., and Zocchi, M. R. (2002). NK cell activation by dendritic cells is dependent on LFA-1-mediated induction of calcium-calmodulin kinase II: inhibition by HIV-1 Tat C-terminal domain. J. Immunol. *168*, 95-101.

Raulet, D. H. (2003a). Natural Killer Cells, In Fundamental Immunology, W. E. Paul, ed. (Philadelphia: Lippincott Williams and Wilkins), pp. 365-391.

Raulet, D. H. (2003b). Roles of the NKG2D immunoreceptor and its ligands. Nat. Rev. Immunol. *3*, 781-790.

Raulet, D. H., Vance, R. E., and McMahon, C. W. (2001). Regulation of the natural killer cell receptor repertoire. Annu. Rev. Immunol. *19*, 291-330.

Ravetch, J. V., and Bolland, S. (2001). IgG Fc receptors. Annu. Rev. Immunol. *19*, 275-290.

Reyburn, H. T., Mandelboim, O., Vales-Gomez, M., Davis, D. M., Pazmany, L., and Strominger, J. L. (1997). The class I MHC homologue of human cytomegalovirus inhibits attack by natural killer cells. Nature *386*, 514-517.

Riteau, B., Barber, D. F., and Long, E. O. (2003). Vav1 phosphorylation is induced by beta2 integrin engagement on natural killer cells upstream of actin cytoskeleton and lipid raft reorganization. J. Exp. Med. *198*, 469-474.

Robertson, M. J. (2002). Role of chemokines in the biology of natural killer cells. J. Leukoc. Biol. *71*, 173-183.

Robinson, D., Shibuya, K., Mui, A., Zonin, F., Murphy, E., Sana, T., Hartley, S. B., Menon, S., Kastelein, R., Bazan, F., and O'Garra, A. (1997). IGIF does not drive Th1 development but synergizes with IL-12 for interferon-gamma production and activates IRAK and NFkappaB. Immunity *7*, 571-581.

Roulston, A., Marcellus, R. C., and Branton, P. E. (1999). Viruses and apoptosis. Annu. Rev. Microbiol. *53*, 577-628.

Saederup, N., Aguirre, S. A., Sparer, T. E., Bouley, D. M., and Mocarski, E. S. (2001). Murine cytomegalovirus CC chemokine homolog MCK-2 (m131-129) is a determinant of dissemination that increases inflammation at initial sites of infection. J. Virol. *75*, 9966-9976.

Salazar-Mather, T. P., Orange, J. S., and Biron, C. A. (1998). Early murine cytomegalovirus (MCMV) infection induces liver natural killer (NK) cell inflammation and protection through macrophage inflammatory protein 1alpha (MIP-1alpha)-dependent pathways. J. Exp. Med. *187*, 1-14.

Samuel, C. E. (2001). Antiviral actions of interferons. Clin. Microbiol. Rev. *14*, 778-809, table of contents.

Sarin, A., Williams, M. S., Alexander-Miller, M. A., Berzofsky, J. A., Zacharchuk, C. M., and Henkart, P. A. (1997). Target cell lysis by CTL granule exocytosis is independent of ICE/Ced-3 family proteases. Immunity *6*, 209-215.

Scalzo, A. A. (2002). Successful control of viruses by NK cells--a balance of opposing forces? Trends. Microbiol. *10*, 470-474.

Scalzo, A. A., Fitzgerald, N. A., Simmons, A., La Vista, A. B., and Shellam, G. R. (1990). Cmv-1, a genetic locus that controls murine cytomegalovirus replication in the spleen. J. Exp. Med. *171*, 1469-1483.

Scalzo, A. A., Fitzgerald, N. A., Wallace, C. R., Gibbons, A. E., Smart, Y. C., Burton, R. C., and Shellam, G. R. (1992). The effect of the Cmv-1 resistance gene, which is linked to the natural killer cell gene complex, is mediated by natural killer cells. J. Immunol. *149*, 581-589.

Schust, D. J., Tortorella, D., Seebach, J., Phan, C., and Ploegh, H. L. (1998). Trophoblast class I major histocompatibility complex (MHC) products are resistant to rapid degradation imposed by the human cytomegalovirus (HCMV) gene products US2 and US11. J. Exp. Med. *188*, 497-503.

Senkevich, T. G., and Moss, B. (1998). Domain structure, intracellular trafficking, and beta2-microglobulin binding of a major histocompatibility complex class I homolog encoded by molluscum contagiosum virus. Virology *250*, 397-407.

Shisler, J., Yang, C., Walter, B., Ware, C. F., and Gooding, L. R. (1997). The adenovirus E3-10.4K/14.5K complex mediates loss of cell surface Fas (CD95) and resistance to Fas-induced apoptosis. J. Virol. *71*, 8299-8306.

Sivori, S., Parolini, S., Falco, M., Marcenaro, E., Biassoni, R., Bottino, C., Moretta, L., and Moretta, A. (2000). 2B4 functions as a co-receptor in human NK cell activation. Eur. J. Immunol. *30*, 787-793.

Smith, H. R., Heusel, J. W., Mehta, I. K., Kim, S., Dorner, B. G., Naidenko, O. V., Iizuka, K., Furukawa, H., Beckman, D. L., Pingel, J. T., et al. (2002a). Recognition of a virus-encoded ligand by a natural killer cell activation receptor. Proc. Natl. Acad. Sci. USA. *99*, 8826-8831.

Smith, H. R., Heusel, J. W., Mehta, I. K., Kim, S., Dorner, B. G., Naidenko, O. V., Iizuka, K., Furukawa, H., Beckman, D. L., Pingel, J. T., et al. (2002b). Recognition of a virus-encoded ligand by a natural killer cell activation receptor. Proc. Natl. Acad. Sci. USA. *99*, 8826-8831.

Tabeta, K., Georgel, P., Janssen, E., Du, X., Hoebe, K., Crozat, K., Mudd, S., Shamel, L., Sovath, S., Goode, J., et al. (2004). Toll-like receptors 9 and 3 as essential components of innate immune defense against mouse cytomegalovirus infection. Proc. Natl. Acad. Sci. USA. *101*, 3516-3521.

Takeda, K., Tsutsui, H., Yoshimoto, T., Adachi, O., Yoshida, N., Kishimoto, T., Okamura, H., Nakanishi, K., and Akira, S. (1998). Defective NK cell activity and Th1 response in IL-18-deficient mice. Immunity *8*, 383-390.

Tewari, M., and Dixit, V. M. (1995). Fas- and tumor necrosis factor-induced apoptosis is inhibited by the poxvirus crmA gene product. J. Biol. Chem. *270*, 3255-3260.

Thäle, R., Lucin, P., Schneider, K., Eggers, M., and Koszinowski, U. H. (1994). Identification and expression of a murine cytomegalovirus early gene coding for an Fc receptor. J. Virol. *68*, 7757-7765.

Tieng, V., Le Bouguenec, C., du Merle, L., Bertheau, P., Desreumaux, P., Janin, A., Charron, D., and Toubert, A. (2002). Binding of *Escherichia coli* adhesin AfaE to CD55 triggers cell-surface expression of the MHC class I-related molecule MICA. Proc. Natl. Acad. Sci. USA. *99*, 2977-2982.

Tollefson, A. E., Hermiston, T. W., Lichtenstein, D. L., Colle, C. F., Tripp, R. A., Dimitrov, T., Toth, K., Wells, C. E., Doherty, P. C., and Wold, W. S. (1998). Forced degradation of Fas inhibits apoptosis in adenovirus-infected cells. Nature *392*, 726-730.

Tomasec, P., Braud, V. M., Rickards, C., Powell, M. B., McSharry, B. P., Gadola, S., Cerundolo, V., Borysiewicz, L. K., McMichael, A. J., and Wilkinson, G. W. (2000). Surface expression of HLA-E, an inhibitor of natural killer cells, enhanced by human cytomegalovirus gpUL40. Science *287*, 1031.

Trapani, J. A., and Smyth, M. J. (2002). Functional significance of the perforin/granzyme cell death pathway. Nat. Rev. Immunol. *2*, 735-747.

Trinchieri, G. (1989). Biology of natural killer cells. Adv. Immunol. *47*, 187-376.

Trinchieri, G. (2003). Interleukin-12 and the regulation of innate resistance and adaptive immunity. Nat. Rev. Immunol. *3*, 133-146.

Trowsdale, J., Barten, R., Haude, A., Stewart, C. A., Beck, S., and Wilson, M. J. (2001). The genomic context of natural killer receptor extended gene families. Immunol. Rev. *181*, 20-38.

Tseng, C. T., and Klimpel, G. R. (2002). Binding of the hepatitis C virus envelope protein E2 to CD81 inhibits natural killer cell functions. J. Exp. Med. *195*, 43-49.

Ulevitch, R. J. (2004). Therapeutics targeting the innate immune system. Nat. Rev. Immunol. *4*, 512-520.

Vance, R. E., Jamieson, A. M., and Raulet, D. H. (1999). Recognition of the class Ib molecule Qa-1(b) by putative activating receptors CD94/NKG2C and CD94/NKG2E on mouse natural killer cells. J. Exp. Med. *190*, 1801-1812.

Vance, R. E., Kraft, J. R., Altman, J. D., Jensen, P. E., and Raulet, D. H. (1998). Mouse CD94/NKG2A is a natural killer cell receptor for the nonclassical major histocompatibility complex (MHC) class I molecule Qa-1(b). J. Exp. Med. *188*, 1841-1848.

Vester, B., Muller, K., Solbach, W., and Laskay, T. (1999). Early gene expression of NK cell-activating chemokines in mice resistant to Leishmania major. Infect. Immun. *67*, 3155-3159.

Vilches, C., and Parham, P. (2002). KIR: diverse, rapidly evolving receptors of innate and adaptive immunity. Annu. Rev. Immunol. *20*, 217-251.

Vivier, E., and Daeron, M. (1997). Immunoreceptor tyrosine-based inhibition motifs. Immunol. Today *18*, 286-291.

Voigt, V., Forbes, C. A., Tonkin, J. N., Degli-Esposti, M. A., Smith, H. R., Yokoyama, W. M., and Scalzo, A. A. (2003). Murine cytomegalovirus m157 mutation and variation leads to immune evasion of natural killer cells. Proc. Natl. Acad. Sci. USA. *100*, 13483-13488.

Wack, A., Soldaini, E., Tseng, C., Nuti, S., Klimpel, G., and Abrignani, S. (2001). Binding of the hepatitis C virus envelope protein E2 to CD81 provides a co-stimulatory signal for human T cells. Eur. J. Immunol. *31*, 166-175.

Wagner, M., Gutermann, A., Podlech, J., Reddehase, M. J., and Koszinowski, U. H. (2002). Major histocompatibility complex class I allele-specific cooperative and competitive interactions between immune evasion proteins of cytomegalovirus. J. Exp. Med. *196*, 805-816.

Wang, E. C., McSharry, B., Retiere, C., Tomasec, P., Williams, S., Borysiewicz, L. K., Braud, V. M., and Wilkinson, G. W. (2002). UL40-mediated NK evasion during productive infection with human cytomegalovirus. Proc. Natl. Acad. Sci. USA. *99*, 7570-7575.

Watzl, C. (2003). The NKG2D receptor and its ligands-recognition beyond the "missing self"? Microbes. Infect. *5*, 31-37.

Welsh, R. M., Brubaker, J. O., Vargas-Cortes, M., and O'Donnell, C. L. (1991). Natural killer (NK) cell response to virus infections in mice with severe combined immunodeficiency. The stimulation of NK cells and the NK cell-dependent control of virus infections occur independently of T and B cell function. J. Exp. Med. *173*, 1053-1063.

Williams, N. S., Kubota, A., Bennett, M., Kumar, V., and Takei, F. (2000). Clonal analysis of NK cell development from bone marrow progenitors *in vitro*: orderly acquisition of receptor gene expression. Eur. J. Immunol. *30*, 2074-2082.

Wu, J., Chalupny, N. J., Manley, T. J., Riddell, S. R., Cosman, D., and Spies, T. (2003). Intracellular retention of the MHC class I-related chain B ligand of NKG2D by the human cytomegalovirus UL16 glycoprotein. J. Immunol. *170*, 4196-4200.

Xiang, Y., and Moss, B. (1999). IL-18 binding and inhibition of interferon gamma induction by human poxvirus-encoded proteins. Proc. Natl. Acad. Sci. USA. *96*, 11537-11542.

Yokoyama, W. M., Ryan, J. C., Hunter, J. J., Smith, H. R., Stark, M., and Seaman, W. E. (1991). cDNA cloning of mouse NKR-P1 and genetic linkage with LY-49. Identification of a natural killer cell gene complex on mouse chromosome 6. J. Immunol. *147*, 3229-3236.

York, I. A., and Johnson, D. C. (1993). Direct contact with herpes simplex virus-infected cells results in inhibition of lymphokine-activated killer cells because of cell-to-cell spread of virus. J. Infect. Dis. *168*, 1127-1132.

Ziegler, H., Thäle, R., Lucin, P., Muranyi, W., Flohr, T., Hengel, H., Farrell, H., Rawlinson, W., and Koszinowski, U. H. (1997). A mouse cytomegalovirus glycoprotein retains MHC class I complexes in the ERGIC/cis-Golgi compartments. Immunity *6*, 57-66.

Zocchi, M. R., Rubartelli, A., Morgavi, P., and Poggi, A. (1998). HIV-1 Tat inhibits human natural killer cell function by blocking L-type calcium channels. J. Immunol. *161*, 2938-2943.

Zou, Z., Nomura, M., Takihara, Y., Yasunaga, T., and Shimada, K. (1996). Isolation and characterization of retinoic acid-inducible cDNA clones in F9 cells: a novel cDNA family encodes cell surface proteins sharing partial homology with MHC class I molecules. J. Biochem. (Tokyo) *119*, 319-328.

3

Mucosal Innate Immune Response

*Barbara Fernie-King**

Abstract

The innate immune system provides the first line of defence against invading microorganisms. Most infectious agents infect their host through a mucosal surface, be it the digestive, respiratory or reproductive tract. The innate immune system comprises cellular activities stimulated by the recognition of pathogens by pattern recognition receptors (PRRs) in response to pathogen-associated molecular patterns (PAMPS) together with a multitude of secreted soluble antimicrobial factors. In order to protect themselves from this array of defences and to promote survival within the host, microorganisms have developed a number of different counter measures. These include modification of bacterial cell membranes or cell walls, mimicry of host cell membrane lipids, secretion of proteins or enzymes that inhibit or degrade antimicrobial factors, and down-regulation of expression of antimicrobial substances. This chapter describes the various strategies adopted by both the host and invader on the battleground of mucosal surfaces.

1. Introduction

The innate immune system is extremely ancient being present not only in vertebrates but also in plants, insects and other invertebrates. It provides the first line of defence against invading microorganisms. The mucosal surfaces of the human body, which include the upper respiratory tract and lungs, the gastro-intestinal system, the eyes and the reproductive tract, are continuously exposed to challenge by both commensal and potentially pathogenic microorganisms. However, there is only a low incidence of infection as the mucosal innate immune system contains an array of different defence mechanisms, both chemical and mechanical, which create a hostile environment for microorganisms until antigen-specific cells are recruited to the site.

As implied by the name, all mucosal surfaces are covered in a layer of mucus, and some are also ciliated. The mucosal innate immune system expresses a wide variety

*For correspondence email baf22@cam.ac.uk

of antimicrobial proteins and peptides (AMPs) that are present in the fluid phase and which include lysozyme (the most abundant), secretory leukocyte proteinase inhibitor (SLPI), elafin, α and β defensins, the cathelicidin LL-37, lactoferrin and lactoferricin, lactoperoxidase (and several other milk proteins), lung surfactant proteins A and D, Group II phospholipase A2 ($SPLA_2$), trefoil factor family (TFF) peptides, complement in gingival crevicular fluid, adrenomedullin, and the salivary histatins, some of which act together synergistically. Some of these are present at most mucosal surfaces, while others are only expressed at one particular site; some are constitutively expressed, and others are expressed in response to microbial challenge. Another important facet of mucosal innate immunity is the recognition of microorganisms as "foreign" and this is achieved through the action of "pattern recognition receptors" (PRRs) in response to "pathogen-associated molecular patterns" (PAMPS).

1.1. Antimicrobial proteins and peptides in the fluid phase

1.1.1. Lysozyme

Lysozyme was the first antimicrobial enzyme to be described, having been discovered by Alexander Fleming in 1922 (Fleming, 1922), and is present at high concentrations on almost all mucosal surfaces. Hen egg lysozyme has since become one of the most studied and best characterised globular proteins and was the first enzyme to have its structure determined by X-ray diffraction (Blake et al., 1965). Lysozyme exerts is antimicrobial activity by hydrolyzing β-1, 4 linkages of *N*-acetylmuramic acid and 2-acetylamino-2-deoxy-D-glucose residues of Gram-positive bacterial cell wall peptidoglycan (PGN). It acts synergistically with lactoferrin (Ellison and Giehl, 1991), $SPLA_2$ (Buckland et al., 2000) and SLPI (Singh et al., 2000) but the combination of lysozyme with human β-defensins or LL-37 is only additive (Singh et al., 2000). Killing of Gram negative bacteria by lysozyme is dependant on the disruption of lipopolysaccharide (LPS) outer membrane by lactoferrin in order for lysozyme to gain access to the inner PGN (Ellison and Giehl, 1991). It has also been suggested that certain sub-units of lysozyme may act non-enzymatically, and are antimicrobial in a manner similar to the small cationic antimicrobial peptides like the defensins (see below and reviewed in Masschalck and Michiels, 2003). Lysozyme has also been shown to inhibit the growth of HIV *in vitro* but the mechanism of antiviral activity is not yet known (Lee-Huang et al., 1999).

1.1.2. Lactoferrin and other milk proteins

Lactoferrin is an avid iron-binding protein found in several mucosal surface fluids, including breast milk, saliva, seminal fluid and the endometrium, but most abundantly in human colostrum where the concentration can be up to 7 mg/ml, and in tears up to 2 mg/ml (reviewed in Farnaud and Evans, 2003). The antimicrobial (bacteriostatic) activity of lactoferrin was originally attributed to its ability to sequester iron and thus deprive microorganisms of an essential nutrient (Arnold et al., 1977), but it was subsequently shown to be directly bactericidal in its iron-free state (apolactoferrin) (Arnold et al., 1982). It is now established that lactoferrin interacts directly with the LPS of Gram-negative bacteria causing release of LPS from the membrane, but that this interaction can be blocked by the addition of $Ca^{2}+$ and $Mg^{2}+$ ions (Appelmelk et al., 1994). Lactoferrin also has anti-fungal activity, and may act synergistically with anti-fungal drugs (Kuipers et al., 1999). It has anti-viral activity and inhibits replication of a wide range of viruses probably by preventing infection of the host cell by direct binding to virus particles, rather than inhibiting virus replication after the target cell has been infected (reviewed in van der Strate et al., 2001). Another mechanism for this activity has been shown to be by binding to host cell heparin sulphate and chondroitin sulphate that the virus uses as a receptor, thus blocking viral entry into the cell (Marchetti et al., 2004). Lactoferrin-derived peptides such as lactoferricin have been shown to have greater antimicrobial activity than the parent protein (Farnaud and Evans, 2003).

Other antimicrobial milk proteins (reviewed in Lönnerdahl, 2003), which probably mainly exert their effect in the gastrointestinal tract of the newborn infant, include:

(a) κ-casein - a minor casein submit that is thought to prevent adhesion/attachment of bacteria to the mucosal lining by acting as a receptor analogue for similar surface-exposed carbohydrate structures on the gastric mucosa;
(b) lactoperoxidase (also produced in the salivary glands) - in the presence of hydrogen peroxide, which is formed in small quantities by cells, lactoperoxidase catalyses the oxidation of thiocyanate (a constituent of saliva), forming hypothiocyanate which can kill both Gram-positive and negative bacteria and may thus contribute to the defence of the mouth and upper gastrointestinal tract.
(c) haptocorrin - this is primarily thought of as a Vitamin B12 binder, but it is very unsaturated in human milk. The antibacterial mechanism involved was originally thought to be by binding and withholding Vitamin B12 from microorganisms, but it has since been found to inhibit the growth of *E. coli* independently of the presence/absence of Vitamin B12. How this occurs is still unknown;
(d) α-lactalbumin - proteolytic fragments generated by proteases present in the gut are active against a variety of bacteria but the mechanism is as yet unknown.

1.1.3. Secretory leukocyte proteinase inhibitor (SLPI) and elafin (elastase-specific inhibitor)

These two related proteins belong to the four-disulphide core (FDC) domain family that, as well as functioning as protease inhibitors, have antimicrobial activity (Hiemstra et al., 1996; Simpson et al., 1999). SLPI and elafin are constitutively expressed at most mucosal sites including the upper and lower respiratory, intestinal, male and female reproductive tracts. SLPI is an 11.7 kDa protein, consisting of 107 amino acids and contains two very similar FDC domains. The N-terminal domain of SLPI contains the antimicrobial function (Hiemstra et al., 1996), and the C-terminal domain the antiproteinase activity (Eisenberg et al., 1990). Elafin contains one FDC domain (the C-terminal domain), which contains both the antiproteinase and antimicrobial functions, together with another unrelated N-terminal domain involved in anchoring the protein to epithelial surfaces via transglutamination (Nara et al., 1994). SLPI and elafin have been shown to have antimicrobial activity against both Gram-positive and Gram-negative bacteria (Hiemstra et al., 1996; Simpson et al., 1999; Fernie-King et al., 2002), fungi (Tomee et al., 1997) and HIV (McNeely et al., 1995) but the exact mode of action is as yet unknown.

1.1.4. Defensins

The defensins are a group of small cationic peptides, a number of which are expressed on many mucosal surfaces, and which are characterised by containing six conserved cysteine residues which form specific disulphide bridges. They are divided into two sub-families, the α and β defensins, based on the arrangement of the disulphide bridges, and both groups have a characteristic β-sheet structure. The amino acid sequences of defensins are highly variable except for the conserved cysteine framework in each sub-family. The α-defensin group consist of six peptides that are found in neutrophils and in Paneth cells of the small intestine, whereas the β-defensin group consists of four peptides which are synthesised by epithelial cells in a wide range of tissues, including the airways (Schutte and McCray, 2002). It is likely that several more defensin peptides exist, that are still uncharacterised, as there are a large number of defensin genes located in a cluster on chromosome 8p23, and which differ in number between individuals. Some defensins are constitutively expressed, whilst others are expressed in response to inflammatory stimuli or microbial challenge. Human neutrophil defensins (α-defensins) are constitutively synthesised by the bone marrow precursors of neutrophils and stored in primary (azurophil) granules prior to release into the blood and tissues. The synthesis of α-defensins HD5 and HD6, is probably also constitutive. They are specific to Paneth cells, from where they are released into the lumen of intestinal crypts after exposure to bacteria etc. Most defensins are microbicidal against bacteria and fungi, especially under low ionic strength conditions, and some are active against certain enveloped viruses. The mode of action of defensins is by permeabilisation of target membranes, but the exact method is as yet unknown. Insertion of the defensin molecule into the negatively charged microbial membrane is certainly facilitated by the cationic nature of defensins. For an extensive review of the defensin family see Ganz (2003) or Lehrer (2004); and for more specific papers on defensins in the gut see Cunliffe (2003) and Dhaliwal et al. (2003).

1.1.5. LL-37

The single human cathelicidin LL-37 (or hCAP18) is a salt-sensitive, cationic, α-helical, 37 amino acid peptide with no disulphide bonds which is expressed on a number of mucosal epithelial surfaces including saliva (Murakami et al., 2002), the upper and lower airways (Bals et al., 1998) and the gastric mucosae (Hase et al., 2002). It has a broad spectrum of antimicrobial activity against a range of organisms including Gram-positive and Gram-negative bacteria, fungi, parasites and enveloped viruses (reviewed by Nizet and Gallo 2003). LL-37 has been shown to act synergistically with lysozyme, but only in an additive manner with hBD-1 and hBD-2 (Singh et al., 2000). Its mode of action is by interaction with, and disruption of, the anionic bacterial cell membrane by a toroidal pore mechanism (Henzler Wildman et al., 2003). Expression is up-regulated in the nasal mucosae of patients with chronic rhinitis (Kim et al., 2003), high concentrations of LL-37 are found in the bronchoalveolar lavage (BAL) fluid of infants with pneumonia (Schaller-Bals et al., 2002) and expression by the gastric epithelium is up-regulated in response to infection with *Helicobacter pylori* (Hase et al., 2003). LL-37 has also been shown to have a number of other functions including up-regulation of epithelial expression of chemokines and chemokine receptor genes, and also acts as a chemoattractant for neutrophils, monocytes, T-lymphocytes and mast cells (reviewed in Nizet and Gallo, 2003).

1.1.6. Histatins

The histatins are a family of twelve salivary proteins, the majority of which are degradation fragments of histatin 1 and 3, that are rich in arginine, histidine and lysine residues. They are small in size, approximately 3-4 kDa, and are constitutively synthesised by the ductal cells of human parotid and submandibular salivary glands (reviewed in van Nieuw Amerongen and Veerman, 2002). They have been shown to be microbicidal against several bacteria, including *Streptococcus mutans* (Mackay et al., 1984), and have potent candidacidal activity. The mechanism by which the histatins kill *Candida albicans* is a multistep process initiated by binding to a specific receptor on the cell surface, followed by internalisation, interaction with mitochondria and release of cellular ATP (Fitzgerald et al., 2003).

1.1.7. Complement

On normal uninflamed mucous surfaces there is no functioning complement system. Indeed, normal saliva and normal intestinal juice contain C3 binding lectins which inhibit C3 function, and which, in turn are easily neutralised by C3. These lectins therefore are a marker for the absence of C3. The lectins circulate as large complexes containing mucin and also IgA (Lachmann and Thompson, 1970; Eggert, 1980a; 1980b; 1980c; 1980d). These complexes also have some activity in agglutinating bacteria. This seems to represent some cross-reaction between carbohydrates on these bacteria, for example *Streptococcus mitis*, and on C3. Their functional importance is not known.

Complement is found in the crevicular fluid around the teeth. This resembles dilute plasma. When there is inflammation around the teeth, complement-mediated reactions may well occur and the same subversion mechanisms may come into play that are described Chapter 1. There are suggestions that where crevicular fluid encounters saliva, some potentiation of the complement system may occur and this maybe helpful for the eradication of microorganisms (Boackle, 1991).

Intestinal juice may contain a further anti-complementary factor. This is the intestinal glycoprotein which was shown to activate complement (Winsnes, 1980). It does so by binding Factor H (Winsnes and Lachmann, 1982). However, unlike the binding of functional Factor H to the surface of bacteria, this binding by the intestinal glycoprotein inactivates the Factor H and produces what is essentially a Factor H depleted serum allowing the uncontrolled activation of the alternative pathway. This is an *in vitro* phenomenon and, again, its physiological or pathophysiological role is unknown.

1.1.8. Lung surfactant proteins A and D (SP-A and SP-D)

SP-A and SP-D are members of the collectin protein family, whose best known function is their ability to directly opsonize pathogens (by acting as pattern recognition receptors for microbial carbohydrates) so that they are phagocytosed by macrophages and monocytes. Collectins are built of structural subunits each composed of three identical, or almost identical, polypeptide chains (the carbohydrate recognition domains) attached to a collagenous region. By electron microscopy, SP-A appears to have a central knob, probably formed by the N-terminal interlinking regions, with the subunits radiating out like a small umbel, whereas SP-D has a cruciform structure. SP-D may form oligomers resembling a cart wheel. Both molecules are produced and secreted onto the epithelial surface of the lung by alveolar type II cells and unciliated bronchial epithelial cells. In humans, SP-A activity is restricted to the lung and serous glands of the proximal trachea, whereas SP-D is also found in some epithelial cells of the gastrointestinal tract (reviewed in Holmskov et al, 2003). SP-A and SP-D bind to bacterial cell wall components like lipoteichoic acid (LTA) of Gram-positive bacteria and LPS, and SP-D expression by the gastric mucosa is up regulated in *H. pylori*-associated gastritis compared to that in the normal mucosa, resulting in agglutination and reduced motility of the bacteria (Murray et al., 2002). SP-D also strongly agglutinates viruses, for instance respiratory syncytial virus (LeVine et al., 2004), which could lead to mucociliary clearance of viruses. Both SP-A and SP-D bind to a number of different fungi that can cause pulmonary infections, and directly inhibit the growth of *Histoplasma capsulatum* (McCormack et al., 2003). SP-A and SP-D have also recently been shown to be directly antimicrobial and are able to inhibit growth of Gram-negative bacteria in the absence of immune effector cells, and independent of collectin-mediated aggregation of bacteria, by increasing permeability of the bacterial cell membrane (Wu et al., 2003). The exact mechanism by which they achieve this is as yet unknown.

1.1.9. Group II phospholipase A2 ($SPLA_2$)

$SPLA_2$ is a highly cationic 13.9 kDa protein which hydrolyses the phosphatidylglycerol of bacterial cell membranes several hundred times better than the phosphatidylcholine of mammalian cells. $SPLA_2$ is secreted from a number of different cells such as macrophages, the Paneth cells of the intestine and the cells of the lacrimal duct. It is also present in seminal plasma, rheumatoid synovial fluid, platelets and neutrophils (reviewed in Buckland and Wilton, 2000). It is thought to be the principal bactericide in tears for Gram-positive bacteria, where it is present at greater than 30 μg/ml (Qu and Lehrer, 1998). For some species of Gram-positive bacteria, $SPLA_2$ requires the presence of lysozyme to disrupt the PGN cell wall and the two enzymes are considered to act synergistically (Buckland et al., 2000). Its action against Gram-negative bacteria is dependent on bactericidal-permeability-increasing protein (which is a 55 kDa basic protein found in azurophilic granules of PMNs and on the surface of neutrophils) first disrupting the LPS so that the cell membrane becomes accessible (Weiss et al., 1994).

1.1.10. Trefoil factor family peptides (TFFs)

TFFs are major secretory products of normal mucus epithelia whose primary function appears to be as healing factors in conjunction with epidermal growth factor. They have been shown *in vitro* to protect the intestinal epithelial barrier from damage via a co-operative interaction with mucin glycoprotein (Kindon et al., 1995). They have also been shown to bind to the surface *of Staphylococcus aureus* and *Pseudomonas aeruginosa* in sections and sputum from infected cystic fibrosis airways, leading the authors to speculate that TFFs may exert influence on adhesion to host cells or phagocytosis and contribute to the antimicrobial mucociliary defence system (dos Santos Silva et al., 2000). *H. pylori* (but not *Campylobacter jejuni*) has also been shown to bind to TFF1 and subsequently to gastric mucin, suggesting that TFF1 may act as a receptor for the organism *in vivo* (Clyne et al., 2004). In contrast, the authors conclude that this interaction may account for the tissue tropism observed for this organism and thus act as an aid to colonisation.

1.1.11. Adrenomedullin

Adrenomedullin (AM) is found in picomolar concentrations in plasma and up to 6-fold higher concentrations in saliva. It is constitutively expressed by a wide variety of tissues and cells (including epithelial), and has been shown to accumulate at the apical regions of normal human bronchial epithelium. It exhibits antimicrobial activity against a wide range of oral, respiratory tract, gut and skin pathogens, including *S. aureus, Porphyromonas gingivalis, S. mutans, Streptococcus pneumoniae, Streptococcus pyogenes, Haemophilus influenzae* and *H. pylori*, some being more sensitive than others (Allaker et al., 1999). AM has 30% homology at the genetic level to the cecropin group of antimicrobial peptides and therefore might have a similar mechanism of action, involving channel formation in membranes and subsequent lysis. It has subsequently been shown to be present on the surface of the colonic mucosa and to be bactericidal against *Escherichia coli.*

1.2. Mechanical means of elimination

1.2.1. Mucus and cilia

The initial process of clearance of pathogens from the upper respiratory tract is largely mechanical, provided by the movement of mucus along the surface of the nasal passages, throat and lungs by beating cilia. There are an estimated 10^9 cilia/cm^2 lining the respiratory tract which beat with a waveform consisting of an effective stroke that penetrates the layer of mucus and a recovery stroke that dives beneath it into the periciliary fluid. The mucus layer is swept towards the oropharynx where is it swallowed and the trapped microbes are eliminated by digestive enzymes (described in Meeks and Bush, 2000). Respiratory mucus is the product of two mucin genes, MUC5AC and MUC5B, and is secreted by surface epithelial goblet cells and the mucous cells of the submucosal glands (reviewed in Jackson, 2001). The main constituents of mucus are the mucin glycoproteins, which consist of a peptide backbone containing alternating glycosylated and non-glycosylated domains covered in oligosaccherides. Mucus acts as a dynamic defensive barrier by entrapping microbes and preventing their adhesion to host epithelial cell surface receptors, probably by binding specific bacterial adhesins and preventing their attachment to similar epitopes on host cell surface receptors. The epithelium of the gastrointestinal tract is also covered by a protective mucus gel which is synthesised and secreted from the apical surface of the goblet cells. Paradoxically, mucus also provides an ecological advantage to bacteria, *i.e.* is a source of carbohydrates and peptides and exogenous nutrients including vitamins and minerals which are present within the mucus matrix. Also, adhesion to specific mucin epitopes could facilitate mucus colonisation by commensals giving growth advantage in that they would avoid rapid expulsion from the gut. Mucus secretion is actually enhanced in response to some bacteria (mainly commensals but also some pathogens), *e.g. Vibrio cholerae*, which triggers massive mucin release via a cAMP-dependent mechanism, and *P. aeruginosa* which activates mucin gene expression in both airway and goblet cell cultures through an NF-κB signalling pathway (reviewed in Deplancke and Gaskins, 2001).

1.3. Recognition of microbes

In innate immunity, recognition of pathogens is a crucial step and is brought about by a number of "pattern recognition receptors" (PRRs) both soluble and membrane bound. The PRRs interact with "pathogen-associated molecular patterns" (PAMPS) on microorganisms which include highly conserved molecules such as lipopolysaccharide (LPS), lipoproteins, PGN, oligosaccharides and flagellin, the structural subunit of bacterial flagellae. PRRs are found on many cells of the innate immune system that are found in and around mucosal surfaces, including macrophages, and on the airway and intestinal epithelium. There are several different families of PRRs, and those involved in the mucosal innate immune response include the membrane-bound and soluble forms of the Toll-like receptors (TLRs) together with the fluid-phase receptors, soluble CD14, and the pulmonary collectins SP-A and SP-D (see above).

The TLRs are an evolutionarily ancient set of conserved proteins that are found in plants, insects and mammals. They consist of an ecto-domain comprising several leucine-rich repeats (LRR), a transmembrane domain, and a cytoplasmic domain. Ten human TLRs have been discovered, and these are expressed at higher levels in tissues exposed to the external environment such as the lung and the gastrointestinal tract (as well as on peripheral blood leukocytes and in the spleen) and have different specificities. In particular, TLR2 binds lipoproteins, PGN, lipoteichoic acid and porins, TLR4 binds LPS, and TLR5 responds to flagellin. Binding of microbial ligands to TLRs have a number of consequences, leading to recruitment of phagocytes, production of pro-inflammatory cytokines *e.g.* IL-8, activation of the NF-κB signalling pathway, up-regulation of co-stimulatory molecules on antigen presenting cells and the maturation of naïve dendritic cells. Activation of TLR2 on human tracheobronchial epithelial cells has also been shown to up-regulate production of the human β defensin-2. For detailed reviews of TLRs see Beutler (2003) and Takeda et al. (2003). A recent study has shown that there is a biologically active soluble form of TLR2 (sTLR2) which is found naturally expressed in breast milk and plasma, as a result of the post-translational modification of the TLR2 protein in an intracellular compartment. Levels of milk sTLR2 mirrored those of the TLR co-receptor CD14 (see below) and co-immunoprecipitation experiments have shown an interaction between the two molecules (LeBouder et al., 2003).

CD14 is a cell surface glycosylphosphatidylinositol (GPI)-linked 55 kDa glycoprotein expressed predominantly on myelomonocytic cells (including macrophages) but, without the GPI anchor, is also present in a soluble form in breast milk at between 20-100 μg/ml where it is derived from mammary epithelial cells and milk-derived macrophages (Labéta et al., 2000). It is also present in seminal plasma at about 1.89 μg/ml, where it is thought to be derived in part from testis and/or seminal vesicles (Harris et al., 2001). It acts as a PRR for LPS and PGN, plus a number of other macrophage activating molecules from a wide variety of microorganisms (Dziarski, 2003). Soluble CD14 forms complexes with LPS and PGN and can then bind and activate both CD14-positive and negative cells, via interactions with membrane bound TLR4, as measured by production of IL-6 and IL-8, and thus contribute to the maintenance of intestinal immune homeostasis in the newborn infant.

2. Evasion by microorganisms

To be successful pathogens, microorganisms need to evade or to subvert the immune mechanisms of the host. Since the components of the mucosal innate immune system are so numerous and varied in mechanism, it is interesting to review how microorganisms are able to resist them. Evasion of the innate immune system by viruses, evasion of the complement system by bacteria, and molecular mimicry are reviewed in detail in separate chapters.

2.1. Paralysis of cilia

Tracheal cytotoxin (a muramyl tetrapeptide bacterial cell wall fragment) of *Bordetella pertussis*, in combination with LPS, paralyses respiratory epithelial cilia, thus inhibiting mucociliatory clearance. In an air-interface organ culture of nasal turbinate and adenoid tissues, *B. pertussis* was observed to adhere to the base of the cilia, disorganising their spatial arrangement, and also adhered to damaged tissue and mucus, but very rarely to unciliated cells (Soane et al., 2000).

2.2. Modification of bacterial cell membranes or cell walls

The role of the composition of the bacterial cell membrane and capsule in the resistance of microorganisms to the action of antimicrobial peptides is the most widely studied mechanism so far and has been observed in a number of different Gram-positive and Gram-negative bacteria. The common theme is to modify the membrane lipids so that they are less negatively charged and thus less attractive to cationic peptides. For instance, *Salmonella enterica Typhimurium, E. coli* and *Yersinia enterocolitica* are able to produce more palmitoylated heptaacylated lipid which confers resistance to the rabbit α-defensins NP-1 and NP-2 and human α-defensins. *P. aeruginosa* is also able to adapt its cell membrane in response to several AMPs, including LL-37, as are *Proteus mirabilis* due to the presence of aminoarabinose in their LPS, and *H. influenzae* which are resistant to hBD-2 due to the presence of hexaacylated lipid A. Wild type *Yersinia pseudotuberculosis* is resistant to polymyxin B and cecropin, whereas mutants lacking 4-aminoarabinose in their LPS were killed. *S. aureus* modifies phosphatidylglycerol, the major phospholipid of *S. aureus*, with L-lysine under the control of the *mprF* gene.

This confers resistance to the α-defensins 1, 2 and 3 by reducing the negative charge. *mprF* analogs also exist in *Mycobacterium tuberculosis*, *P. aeruginosa* and *Enterococcus faecalis*. In addition, *S. aureus* can modify the structure of teichoic acid in the cell wall by the incorporating D-alanine, causing the surface to have a stronger positive charge. *S. aureus* are also resistant to the β-defensins hBD-1 and hBD-2, but are vulnerable to hBD-3. For a more detailed review of this and other methods of bacterial evasion see Hiemstra et al. (2004).

2.3. Inherent structural resistance

Bordetella species have been shown to be protected from the bactericidal functions of SP-A and SP-D by the presence of the non-repeating terminal trisaccharide of their LPS. Mutants of *B. pertussis* and *B. brochiseptica* lacking the terminal trisaccharide were susceptible to aggregation and permeabilisation by both SP-A and SP-D. Interestingly, the O antigen and palmitoylated lipid A of *B. brochiseptica* plays no role in this resistance (Schaeffer et al., 2004). *Klebsiella pneumoniae* has recently been shown to be resistant to a number of different antimicrobial peptides due to the presence of capsule polysaccharide which limits the interaction of AMPS with the bacterial surface. The amount of capsule polysaccharide produced was increased when grown in the presence of polymyxin B and lactoferrin (Campos et al., 2004).

2.4. Secretion of an antimicrobial peptide binding/inactivating protein

Streptococcal inhibitor of complement (SIC) is a 31 kDa anionic extracellular protein produced by a few highly virulent strains of *S. pyogenes* (predominantly by the M1 strain), and was initially identified as an inhibitor of the membrane attack complex of complement (Åkesson et al., 1996; Fernie-King et al., 2001). SIC was subsequently reported to inhibit the antimicrobial properties of five more proteins of the innate immune system found on mucosal surfaces, namely SLPI, lysozyme, (Fernie-King et al., 2002), LL-37, α-defensin 1 (HNP1) (Frick et al., 2003) and the β-defensins 1, 2, and 3 (Fernie-King et al., 2004). Interestingly, SIC had no effect on the anti-proteinase activity of SLPI implying that it interacts specifically with the antimicrobial domain of SLPI. The M12 and M55 strains of *S. pyogenes* produce a similar protein termed 'distantly related to SIC' or drs, for which complement inhibitory activity has also been demonstrated (Binks and Sriprakash 2004). Staphylokinase, a protein secreted by *S. aureus,* has been shown to induce both the extracellular release of neutrophil α-defensins, and subsequently to complex with them, resulting in almost complete inhibition of bactericidal activity (Jin et al., 2004). The product of the *E. coli* ORFan gene *ykfE* has been shown to be a potent *in vitro* inhibitor of vertebrate lysozyme (Ivy) (Monchois et al., 2001). A more recent study has concluded that Ivy (a periplasmic protein) is one of several factors involved in lysozyme resistance in *E. coli*, and this property would mainly be utilised when the integrity of the outer membrane had already been breached by lactoferrin (Deckers et al., 2004).

2.5. Degradation of antimicrobial substances by bacteria

Many bacteria (both commensals and pathogens) degrade mucus enzymatically thus releasing sources of energy to the bacteria, and enabling them to reach the epithelial surface. Degradation is a multistep process involving firstly proteolysis of the non-glycosylated regions of the mucin glycoproteins by proteases and secondly degradation of the glycylated sub-units by enzymes like sialidases, glycosulphatases and glycosidases. Microorganisms known to do this include *Entamoeba histolytica, S. typhimurium, H. pylori, C. albicans*, a mammalian reovirus (Bisaillon et al., 1999) and ocular commensals (Berry et al., 2002).

Most bacteria also secrete proteinases and these are thought to act as virulence factors. For example, *S. typhimurium* expresses an outer membrane endopeptidase, PgtE. *pgtE* mutants were found to be more sensitive to LL-37 and other α-helical AMPs, and this phenotype could be reversed by the introduction of a high copy number plasmid expressing pgtE. Interestingly, *pgtE* mutants were not more sensitive or resistant to killing by HNP-1 or other β-sheet AMPs (Guina et al., 2000). LL-37 is progressively degraded by purified *P. aeruginosa* elastase, *S. pyogenes* cysteine proteinase, a *S. aureus* metalloproteinase (aureolysin), and the growth medium from *E. faecalis* (which contained gelatinase). Growth medium from *P. mirabilis* culture totally degraded LL-37 with no evidence of intermediate fragments. This degradation of LL-37 could be inhibited by the appropriate proteinase inhibitors. A more recent study (Belas et al., 2004) shows that *P. mirabilis* ZapA metalloproteinase degrades a wide range of substrates, including hBD-1, LL-37, and the complement components C1q and C3. The authors postulate that a function of ZapA during urinary tract infections is the proteolysis of antimicrobial peptides associated with the innate immune response. *P. aeruginosa* elastase is also able to degrade human α-defensin, SP-A and SP-D (Mariencheck et al., 2003), and to inactivate SLPI.

2.6. Degradation, stimulation or release of host proteoglycans

The extra-cellular proteinases of *P. aeruginosa*, *E. faecalis* and *S. pyogenes* degrade tissue proteoglycans, especially decorin, causing release of glycoaminoglycans (GAG) (Schmidtchen et al., 2001). *P. aeruginosa* also stimulated the shedding of syndecan-1 (the main cell-surface heparan sulphate proteoglycans of epithelia) both *in vitro* and *in vivo* and both processes seem to enhance bacterial virulence (Park et al., 2001). It was shown that dermatan sulphate, the GAG of human decorin, bound to and inhibited the bactericidal action of HNP-1 whilst syndecan-1 binds to cathelicidin-type AMPS. This inhibition probably occurs via ionic binding in that both syndecan-1 and dermatan sulphate are anionic in nature.

2.7. Down-regulation of expression of antimicrobial substances

H. pylori initially stimulates mucus secretion but eventually prolonged exposure results in a decrease in mucin synthesis and secretion by suppression of MUC1 and MUC5AC genes (Deplancke and Gaskins 2002). *Shigella spp.* evade killing by LL-37 and hBD-1 in the epithelial cells of the gut by downregulating expression of LL-37 and hBD-1 (Islam et al., 2001). This is effected by plasmid DNA from *S. dysenteriae*, *S. flexneri* and *S. boyndii* in some way. Interestingly, downregulation of LL-37 and hBD-1 expression was also observed in cases of watery diarrhoea of unidentified cause, but not due to *Shigella* or *Salmonella* infection, and only early in the infection. This was assumed not to be due to *E. coli* either, as hBD-1 expression had earlier been shown to be unaffected by either *E. coli* or *Salmonella* so evidently other types of enteric bacteria are able to produce the same effect.

2.8. Mimicry of host cell membrane lipids

Antimicrobial peptides are thought to be less attracted to host cell membranes due to the more positive surface charge on eukaryotic cell membranes caused by the presence of phosphatidylcholine. However, a number of laboratories have observed surface structures containing choline, in the form of phosphorylcholine (ChoP), on a various bacteria including oral streptococci, pneumococci, haemophili, mycoplasmas and neisseriae (for example, Lysenko et al., 2000). This property has been studied in detail in *H. influenzae* in relation to its resistance to killing by LL-37 in which expression of ChoP appears to be inducible in response to exposure to LL-37.

2.9. Energy-dependent efflux systems

Gonococci possess an energy-dependent efflux pump termed mtr (multiple transferable resistance), encoded by a three-gene, single transcriptional unit (*mtrCDR*) which are highly similar to efflux pump proteins in *E. coli* and *P. aeruginosa.* This system has been shown in gonococci to enhance resistance to antimicrobial cationic compounds which affect the cytoplasmic membrane such as LL-37 but, interestingly, has no effect at all on susceptibility to hNP-2 so some other mechanism is presumed to be responsible for resistance to defensins (Hagman et al., 1997; Schafer et al., 1998). *Y. enterocolitica* possess a temperature-regulated efflux pump/potassium antiporter system which is also induced by the presence of antimicrobial peptides and it has been proposed that this system protects the bacteria by both acidifying the cytoplasm to prevent the action of AMPs and by pumping the AMPs out of the cell (Bengoechea, 2000).

2.10. Induced resistance of unknown cause

Histatin resistant derivatives of *C. albicans* can be generated by repeated exposure to increasing amounts of histatin. It is not yet known how they become resistant, but apart from the resistant phenotype they are no different from the parent strains (Fitzgerald et al., 2003).

References

Åkesson, P., Sjöholm, A.G., and Björck, L. (1996). Protein SIC, a novel extracellular protein of *Streptococcus pyogenes* interfering with complement function. J. Biol. Chem. *271*, 1081-1088.

Allaker, R.P., Zihni, C., and Kapas, S. (1999). An investigation into the antimicrobial effects of adrenomedullin on members of the skin, oral, respiratory tract and gut microflora. FEMS Immunol. Med. Microbiol. *23*, 289-293.

Appelmelk, B.J., An, Y.-Q., Geerts, M., Thijs, B.D., de Boer, H.A., MacLaren, D.M., de Graaff, J., and Nuijens, J.H. (1994). Lactoferrin is a lipid A-binding protein. Infect. Immun. *62*, 2628-2632.

Arnold, R.R., Cole, R.M., and McGhee, J.R. (1977). A bactericidal effect for human lactoferrin. Science (Washington, DC). *197*, 263-265.

Arnold, R.R., Russell, J.E., Champion, W.J., Brewer, M., and Gauthier, J.J. (1982). Bactericidal activity of human lactoferrin: differentiation from the stasis of iron deprivation. Infect. Immun. *35*, 792-799.

Bals, R., Wang, X., Zasloff, M., and Wilson, J.M. (1998). The peptide antibiotic LL-37/ hCAP-18 is expressed in epithelia of the human lung where it has broad antimicrobial activity at the airway surface. Proc. Natl. Acad. Sci. USA. *95*, 9541-9546.

Belas, R., Manos, J., and Suvanasuthi, R. (2004). *Proteus mirabilis* ZapA metalloprotease degrades a broad spectrum of substrates, including antimicrobial peptides. Inf. Imm. *72*, 5159-5167.

Bengoechea, J.A., and Skurnik, M. (2000). Temperature-regulated efflux pump/potassium antiporter system mediates resistance to cationic antimicrobial peptides in *Yersinia*. Mol. Microbiol. *37*, 67-80.

Berry, M., Harris, A., Lumb, R., and Powell, K. (2002). Commensal ocular bacteria degrade mucins. Br. J. Ophthalmol. *86*, 1412-1416.

Beutler, B. (2003). Innate immune responses to microbial poisons: discovery and function of the Toll-like receptors. Ann. Rev. Pharmacol. Toxicol. *43*, 609-628.

Binks, M., and Sriprakash, K.S. (2004). Characterization of a complement-binding protein drs, from strains of *Streptococcus pyogenes* containing the *emm12* and *emm55* genes. Inf. Imm. *72*, 3981-3986.

Bisaillon, M., Sénéchal, S., Bernier, L., and Lemay, G. (1999). A glycosyl hydrolase activity of mammalian reovirus σ1 protein can contribute to viral infection through a mucus layer. J. Mol. Biol. *286*, 759-773.

Blake, C.C., Koenig, D.F., Mair, G.A., North, A.C., Phillips, D.C., and Sarma, V.R. (1965). Structure of hen egg-white lysozyme. A three-dimensional Fourier synthesis at 2 Ångstrom resolution. Nature *206*, 757-761.

Boackle, R.J. (1991). The interaction of salivary secretions with the human complement system - a model for the study of host defense systems on inflamed mucosal surfaces. Crit. Rev. Oral Biol. Med. *2*, 355-367,

Buckland, A.G., Heeley, E.L., and Wilton, D.C. (2000). Bacterial cell membrane hydrolysis by secreted phospholipases A_2: a major physiological role of human group IIa $sPLA_2$ involving both bacterial cell wall penetration and interfacial catalysis. Biochim. Biophys. Acta. *1484*, 195-206.

Buckland, A.G., and Wilton, D.C. (2000). The antibacterial properties of secreted phospholipases A_2. Biochim. Biophys. Acta. *1488*, 71-82.

Campos, M.A., Vargas, M.A., Regueiro, V., Llompart, C.M., Albertí, S., and Bengoechea, J.A. (2004). Capsule polysaccharide mediates bacterial resistance to antimicrobial peptides. Inf. Imm. *72*, 7107-7114.

Clyne, M., Dillon, P., Daly, S., O'Kennedy, R., May, F.E.B., Westley, B.R., and Drumm, B. (2004). *Helicobacter pylori* interacts with the human single-domain trefoil protein TFF1. Proc. Nat. Acad. Sci. *101*, 7409-7414.

Cunliffe, R.N. (2003). α-defensins in the gastrointestinal tract. Mol. Immunol. *40*, 463-467.

Deckers, D., Masschalck, B., Aertsen, A., Callewaert, L., Van Tiggelen, C.G., Atanassova, M., and Michiels, C.W. (2004). Periplasmic lysozyme inhibitor contributes to lysozyme resistance in *Escherichia coli*. Cell Mol. Life Sci. *61*, 1229-1237.

Deplancke, B., and Gaskins, H.R. (2001). Microbial modulation of innate defense: goblet cells and the intestinal mucus layer. Am. J. Clin. Nutr. *73(suppl)*, 1131S-1141S.

Dhaliwal, W., Bajaj-Elliott, M., and Kelly, P. (2003). Intestinal defensin gene expression in human populations. Mol. Immunol. *40*, 469-475.

dos Santos Silva, E., Ulrich, M., Döring, G., Botzenhart, K., and Gött, P. (2000). Trefoil factor family domain peptides in the human respiratory tract. J. Pathol. *190*, 133-142.

Dziarski, R. (2003). Recognition of bacterial peptidoglycan by the innate immune system. Cell. Mol. Life Sci. *60*, 1793-1804.

Eggert, F.M. (1980a). The nature of secretory agglutinins and aggregating factors. I. Secretory conglutinin-like factor, secretory bacterial aggregating factors and secretory IgA antibody in human saliva and amniotic fluid. Int. Arch. Allergy Appl. Immunol. *61*, 192-202.

Eggert, F.M. (1980b). The nature of secretory agglutinins and aggregating factors. II. Biochemical and immunochemical properties of factors in human saliva and amniotic fluid. Int. Arch. Allergy Appl. Immunol. *61*, 203-212.

Eggert, F.M. (1980c). The nature of secretory agglutinins and aggregating factors. III. Secretory conglutinin-like factor SKF detects a cross-reaction between bacteria and complement component C3. Int. Arch. Allergy Appl. Immunol. *62*, 34-45.

Eggert, F.M. (1980d). The nature of secretory agglutinins and aggregating factors. IV. Complexing bvetween non-mucin glycoproteins, immunoglobulins and mucins in human saliva and amniotic fluid. Int. Arch. Allergy Appl. Immunol. *62*, 46-58.

Eisenberg, S.P., Hale, K.K., Heimdal, P., and Thompson, R.C. (1990). Location of the protease-inhibitory region of secretory leukocyte protease inhibitor. J. Biol. Chem. *265*, 7976-7981.

Ellison, R.T., and Giehl, T.J. (1991). Killing of Gram-negative bacteria by lactoferrin and lysozyme. J. Clin. Invest. *88*, 1080-1891.

Farnaud, S., and Evans, R.W. (2003). Lactoferrin - a multifunctional protein with antimicrobial properties. Mol. Immunol. *40*, 395-405.

Fernie-King, B.A., Seilly, D.J., and Lachmann, P.J. (2004). The interaction of streptococcal inhibitor of complement (SIC) and its proteolytic fragments with the human beta defensins. Immunol. *111*, 444-452.

Fernie-King, B.A., Seilly, D.J., Davies, A., and Lachmann, P.J. (2002). Streptococcal Inhibitor of Complement inhibits two additional components of the mucosal innate immune system: secretory leukocyte proteinase inhibitor and lysozyme. Infect. Immun. *70*, 4908-4916.

Fernie-King, B.A., Seilly, D.J., Willers, C., Würzner, R., Davies, A., and Lachmann, P.J. (2001). Streptococcal inhibitor of complement (SIC) inhibits the membrane attack complex by preventing uptake of C567 onto cell membranes. Immunol. *103*, 390-398.

Fitzgerald, D.H., Coleman, D.C., and O'Connell, B.C. (2003). Binding, internalization and degradation of histatin 3 in histatin-resistant derivatives of *Candida albicans*. FEMS Microbiol. Letts. *220*, 247-253.

Fleming, A. (1922). On a remarkable bacteriolytic element found in tissues and secretions. Proc. R. Soc. Lond. (Biol.) *93*, 306-317.

Frick. I.-M., Åkesson, P., Rasmussen, M., Schmidtchen, A., and Björck, L. (2003). SIC - a secreted protein of *Streptococcus pyogenes* that inactivates antibacterial peptides. J. Biol. Chem. *278*, 16561-16566

Ganz, T. (2003). Defensins: antimicrobial peptides of innate immunity. Nat. Rev. Imm. *3*, 710-720.

Guina, T., Yi, E.C., Wang, H., Hackett, M., and Miller, S. (2000). A PhoP-regulated outer membrane protease of *Salmonella enterica* serovar *typhimurium* promotes resistance to alpha-helical antimicrobial peptides. J. Bacteriol. *182*, 4077-4086

Hagman, K.E., Lucas, C.E., Balthazer, J.T., Snyder, L., Nilles, M., Judd, R.C., and Shafer, W.M. (1997). The MtrD protein of *Neisseria gonorrhoeae* is a member of the resistance/modulation/division protein family constituting part of an efflux system. Microbiol. *143*, 2117-2125.

Harris, C.L., Vigar, M.A., Rey Nores, J.E., Horejsi, V., Labeta, M.O., and Morgan, B.P. (2001). The lipopolysaccharide co-receptor CD14 is present and functional in seminal plasma and expressed on spermatozoa. Immunology *104*, 317-323.

Hase, K., Eckmann, L., Leopard, J.D., Varki, N., and Kagnoff, M.F. (2002). Cell differentiation is a key determinant of cathelicidin LL-37/human cationic antimicrobial protein 18 expression by human colon epithelium. Infect. Immun. *70*, 953-963.

Hase, K., Murakami, M., Iimura, M., Cole, S.P., Horibe, Y., Ohtake, T., Obonyo, M., Gallo, R.L., Eckmann, L., and Kagnoff, M.F. (2003). Expression of LL-37 by human gastric epithelial cells as a potential host defense mechanism against *Helicobacter pylori. 125*, 1613-1625.

Henzler Wildman, K.A., Lee, D.-K., and Ramamoorthy, A. (2003). Mechanism of lipid bilayer disruption by the human antimicrobial peptide, LL-37. Biochem. *42*, 6545-6558.

Hiemstra, P.S., Fernie-King, B.A., McMichael, J., Lachmann, P.J., and Sallenave, J.-M. (2004). Antimicrobial peptides: mediators of innate immunity as templates for the development of novel anti-infective and immune therapeutics. Curr. Pharm. Des. *10*, 2891-2905.

Hiemstra, P.S., Maassen, R.J., Stolk, J., Heinzel-Wieland, R., Steffens, G.J., and Dijkman, J.H. (1996). Antibacterial activity of antileukoprotease. Inf. Imm. *64*, 4520-4524.

Holmskov, U., Thiel, S., and Jensenius, J.C. (2003). Collectins and Ficolins: humoral lectins of the innate immune defense. Ann. Rev. Immunol. *21*, 547-578.

Islam, D., Bandholtz, L., Nilsson, J., Wigzell, H., Christensson, B., Agerberth. B., and Gudmundsson, G.H. (2001). Downregulation of bactericidal peptides in enteric infections: a novel immune escape mechanism with bacterial DNA as a potential regulator. Nat. Med. *7*, 180-185.

Jackson, A.D. (2001). Airway goblet-cell mucus secretion. Trends in Pharmacol. Sci. *22*, 39-45.

Jin, T., Bokarewa, M., Foster, T., Mitchell, J., Higgins, J., and Tarkowski, A. (2004). *Staphylococcus aureus* resists human defensins by production of staphylokinase, a novel bacterial evasion mechanism. J. Immunol. *172*, 1169-1176.

Kindon, H., Pothoulakis, C., Thim, L., Lynch-Devaney, K., and Podolsky, D.K. (1995). Trefoil peptide protection of intestinal epithelial barrier function: cooperative interaction with mucin glycoprotein. Gastroenterol. *109*, 516-523.

Kim, S.T., Cha, H.E., Kim, K.Y., Han G.C., Chung, Y.S., Lee, Y.J., Hwang, Y.J., and Lee, H.M. (2003). Antimicrobial peptide LL-37 is upregulated in chronic nasal inflammatory disease. Acta Otolaryngol. *123*, 81-85.

Kuipers, M.E., de Vries, H.G., Eikelboom, M.C., Meijer, D.K.F., and Swart, P.J. (1999). Synergistic fungistatic effects of lactoferrin in combination with antifungal drugs against clinical *Candida* isolates. Antimicrob. Agents Chemother. *43*, 2635-2641.

Labéta M.O., Vidal, K., Rey Nores, J.E., Arias, M., Vita, N., Morgan, B.P., Guillemot, J.C., Loyaux, D., Ferrera, P., Schmid, D., Affolter, M., Borysiewicz, L.K., Donnet-Hughes, A., and Schiffrin, E.J. (2000). Innate recognition of bacteria in human milk is mediated by a milk-derived highly expressed pattern recognition receptor, soluble CD14. J. Exp. Med. *191*, 1807-1812.

Lachmann, P.J., and Thomson, R.A. (1970). Immunoconglutinins in human saliva--a group of unusual IgA antibodies. Immunol. *18*, 157-69.

LeBouder, E., Rey-Nores, J.E., N.K., Grigorov, M., Lawn, S.D., Affolter, M., Griffin, G.E., Ferrara, P., Schiffrin, E.J., Morgan, B.P., and Labéta, M.O. (2003). Soluble forms of Toll-like receptor (TLR) 2 capable of modulating TLR2 signalling are present in human plasma and breast milk. J. Immunol. *171*, 6680-0.

Lehrer, R.I. (2004) Primate defensins. Nat. Rev. Immunol. *2*, 727-738.

Lee-Huang, S., Huang, P.L., Sun, Y., Kung, H.F., Blithe, D.L., and Chen, H.C. (1999). Lysozyme and RNAses as anti-HIV components in beta-core preparations of human chorionic gonadotrophin. Proc. Natl. Acad. Sci. USA. *96*, 2678-1681.

LeVine, A.M., Elliott, J., Whitsett, J., Srikiatkhachorn, A., Crouch, E., DeSilva, N., and Korfhagen, T. (2004). Surfactant protein-D enhances phagocytosis and pulmonary clearance of respiratory syncytial virus. Am. J. Respir. Cell Mol. Biol. *31*, 193-199.

Lönnerdahl, B. (2003). Nutritional and physiologic significance of human milk proteins. Am. J. Clin. Nutr. *77* (suppl), 1537S-1543S.

Lysenko, E.S., Gould, J., Bals, R., Wilson, J.M., and Weisner, J.N. (2000). Bacterial phosphorylcholine decreases susceptibility to the antimicrobial peptide LL-37/hCAP18 expressed in the upper respiratory tract. Inf. Imm. *68*, 1664-1671.

Mackay, B.J., Denepitiya, L., Iacono, V.J., Krost, S.B., and Pollock, J.J. (1984). Growth-inhibitory and bactericidal effects of human parotid salivary histidine-rich polypeptides on *Streptococcus mutans*. *44*, 695-701.

Marchetti, M., Trybala, E., Superti, F., Johansson, M., and Bergström, T. (2004). Inhibition of herpes simplex virus infection by lactoferrin is dependent on interference with the virus binding to glycosaminoglycans. Virology *318*, 405-413.

Mariencheck, W.I., Alcorn, J.F., Palmer, S.M., and Wright, J.R. (2003). *Pseudomonas aeruginosa* elastase degrades Surfactant Proteins A and D. Am. J. Respir. Cell. Mol. Biol. *28*, 528-537.

Marutsuka, K., Nawa, Y., Asada Y., Hara, S., Kitamura, K., Eto, T., and Sumiyoshi, A. (2001). Adrenomedullin and proadrenomedullin N-terminal 20 peptide (PAMP) are present in human colonic epithelia and exert an antimicrobial effect. Exp. Physiol. *86*, 543-545.

Masschalck, B., and Michiels, C.W. (2003). Antimicrobial properties of lysozyme in relation to foodborne vegetative bacteria. Crit. Rev. Microbiol. *29*, 191-214.

McCormack, F.X., Gibbons, R., Ward, S.R., Kuzmenko, A., Wu, H., and Deepe Jr., G.S. (2003). Macrophage-independent fungicidal action of the pulmonary collectins. J. Biol. Chem. *278*, 36250-36256.

McNeely, T.B., Dealy, M., Dripps, D.J., Orenstein, J.M., Eisenberg, S.P., and Wahl, S.M. (1995). Secretory leukocyte protease inhibitor: a human saliva protein exhibiting anti-human immunodeficiency virus 1 activity *in vitro*. J. Clin. Invest. *96*, 456-464.

Meeks, M., and Bush, A. (2000). Primary ciliary dyskinesia (PCD). Ped. Pulmonol. *29*, 307-316.

Monchois, V., Abergel, C., Sturgis, J., Jeudy, S., and Claverie, J-M. (2001). *Escherichia coli ykfE* ORFan gene encodes a potent inhibitor of C-type lysozyme. J. Biol. Chem. *276*, 18437-18441.

Murakami, M., Ohtake, T., Dorschner, R.A., and Gallo, R.L. (2002). Cathelicidin antimicrobial peptides are expressed in salivary glands and saliva. J. Dent. Res. *81*, 845-850.

Murray, E., Khamri, W., Walker, M.M., Eggleton, P., Moray, A.P., Ferris, J.A., Knapp, S., Karim, Q.N., Worku, M., Strong, P., Reid, K.B., and Thursz, M.R. (2002). Expression of surfactant protein D in the human gastric mucosa and during *Helicobacter pylori* infection. Infect. Immun. *70*, 1481-1487.

Nara, K., Ito, S., Ito, T., Suzuki, Y., Ghoneim, M.A., Tachibana, S., and Hirose, S. (1994). Elastase inhibitor elafin is a new type of proteinase inhibitor which has a transglutaminase-mediated anchoring sequence termed "cementoin". J. Biochem. (Tokyo) *115*, 441-448.

van Nieuw Amerongen, A., and Veerman, E.C.I. (2002). Saliva - the defender of the oral cavity. Oral. Dis. *8*, 12-22.

Nizet, V., and Gallo, R.L. (2003). Cathelicidins and innate defense against invasive bacterial infection. Scand. J. Infect. Dis. *35*, 670-676.

Park, P.W., Pier, G.B., Hinkes, M.T., and Bernfield, M. (2001). Exploitation of syndecan-1 shedding by *Pseudomonas aeruginosa* enhances virulence. Nature *411*, 98-102.

Qu, X-D., and Lehrer, R.I. (1998). Secretory phospholipase A2 is the principal bactericide for Staphylococci and other Gram-positive bacteria in human tears. Inf. Imm. *66*, 2791-1979.

Schaeffer, L.M., McCormack, F.X., Wu, H., and Weiss, A.A. (2004). Interactions of pulmonary collectins with *Bordetella bronchiseptica* and *Bordetella pertussis* lipopolysaccharide elucidate the structural basis of their antimicrobial activities. Inf. Imm. *72*, 7124-7130.

Shafer, W.M., Qu, X.-D., Waring, A.J., and Lehrer, R.I. (1998). Modulation of *Neisseria gonorrhoea* susceptibility to vertebrate antibacterial peptides due to a member of the resistance/modulation/division efflux pump family. Proc. Nat. Acad. Sci. USA. *95*, 1829-1833.

Schaller-Bals, S., Schulze, A., and Bals, R. (2002). Increased levels of antimicrobial peptides in tracheal aspirates of newborn infants during infection. Am. J. Respir. Crit. Care Med. *165*, 992-995.

Schmidtchen, A., Frick, I.M., Andersson, E., Tapper, H., and Björck, L. (2002). Proteinases of common pathogenic bacteria degrade and inactivate the antibacterial peptide LL-37. Mol. Microbiol. *46*, 157-168.

Schmidtchen, A., Frick, I.-M., and Björck, L. (2001). Dermatan sulphate is released by proteinases of common pathogenic bacteria and inactivates antibacterial α-defensin. Mol. Microbiol. *39*, 708-713.

Schutte, B.S., and McCray Jr., P.B. (2002). β-defensins in lung host defense. Ann. Rev. Physiol. *64*, 709-48.

Sieprawska-Lupa, M., Mydel, P., Krawczyk, K., Wojcik, K., Puklo, M., Lupa, B., Suder, P., Silberring, J., Reed, M., Pohl, J., Shafer, W., McAleese, F., Foster, T., Travis, J., and Potempa, J. (2004). Degradation of human antimicrobial peptide LL-37 by *Staphylococcus aureus*-derived proteinases. Antimicrob. Agents Chemother. *48*, 4673-4679.

Simpson, A.J., Maxwell, A.I., Govan, J.R.W., Haslett, C., and Sallenave, J.-M. (1999). Elafin (elastase-specific inhibitor) has anti-microbial activity against Gram-positive and Gram-negative respiratory pathogens. FEBS Letts. *452*, 309-313.

Singh, P.K., Tack, B.F., McCray Jr., P.B., and Welsh, M.J. (2000). Synergistic and additive killing by antimicrobial factors found in human airway surface liquid. Am. J. Physiol. Lung Cell. Mol. Physiol. *279*, L799-L805.

Soane, M.C., Jackson, A., Maskell, D., Allen, A., Keig, P., Dewar, A., Dougan, G., and Wilson, R. (2000). Interaction of *Bordetella pertussis* with human respiratory mucosa *in vitro*. Respir. Med. *94*, 791-799.

Takeda, K., Kaisho, T., and Akira, S. (2003). Toll-like receptors. Ann. Rev. Immunol. *21*, 335-376.

Tomee, J.F.C., Hiemstra, P.S., Heinzel-Wieland, R., and Kauffman, H.F. (1997). Antileukoprotease: an endogenous protein in the innate mucosal defense against fungi. J. Infect. Dis. *176*, 740-747.

van der Strate, B.W.A., Beljaars, L., Molema, G., Harmsen, M.C., and Meijer, D.K.F. (2002). Antiviral activities of lactoferrin. Antivir. Res. *52*, 225-239.

Weiss, J., Inada, M., Elsbach, P., and Crowl, R.M. (1994). Structural determinants of the action against *Escherichia coli* of a human inflammatory fluid phospholipase A(2) in concert with polymorphonuclear leukocytes. J. Biol. Chem. *269*, 26331-26337.

Winsnes, R. (1980). Complement (C3) conversion by rat intestinal glycoprotein and its degradation products. Scand. J. Immunol. *12*, 67-76.

Winsnes, R., and Lachmann, P.J. (1982). Intestinal glycoprotein activates the alternative complement pathway by reacting with factor H. Scand. J. Immunol. *15*, 371-378.

Wu, H., Kuzmenk, A., Wan, S., Schaffer, L., Weiss, A., Fisher, J.H., Kim, K.S., and McCormack, F.X. (2003). Surfactant proteins A and D inhibit the growth of Gram-negative bacteria by increasing membrane permeability. J. Clin. Invest. *111*, 1589-1602.

4

Viral Immune Evasion: An Overview

*Antonio Alcami**

Abstract

The presence in the immunocompetent host of numerous mechanisms to prevent the infection by pathogens has forced successful viral pathogens to acquire during evolution strategies to evade the host defense mechanisms. Our knowledge of viral immune evasion strategies has expanded enormously during the last ten years, with viral evasion mechanisms identified for most of the known immune pathways. These studies are relevant to understand viral pathogenesis and shed light into basic mechanisms of immunity.

1. Introduction

Viral replication occurs within an animal host that is well prepared with a variety of defense mechanisms to prevent viral infection. In turn, a variety of virus-encoded strategies to counteract host defenses have evolved to allow viruses to replicate and propagate (Alcami and Koszinowski, 2000; Tortorella et al., 2000).

The immune evasion strategies identified in different viruses are influenced by the nature and replication of the viral genetic material, cell tropism, mechanism of transmission and ability to persist in the infected host. Some viruses, such as human immunodeficiency virus (HIV), infect immune cells and may cause immunosuppression. RNA viruses, such as influenza virus and HIV, mutate rapidly and generate variants that escape recognition by the host immune system. Other viruses, such as herpesviruses, maintain their genome in a latent stage with minimal expression of viral proteins, avoiding recognition by the immune system and allowing viral persistence. Intermitent phases of virus reactivation ensure transmission to other individuals and virus persistence in the host population. Investigations during the last ten years have shown that viruses not only hide to avoid immune recognition but they also play an active role to inhibit mechanisms of immunity. Large DNA viruses, such as poxviruses and herpesviruses, may

*For correspondence email aalcami@cnb.uam.es

accommodate between 100 and 200 genes in their genomes allowing them to encode a large number of proteins that interfere with the immune anti-viral response.

Viral strategies of immune evasion block innate (non-specific) or adaptive (antigen-specific) immune responses. The innate response constitutes the first line of immune defense that is rapidly activated and does not require previous encounters with the viral agent. Some components of the innate response are interferons (IFNs), the complement system or natural killer (NK) cells. The adaptive immune response consists of antibodies and cytotoxic T cells (CTLs) directed against specific virus-encoded antigens. This response requires days or even weeks to develop and induces immunological memory, allowing the host to respond more efficiently in the event of re-infection. The inflammatory response illustrates the close interaction between innate and adaptive immune responses. The inflammatory response is triggered after viral infection and causes an increased permeability of blood vessels and the production of cytokines and chemokines which induce a massive infiltration of cells and soluble factors into areas of infection that may protect the host from viral infection and trigger antigen-specific cellular and humoral responses that will offer further protection mechanisms.

The immune response requires the co-operation of T helper (Th) lymphocytes which differenciate into Th1 and Th2 cells that produce specific sets of cytokines. Th1 cells produce cytokines such as interleukin 2 (IL-2) and IFN-γ that promote cellular pro-inflammatory responses and the maturation of macrophages and CTLs. In contrast, Th2 cells promote humoral immunity inducing the differentiation of B cells to produce antibodies, and the production of IL-4 and IL-10 which may reduce cellular inflammatory responses. Whereas cellular immune responses are thought to be important in viral elimination once viral infection has been established, the antibody response plays an important role in preventing the early stages of infection.

We have just started to understand the complex interaction of viruses with their animal hosts and how viruses achieve an equilibrium leading to the survival of both virus and host. Viruses are fully dependent on the host for their replication, and thus they cannot afford causing extensive damage that may compromise the survival of the host. Much of the pathology associated with viral diseases may be the result of an unbalanced virus-host interaction, and in some cases it represents uncontrolled attempts of the immune system to destroy the virus that lead to damage of the host. A more profound knowledge of virus-host interaction will increase our understanding of the molecular mechanisms of viral pathogenesis.

The study of the interaction of viruses with the immune system is a research area that has expanded enormously in recent years and it is not possible to cover all aspects in depth in a chapter. Here I provide an overview of the different viral immune evasion strategies and basic concepts are discussed and illustrated with representative examples. More information can be found in review articles or other chapters in this book (Alcami and Koszinowski, 2000; Johnson and Hill, 1998; Katze et al., 2002; Seet et al., 2003; Tortorella et al., 2000).

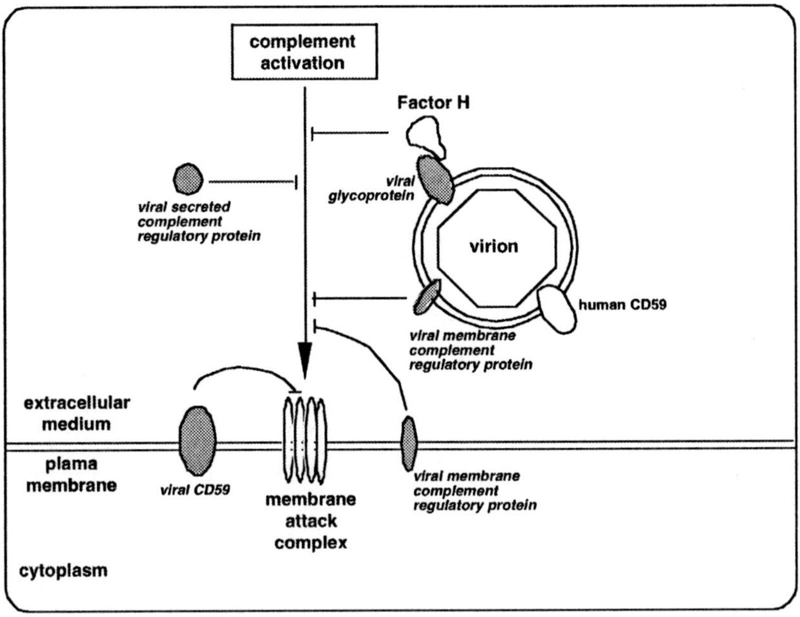

Figure 1. Viral mechanisms to evade the complement response.

2. Inhibition of humoral immunity

2.1. Genetic variability

This is one of the first mechanisms of immune evasion identified in viruses (Flint et al., 2000). In contrast to DNA polymerases, RNA polymerases introduce a high rate of mutations that translate into high amino acid sequence variability. This may generate viral epitopes recognized with lower affinity by neutralizing antibodies and variant peptide sequences that do not bind to major histocompatibility complex (MHC) molecules, thus evading recognition by the cellular immune system as well. This strategy has been well documented in RNA viruses such as rhinoviruses, influenza virus and HIV.

2.2. Fc receptors

Some viruses encode receptors for the Fc domain of immunoglobulin G (IgG), such as the gE-gI complex from herpes simplex virus (HSV), to evade the action of antibodies (Johnson and Hill, 1998).

2.3. Complement system

The complement system constitutes one of the major early defense systems against pathogens. Viral antigens or antigen-antibody complexes may activate the complement cascade which culminates with the formation of a membrane attack complex leading to the destruction of infected cells or the lipid envelope of virus particles. Herpesviruses and poxviruses encode proteins with amino acid sequence similarity to complement regulatory proteins that bind critical factors in the complement activation pathways and inhibit complement activity (Favoreel et al., 2003; Johnson and Hill, 1998; Kotwal et al., 1990) (Figure 1). The HSV membrane

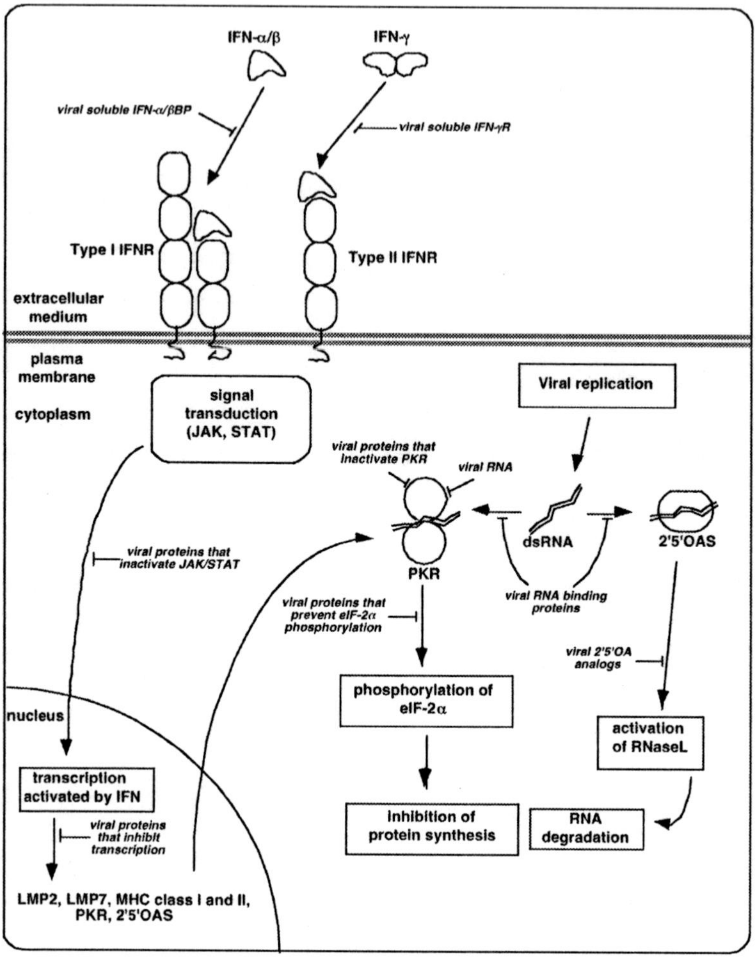

Figure 2. Viral mechanisms to evade IFN activity.

glycoprotein gC binds the complement factor C3 an protects both cell-free virus and HSV-infected cells from the action of complement (Lubinski et al., 1999). The poxviruses variola virus, vaccinia virus (VV) and cowpox virus (CPV) encode a secreted protein, known as viral complement control protein (VCP), that blocks the classical and alternative pathways of complement activation (Kotwal et al., 1990; Rosengard et al., 2002). Herpesvirus saimiri (HVS) encodes a homologue of the inhibitor of the membrane attack complex CD59 that protects the infected cells from complement-mediated lysis (Rother et al., 1994). An alternative strategy is adopted by human retroviruses, VV and human cytomegalovirus (HCMV). Instead of encoding homologues of host proteins, these viruses incorporate into the virion envelope host complement regulatory proteins, such as CD56 and CD59, which confer complement resistance to the virus particle (Saifuddin et al., 1995; Spear et al., 1995; Vanderplasschen et al., 1998).

3. Modulation of IFN functions

IFNs are classified as type I (α and β) and type II (γ), and represent one of the first anti-viral defense mechanisms to be activated. IFNs are produced in response to viral infection and bind specific receptors triggering intracellular pathways that activate anti-viral mechanisms. Two major pathways are the double-stranded RNA (dsRNA)-dependent protein kinase (PKR) and the 2'-5' oligoadenylate system that activates RNase L. In addition, IFNs have immunomodulatory activity. Viruses block the activity of IFN at three different levels: (i) blockade of IFN binding to specific receptors; (ii) inhibition of signal transduction pathways; and (iii) inhibition of anti-viral effector functions (Goodbourn et al., 2000; Katze et al., 2002) (Figure 2).

3.1. Secreted viral IFN receptors (IFNRs) and IFN binding proteins (IFNBPs)

Poxviruses encode secreted proteins that bind with high affinity type I or type II IFN and prevent their interaction with IFNRs (Alcami, 2003; McFadden and Murphy, 2000) (Figure 2). The viral IFN-γR (vIFN-γR), with sequence similarity to the IFN binding domain of their cellular counterpart, has been identified in several poxviruses including myxoma virus (MV), VV and CPV (Alcami and Smith, 1995; Mossman et al., 1995; Upton et al., 1992). A recent report describes a secreted IFN-γBP encoded by fowlpox virus that has no sequence similarity to known cellular IFN-γRs (Puehler et al., 2003), raising the possibility that this virus may have hijacked an avian IFN-γR of structure different to the mammalian counterparts. VV and other poxviruses encode a secreted vIFN-α/βBP that has very limited sequence similarity to the cellular counterparts (Colamonici et al., 1995; Symons et al., 1995).

3.2. Interference with signal transduction induced by IFNs

The binding of IFNs to specific receptors triggers a complex cascade of intracellular pathways through janus kinase (JAK) and signal transducers and activators of transcription (STAT) pathways (Goodbourn et al., 2000; Katze et al., 2002). These signaling events result in the upregulation of cellular proteins that limit viral replication, such as the low-molecular-weight protein 2 (LMP2) and LMP7 subunits of the proteasome, involved in the generation of viral peptides presented MHC class I and class II molecules, or PKR and 2'5'OAS which confer an anti-viral state in cells.

Some viral proteins interfere with the IFN signaling cascade (Goodbourn et al., 2000; Katze et al., 2002) (Figure 2). The protein V from simian virus 5 (SV5) interacts with STAT1 and induces its degradation by the proteasome (Didcock et al., 1999). Other paramyxovirus V proteins target different components of the IFN signaling pathway. Type II human parainfluenza virus V protein targets STAT2, mumps virus V protein targets both STAT1 and STAT3, and measles virus (MeV) V protein causes a defect in STAT nuclear accumulation and STAT-inducible transcription (Nishio et al., 2002; Palosaari et al., 2003; Parisien et al., 2001; Ulane et al., 2003). Kaposi's sarcoma-associated herpesvirus (KSHV) uses a different mechanism by encoding a homologue of the IFN regulatory factor (IRF) that represses the IFN-mediated transcriptional activation of host genes (Zimring et al., 1998).

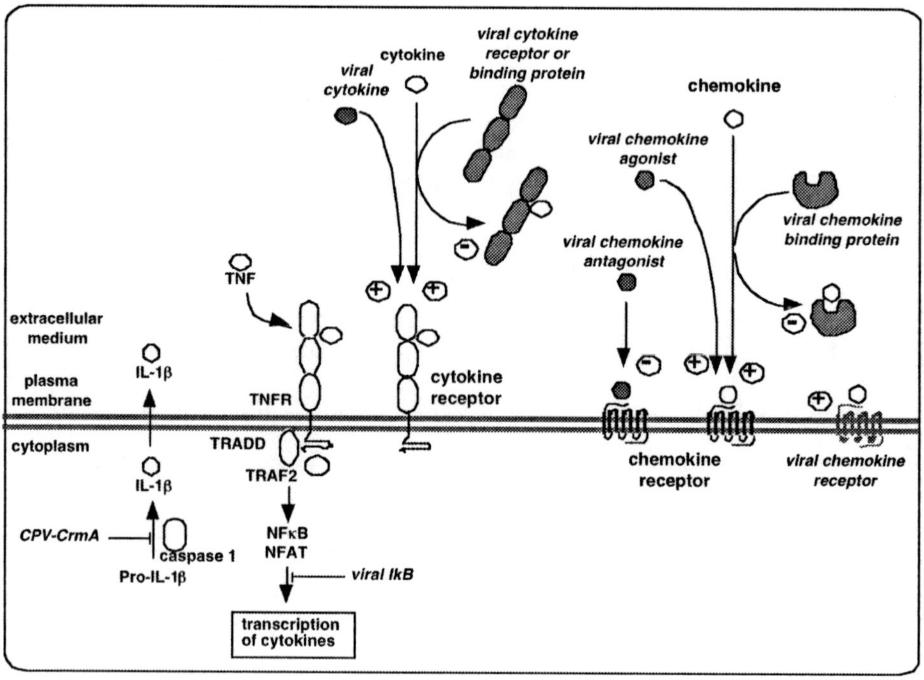

Figure 3. Viral modulation of cytokine networks.

3.3. Viral modulation of IFN effector functions

One of the best-characterized anti-viral effects of IFN is the induction of the PKR and 2'5'OAS pathways that lead to the blockade of viral gene expression (Gale and Katze, 1998; Goodbourn et al., 2000; Katze et al., 2002). Double stranded RNA (dsRNA)-activated PKR phosphorylates the translation initiation factor 2α (eIF-2α) causing the arrest of protein synthesis. Viral mechanisms that prevent the activation of PKR include: (i) proteins encoded by reovirus (σ3) and VV (E3L) that bind dsRNA; (ii) RNAs encoded by adenovirus (VAI) and HIV (TAR) that bind PKR but do not activate the enzyme; and (iii) proteins encoded by HSV (US11) and HIV (Tat) that bind directly to PKR and inactivate the enzyme (Figure 2). An alternative strategy used by VV (K3L) and HSV (IC34.5) to block the PKR pathway is to prevent the phosphorylation of eIF-2α.

The formation of dsRNA during viral replication also activates the enzyme 2'5'OAS, producing 2'5'OA which in turn activates RNase L and the degradation of RNA within the infected cell, causing the blockade of viral replication. HSV produces analogs of 2'5'OA which bind RNaseL and inhibit this pathway (Goodbourn et al., 2000). The proteins σ3 from reovirus and E3L from VV bind dsRNA and thus inhibit the activation of 2'5'OAS as well as PKR (see above).

4. Modulation of cytokine networks

The immune response is orchestrated by a complex network of cytokines and cytokine receptors. In addition, some cytokines such as IFN and tumour necrosis factor (TNF) may act directly on infected cells and restrict virus replication. Viral anti-cytokine strategies often downregulate the immune responses to infection. However, viruses may take advantage of the effects that cytokines have on cell physiology to enhance the ability of immune cells to support viral replication, thus promoting viral replication rather than evading the immune system. Viruses modulate the activity of cytokines at different levels (Alcami, 2003; McFadden and Murphy, 2000; Seet et al., 2003) (Figure 3).

4.1. Viral modulation of cytokine expression and activation

African swine fever virus (ASFV) replicates in macrophages, one of the major cytokine producers, and illustrates the control of cytokine production by viruses. The ASFV IkB homologue interacts with nuclear factor kB (NF-kB) and nuclear factor activated T cell (NFAT) and prevents the transcription of cytokine genes (Miskin et al., 1998) (Figure 3). The CPV protein cytokine response modifier A (CrmA) inhibits the IL-1β converting enzyme (ICE or caspase 1), which proteolytically cleaves the precursor form of pro-inflammatory cytokine IL-1β (pro-IL-1β) to generate mature, active IL-1β (Ray et al., 1992).

4.2. Viral modulation of cytokine signal transduction

In addition to the various anti-IFN mechanisms discussed above, adenoviruses interfere with signaling induced by TNF at different levels (Mahr and Gooding, 1999). The poxvirus VV encodes two intracellular proteins (A52R and A46R) that block signaling mediated by Toll-like and IL-1 receptors by mimicking intracellular regulatory domains of these receptors (Bowie et al., 2000; Harte et al., 2003). The Toll-like receptor pathway is important in NF-kB activation and amplification of inflammatory signals. Some viruses subvert cytokine-mediated signaling for their own benefit. The latent membrane protein 1 (LMP1) of Epstein Barr virus (EBV), a virus that establishes latency in B cells, best illustrates this. LMP1 is a membrane protein that lacks sequence similarity to TNF receptors (TNFRs) but binds TNFR activating factors (TRAFs). Trimeric TNF induces recruitment of TRAFs followed by a signalling cascade leading to either B cell proliferation or apoptosis. In EBV-infected cells, the recruitment of TRAFs by LMP1 induces cell proliferation and ensures survival of infected cells that carry the latent viral genome (Farrell, 1998).

4.3. Molecular mimicry: viral cytokines and cytokine receptors

The genome sequence of large DNA viruses (poxviruses and herpesviruses) identified viral proteins with sequence similarity to ligands and receptors of the cytokine family (Alcami, 2003; McFadden and Murphy, 2000; Seet et al., 2003) (Figure 3). This suggested that, during evolution, viruses may have 'stolen' host immune genes which in some instances have been modified for the benefit of the virus.

4.3.1. Viral cytokines

Viral cytokines act normally as agonists (Figure 3). The EBV IL-10 homologue (vIL-10) was the first viral cytokine identified with immunomodulatory activity, suppresing cellular immunity by inhibiting IFN-γ production (Hsu et al., 1990). Other viral cytokines are the KSHV IL-6 homologue (Moore et al., 1996) and the

HVS IL-17 homologue (Yao et al., 1995). In addition to their immunomodulatory activity, these cytokines promote the proliferation of B and T cells, the major host cell types for these viruses, and may favor viral replication.

4.3.2. Secreted viral cytokine receptors

Viral genes encoding soluble cytokine decoy receptors have been identified mainly in the poxvirus family and in some herpesviruses (Figure 3). The poxvirus soluble cytokine receptors were initially identified because of their sequence similarity to the extracellular binding domain of the host counterparts. A second class of cytokine receptors, known as binding proteins, have very limited sequence similarity to the host cytokine receptors and were identified in cytokine binding or activity assays.

Soluble TNFRs (vTNFRs) are encoded by the M-T2 gene of MV and four different genes, designated CrmB, CrmC, CrmD and CrmE, from other poxviruses including CPV, VV and EV (Hu et al., 1994; Loparev et al., 1998; Saraiva and Alcami, 2001; Smith et al., 1991; 1996). The reasons for the existence in poxvirus genomes of such a variety of vTNFRs are unknown. Further complexity has been added by the recent description of a secreted protein encoded by tanapox virus that binds TNF but has no sequence similarity to TNFRs (Brunetti et al., 2003).

Another member of the TNFR superfamily has been identified in CPV and EV: a secreted CD30 (vCD30) (Panus et al., 2002; Saraiva et al., 2002). vCD30 blocks the interaction of CD30 to CD30 ligand (CD30L) and also induces reverse signaling in CD30L-expressing cells. EV vCD30 down-regulates type 1 cytokine-mediated inflammatory responses in mice, a property that may help the virus to evade an efficient anti-viral response (Saraiva et al., 2002).

The viral IL-1β receptor (vIL-1βR) acts as a decoy receptor (Alcami and Smith, 1992; Spriggs et al., 1992) and, in contrast to the host membrane IL-1Rs, binds exclusively IL-1β and not IL-1α or IL-1 receptor antagonist. Deletion of this gene in VV causes attenuation when the virus is administered intracranially (Spriggs et al., 1992) but leads to enhanced virulence associated with fever and accelerated death in mice infected through the respiratory route, a natural route of poxvirus transmission (Alcami and Smith, 1992; 1996). The enhanced virulence observed illustrates that some viral immunomodulatory proteins may reduce pathology induced by excess production of cytokines, in this case IL-1β, during infection rather than cause immunosuppression.

4.3.3. Viral cytokine binding proteins

IL-18 is a pro-inflammatory cytokine that is required, together with IL-12, for induction of IFN-γ and the generation of an efficient cellular response. The viral IL-18 binding protein (vIL-18BP) encoded by several poxviruses is homologue of the human IL-18BP but has no sequence similarity to membrane IL-18 receptors (Born et al., 2000; Smith et al., 2000; Xiang and Moss, 1999). The vIL-18BP down-regulates IFN-γ production and NK responses in EV-infected mice. An interesting example of a viral protein that binds ligands that interact with different host receptors is the orf virus protein that binds granulocyte macrophage-colony stimulating factor and IL-2 (Deane et al., 2000).

5. Modulation of chemokine networks

Chemokines are a family of small molecular size cytokines that induce cell migration and other biological effects, such as cell differentiation and angiogenesis (Baggiolini, 1998). There are more than 40 chemokines identified that are structurally classified as CC, CXC, C or CX3C. Chemokines mediate their biological effects through nearly 20 seven-transmembrane-domain G-protein coupled receptors. Viruses, mainly herpesviruses and poxviruses, utilize three mechanisms to modulate chemokine activity: expression of chemokine homologues, of chemokine receptor homologues, and of secreted chemokine binding proteins (Alcami, 2003; Lalani et al., 2000; McFadden and Murphy, 2000; Murphy, 2001) (Figure 3).

5.1. Viral chemokine homologues

The chemokine homologues vMIP-II from KSHV and MC148 from molluscum contagiosum virus (MCV) bind to chemokine receptors but do not transduce signals, acting as antagonists (Boshoff et al., 1997; Damon et al., 1998; Kledal et al., 1997; Krathwohl et al., 1997) (Figure 3). By contrast, viruses encoding chemokine agonists take advantage of the chemokine system to recruit to areas of infection immune cells that support viral replication in order to enhance viral dissemination. HCMV UL146, known as vCXC-1, induces neutrophil chemotaxis and may account for the association of neutrophils with HCMV infections (Penfold et al., 1999). Human herpesvirus 6 (HHV-6) U83 induces monocyte migration that may favour the establishment of viral latency (Zou et al., 1999). The KSHV chemokines may play a role in pathogenesis due to their angiogenic properties that may cause the enhanced vascularization observed in Kaposi's sarcoma lesions (Boshoff et al., 1997). In addition, the KSHV chemokines are chemoattractants for Th2 polarized T cells and modulate the anti-viral immune response (Dairaghi et al., 1999; Endres et al., 1999; Sozzani et al., 1998; Stine et al., 2000).

5.2. Viral chemokine receptor homologues

Chemokine receptor homologues have been described in herpesviruses and poxviruses. These proteins are expressed at the surface of infected cells or retained intracellularly, and their ability to neutralize chemokines may be limited (Figure 3). Some viral chemokine receptors bind chemokines and modulate the physiological state of the cell by activating chemokine signaling pathways. KSHV ORF74 is constitutively active and binding of chemokines modulates the activation status of the receptor (Arvanitakis et al., 1997; Rosenkilde et al., 1999). Transgenic mice expressing ORF74 develop Kaposi's sarcoma-like lesions (Arvanitakis et al., 1997; Yang et al., 2000), suggesting that KSHV ORF74 may contribute to neoplasia through a paracrine effect. HCMV US28 modifies the chemokine environment of infected cells by sequestering and internalizing chemokines and depleting them from the medium (Bodaghi et al., 1998). A role of viral chemokine receptors in pathology is also illustrated by HCMV US28, which causes vascular smooth muscle cell migration and may contribute to vascular disease (Streblow et al., 1999).

5.3. Viral chemokine binding proteins (vCKBPs)

The vCKBPs encoded by poxviruses and herpesviruses sequester chemokines without apparent sequence similarity to host chemokine receptors (Seet and McFadden, 2002; Smith et al., 2003) (Figure 3). Most of these viral chemokine inhibitors have been shown to bind with high affinity a broad range of chemokines and to prevent the interaction of chemokines with their specific cellular receptors. Several vCKBPs have been identified: (i) the MV M-T7 protein (Lalani et al., 1997); (ii) the 35 kDa identified in cultures infected with MV, VV or CPV (Alcami et al., 1998; Graham et al., 1997; Smith et al., 1997); (iii) the protein M3 encoded by murine gammaherpesvirus 68 (MHV-68) (Parry et al., 2000; van Berkel et al., 2000); and (iv) the glycoprotein G (gG) encoded by some alphaherpesviruses (Bryant et al., 2003). All vCKBPs are secreted proteins with the exception of the alphaherpesvirus gG that is a membrane protein, present in the virions of some viruses, which is proteolytically cleaved and released to the extracellular space. vCKBPs mimic the interaction of chemokines with the host receptors, in spite of their lack of amino acid sequence similarity (Alexander et al., 2002; Beck et al., 2001; Seet et al., 2001; Webb et al., 2003). The crystal structure of the poxvirus 35 kDa and MHV-68 M3 vCKBPs have been reported (Alexander et al., 2002; Carfi et al., 1999).

6. Evasion of antigen presentation and the cellular response

The variability of antigens from RNA viruses, discussed above, is relevant for evasion of CTL recognition. However, those viruses such as herpesviruses that establish infections for the life of the host but have a very low mutation rate must identify alternative mechanisms to persist in the presence of a vigorous immune response (Fruh et al., 1999; Johnson and Hill, 1998; Yewdell and Hill, 2002) (Figure 4).

6.1. Blockade of antigen presentation mediated by MHC class I molecules

Viral proteins synthesized in the infected cell are degraded in the cytosol mostly by the proteasome, the major intracellular proteolytic complex, and the peptides generated are translocated to the lumen of the endoplasmic reticulum (ER) by transporters associated with antigen processing (TAP). The peptides associate with MHC class I molecules and β2-microglobulin to form a complex that is transported to the cell surface through the Golgi apparatus. The presentation of peptides derived from viral proteins activates specific CTLs that may destroy the infected cell (Figure 4).

The nuclear antigen EBNA1 encoded by EBV has repeated glycine and alanine residues that prevent protein degradation by the proteasome (Levitskaya et al., 1995). The HSV protein ICP47 and the HCMV US6 bind to TAP and inhibit peptide translocation from the cytosol to the ER lumen (Ahn et al., 1997; Hengel et al., 1997; Lehner et al., 1997; York et al., 1994).

The degradation of newly synthesized MHC class I molecules is an alternative mechanism. The HCMV US2 and US11 proteins are membrane glycoproteins that interact with class I molecules and cause them to be transported back to the cytosol,

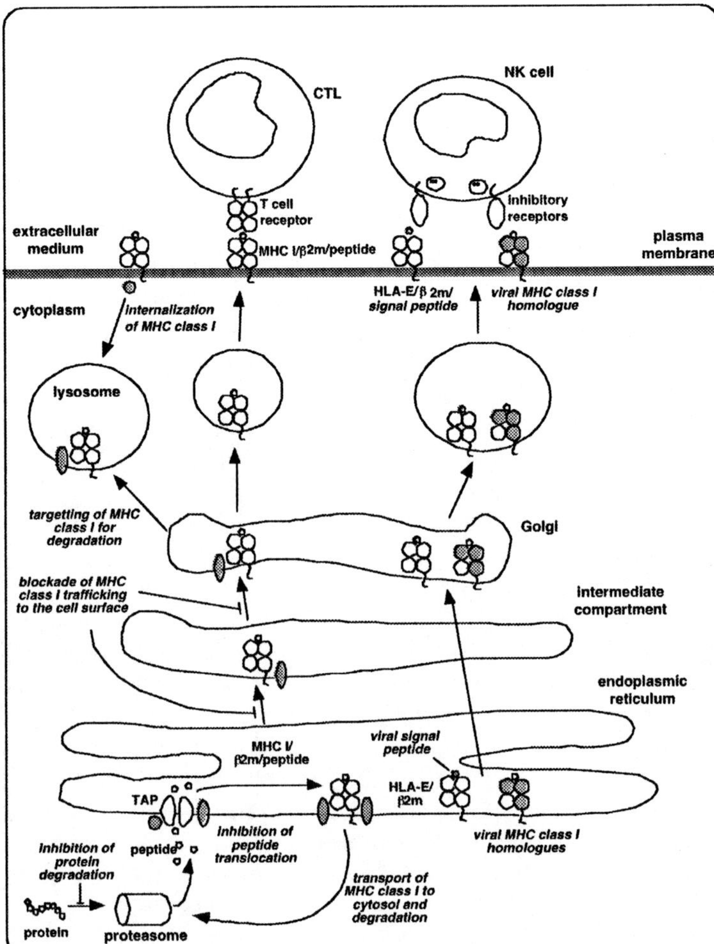

Figure 4. Viral inhibition of antigen presentation and NK cell response.

where class I molecules are ubiquitinylated and destroyed by the proteasome (Wiertz et al., 1996a; 1996b). The molecular mechanism behind this immune evasion strategy has attracted much interest because it sheds light on an important transport system from the ER to the cytosol (Lilley and Ploegh, 2004; Ye et al., 2004).

Several viral proteins, such as the adenovirus E3/19K and MCMV m152, block trafficking of class I molecules to the cell surface (Jefferies and Burgert, 1990; Ziegler et al., 1997). MCMV m06 binds MHC class I molecules and directs them to the lysosome for degradation (Reusch et al., 1999). The KSHV K3 and K5 proteins ubiquitylate MHC class I and induce class I internalization and endosomal degradation (Hewitt et al., 2002).

6.2. Blockade of antigen presentation by class II MHC molecules

MHC class II molecules present peptides derived from exogenous antigens that are internalized and degraded in the lysosome. Only a few viral proteins have been found to block this pathway of antigen presentation. HCMV US2 induces the degradation of MHC class II molecules in the proteasome and HIV Nef interferes with the transport of MHC class II molecules to compartments where they associate with peptides (Tomazin et al., 1999).

6.3. Evasion of the lysis of virus-infected cells by NK cells

NK cells constitute one of the first lines of defense against viruses. These cells possess specific activating receptors that recognize carbohydrate and induce their cytolytic activity. NK cells also express inhibitory receptors that recognize MHC class I molecules and block the NK cell cytotoxic machinery. In this way, NK cells recognize healthy cells expressing normal levels of MHC class I molecules and destroy cells that do not express class I molecules. The down-regulation of MHC class I molecules by some viruses prevents the specific recognition of the infected cell by CTLs, but may have dramatic consequences by increasing the susceptibility of virus-infected cells to NK cell lysis (Orange et al., 2002).

To solve this problem, some viruses encode homologues of MHC class I molecules (Figure 4). The MCMV protein m144 forms a complex with β2-microglobulin and is expressed at the cell surface where it induces inhibitory signals in NK cells (Farrell et al., 1997). HLA-E is an MHC class I molecule that presents peptides derived from signal sequences from other MHC class I molecules, interacts with an inhibitory receptor in NK cells, NKG2A, and prevents lysis mediated by NK cells. The signal peptide of HCMV UL40 binds HLA-E and protects the infected cell from NK cell attack (Tomasec et al., 2000). MCMV m152 interferes with class I antigen presentation (see above) and it was also found to reduce the cell surface levels of the class I-like molecules RAE and H60, which serve as ligands for the NK cell activating receptor NKG2D (Krmpotic et al., 2002).

6.4. Viral inhibition of dendritic cell (DC) functions

Antigen presenting cells (APCs), usually DCs, capture antigen at the site of infection and migrate to the lymph nodes where they interact and activate naïve T cells. The T cells acquire a mature phenotype and migrate to infected tissues where they can activate anti-viral effector functions, such as cytolytic activity or the release of anti-viral cytokines. Inhibition of DC maturation has been reported for some viruses including HSV, MCMV and HCMV (Andrews et al., 2001; Moutaftsi et al., 2002; Salio et al., 1999).

7. Apoptosis

Apoptosis or programmed cell death can be activated by TNF superfamily members, iradiation, inhibitors of cell cycle or infectious agents such as viruses (Figure 5). Apoptosis can be considered an innate mechanism of defence against viral infection that limits virus replication. In addition, CTLs and NK cells destroy virus-infected cells by inducing apoptosis as a result of the production

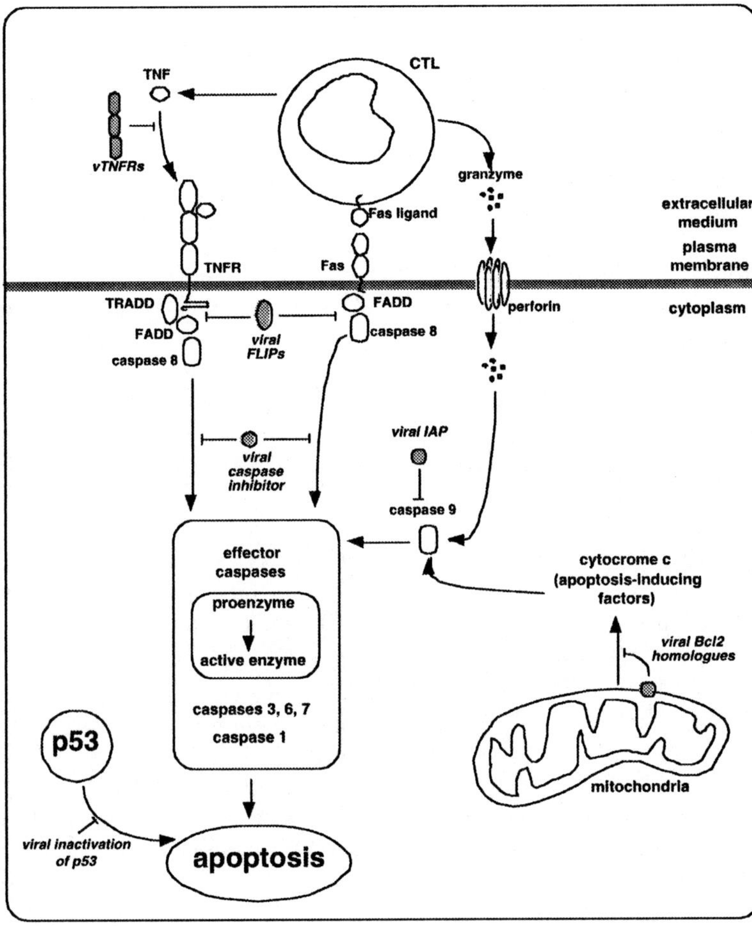

Figure 5. Viral inhibition of apoptosis.

of cytokines such as TNF, the activation of Fas at the cell surface by FasL or the secretion of perforin that enables other secreted proteins (granzymes) to enter the cell. Although the apoptotic response limits virus replication when activated at early times of infection, it may also favour the release of virus particles and viral dissemination under some circumstances. Many cellular proteins involved in the control of apoptosis are targets of viral anti-apoptotic mechanisms (Benedict et al., 2002; Everett and McFadden, 1999) (Figure 5).

7.1. Blockade of signaling through TNFR and Fas

The poxvirus decoy vTNFRs, discussed above, inhibit TNF-mediated apoptosis (Alcami, 2003; McFadden and Murphy, 2000; Seet et al., 2003) (Figure 5). The interaction of TNFRs and Fas with their respective ligands induces the binding of cytoplasmic factors that interact through death domains (DD) and death effector domains (DED) present in adaptor proteins and some caspases. These factors initiate a cascade of interactions that culminate with the activation of caspases and the induction of apoptosis. MCV and KSHV encode vFLIPs (FLICE or caspase 8-inhibitory proteins) that prevent signal transduction through receptors for TNF and FasL (Bertin et al., 1997; Thome and Tschopp, 2001). Adenovirus proteins encoded by the E3 region inhibit TNF- and Fas-mediated apoptosis (Wold, 1999). The adenovirus receptor internalization and degradation (RID) complex (E3/10.4K and E3/14.5K proteins) induces internalization and lysosomal degradation of TNFR and Fas (Burgert et al., 2002; McNees and Gooding, 2002).

7.2. Inhibition of caspases

Caspases are proteolytic enzymes critical for the induction of apoptosis, and many viral proteins inhibit their activity (Figure 5). The CrmA protein encoded by CPV, VV and other poxviruses blocks caspase 1 and caspase 8, and prevents apoptosis initiated by several stimuli including TNF, FasL and granzyme B (Miura et al., 1995; Talley et al., 1995; Tewari et al., 1995). Caspase activity is regulated by a family of inhibitor of apoptosis proteins (IAPs) that have been identified in baculoviruses, a family of insect viruses, and ASFV (Clem and Miller, 1994; Neilan et al., 1993).

7.3. Inhibition of mitochondrial pathways of caspase activation

Apoptosis can also be controlled at the level of mitochondria, as a result of loss of extracellular viral signals or by signals affecting the cell cycle (Rathmell and Thompson, 2002) (Figure 5). The release of cytochrome c from mitochondria activates caspase 9 leading to apoptosis. A number of viruses, including adenovirus, ASFV, EBV, KSHV and MHV-68, encode homologues of Bcl-2, a cellular protein that inhibits the release of cytochrome C, to prevent the induction of apoptosis (Afonso et al., 1996; Cheng et al., 1997; Gangappa et al., 2002; Henderson et al., 1993; Sarid et al., 1997). MV M11L associates with a component of the mitochondrial permeability transition pore and prevents apoptosis (Everett et al., 2002).

7.4. Other anti-apoptotic mechanisms

The tumour suppresor p53 limits cell proliferation by inducing apoptosis or cell cycle arrest. p53 stimulates transcription of genes involved in apoptotic pathways (Figure 5). Viruses such as human papillomavirus (E6 protein), adenovirus (E1B-55K protein) and hepatitis B virus (pX protein) disrupt apoptosis by inactivating p53 (Scheffner et al., 1990; Steegenga et al., 1998; Wang et al., 1995; Werness et al., 1990). An alternative mechanism is encoded by MCV MC66, a glutathione superoxide that protects infected cells from peroxide radicals and ultraviolet radiation, and perhaps from peroxide radicals induced by TNF, macrophages and neutrophils (Shisler et al., 1998).

8. Conclusions and future perspectives

In the past, viral immunology has characterized the host mechanisms that prevent viral infections. Antigenic variability was identified many yeas ago as a viral immune evasion strategy. However, it is in the last ten years that we have become more aware of the large number of virus-encoded proteins that block specific mechanisms of anti-viral immunity. We do not know the function of many viral proteins and it is likely that they may encode new viral immune evasion strategies. The fact that we have not identified these viral functions may reflect our limited knowledge of immunity.

The reasons why a virus blocks a particular immune mechanism may depend on the replication and transmission mechanisms adopted by the virus. Many questions need to be addressed in the future. We need to determine which viral genes are required for the initial steps of the infection or for efficient transmission to another host in the presence of an active immune system. The apparent complexity and redundancy of these viral proteins may only be understood in the context of infection in the animal host. The lack of good animal models of infection for many viruses is limiting our understanding of the function of viral immune evasion strategies *in vivo*.

A better knowledge of virus-immune system interactions will help us to understand the molecular mechanisms of viral pathogenesis. Very often, the pathology observed in viral infections is not caused directly by viral replication but it is a consequence of the immune and inflammatory responses to viral infection that may damage the host. Moreover, some viral proteins may directly cause the pathological effects observed during infection. Some human inflammatory and autoimmune diseases are the result of an uncontrolled production of immune mediators such as TNF and IL-1 that activate the immune response. More examples of viral proteins that enhance these immunopathological responses in the infected host are being identified.

The deletion of viral immune evasion proteins from the viral genome causes viral attenuation in most cases due to increased immune responses. However, enhanced viral virulence has been described for VV and CPV when some of these viral functions have been deleted from the genome. These include the soluble vIL-1βR, IMP and the 35 kDa vCKBP (Alcami and Smith, 1992; Miller et al., 1997; Reading et al., 2003). These surprising observations suggest that some viral immune evasion proteins may not function to promote viral replication but to 'protect' the host from the detrimental effects of an over-reactive immune system, mediated in these examples by IL-1β, complement or chemokines, respectively. An enhanced host survival will be beneficial since the virus fully depends on the host for its replication. The complex interaction of each virus with the immune system is unique, and a better understanding of this interaction will lead to new strategies to control viral replication and the design of more efficient viral vaccines.

The fact that viruses have co-existed with their hosts during evolution and are able to survive in the presence of potent host defence mechanisms suggests that viruses have a better knowledge than us of the mechanisms of immunity. Viruses point at those molecules that play a critical role in defence against pathogens, one of the primary aims of the immune system, and thus viruses offer us a unique opportunity to identify basic mechanisms of immunity. For example, the importance of cytokines and IFNs in anti-viral defense is highlighted by the number of mechanisms that viruses encode to block their activity (Alcami and Koszinowski, 2000; Tortorella et al., 2000). The study of viral mechanisms of immune evasion is fascinating and will increase our understanding of the complex interaction of viruses with the host immune system.

Acknowledgments

Research in the author's laboratory is funded by grants from the Wellcome Trust, the European Union and the Spanish Ministry of Education and Science.

References

Afonso, C. L., Neilan, J. G., Kutish, G. F., and Rock, D. L. (1996). An African swine fever virus Bcl-2 homolog, 5-HL, suppresses apoptotic cell death. J. Virol. *70*, 4858-4863.

Ahn, K., Gruhler, A., Galocha, B., Jones, T. R., Wiertz, E. J., Ploegh, H. L., Peterson, P. A., Yang, Y., and Fruh, K. (1997). The ER-luminal domain of the HCMV glycoprotein US6 inhibits peptide translocation by TAP. Immunity *6*, 613-621.

Alcami, A. (2003). Viral mimicry of cytokines, chemokines and their receptors. Nature Rev. Immunol. *3*, 36-50.

Alcami, A., and Koszinowski, U. H. (2000). Viral mechanisms of immune evasion. Immunol. Today *21*, 447-455.

Alcami, A., and Smith, G. L. (1992). A soluble receptor for interleukin-1 beta encoded by vaccinia virus: a novel mechanism of virus modulation of the host response to infection. Cell *71*, 153-167.

Alcami, A., and Smith, G. L. (1995). Vaccinia, cowpox, and camelpox viruses encode soluble gamma interferon receptors with novel broad species specificity. J. Virol. *69*, 4633-4639.

Alcami, A., and Smith, G. L. (1996). A mechanism for the inhibition of fever by a virus. Proc. Natl. Acad. Sci. USA. *93*, 11029-11034.

Alcami, A., Symons, J. A., Collins, P. D., Williams, T. J., and Smith, G. L. (1998). Blockade of chemokine activity by a soluble chemokine binding protein from vaccinia virus. J. Immunol. *160*, 624-633.

Alexander, J. M., Nelson, C. A., van Berkel, V., Lau, E. K., Studts, J. M., Brett, T. J., Speck, S. H., Handel, T. M., Virgin, H. W., and Fremont, D. H. (2002). Structural basis of chemokine sequestration by a herpesvirus decoy receptor. Cell *111*, 343-356.

Andrews, D. M., Andoniou, C. E., Granucci, F., Ricciardi-Castagnoli, P., and Degli-Esposti, M. A. (2001). Infection of dendritic cells by murine cytomegalovirus induces functional paralysis. Nat. Immunol. *2*, 1077-1084.

Arvanitakis, L., Geras-Raaka, E., Varma, A., Gershengorn, M. C., and Cesarman, E. (1997). Human herpesvirus KSHV encodes a constitutively active G-protein-coupled receptor linked to cell proliferation. Nature *385*, 347-350.

Baggiolini, M. (1998). Chemokines and leukocyte traffic. Nature *392*, 565-568.

Beck, C. G., Studer, C., Zuber, J. F., Demange, B. J., Manning, U., and Urfer, R. (2001). The viral CC chemokine-binding protein vCCI inhibits monocyte chemoattractant protein-1 activity by masking its CCR2B-binding site. J. Biol. Chem. *276*, 43270-43276.

Benedict, C. A., Norris, P. S., and Ware, C. F. (2002). To kill or be killed: viral evasion of apoptosis. Nat. Immunol. *3*, 1013-1018.

Bertin, J., Armstrong, R. C., Ottilie, S., Martin, D. A., Wang, Y., Banks, S., Wang, G. H., Senkevich, T. G., Alnemri, E. S., Moss, B., et al. (1997). Death effector domain-containing herpesvirus and poxvirus proteins inhibit both Fas- and TNFR1-induced apoptosis. Proc. Natl. Acad. Sci. USA. *94*, 1172-1176.

Bodaghi, B., Jones, T. R., Zipeto, D., Vita, C., Sun, L., Laurent, L., Arenzana-Seisdedos, F., Virelizier, J. L., and Michelson, S. (1998). Chemokine sequestration by viral chemoreceptors as a novel viral escape strategy: withdrawal of chemokines from the environment of cytomegalovirus-infected cells. J. Exp. Med. *188*, 855-866.

Born, T. L., Morrison, L. A., Esteban, D. J., VandenBos, T., Thebeau, L. G., Chen, N., Spriggs, M. K., Sims, J. E., and Buller, R. M. (2000). A poxvirus protein that binds to and inactivates IL-18, and inhibits NK cell response. J. Immunol. *164*, 3246-3254.

Boshoff, C., Endo, Y., Collins, P. D., Takeuchi, Y., Reeves, J. D., Schweickart, V. L., Siani, M. A., Sasaki, T., Williams, T. J., Gray, P. W., et al. (1997). Angiogenic and HIV-inhibitory functions of KSHV-encoded chemokines. Science *278*, 290-294.

Bowie, A., Kiss-Toth, E., Symons, J. A., Smith, G. L., Dower, S. K., and O'Neill, L. A. (2000). A46R and A52R from vaccinia virus are antagonists of host IL-1 and toll-like receptor signaling. Proc. Natl. Acad. Sci. USA. *97*, 10162-10167.

Brunetti, C. R., Paulose-Murphy, M., Singh, R., Qin, J., Barrett, J. W., Tardivel, A., Schneider, P., Essani, K., and McFadden, G. (2003). A secreted high-affinity inhibitor of human TNF from Tanapox virus. Proc. Natl. Acad. Sci. USA. *100*, 4831-4836.

Bryant, N. A., Davis-Poynter, N., Vanderplasschen, A., and Alcami, A. (2003). Glycoprotein G isoforms from some alphaherpesviruses function as broad-spectrum chemokine binding proteins. Embo J. *22*, 833-846.

Burgert, H. G., Ruzsics, Z., Obermeier, S., Hilgendorf, A., Windheim, M., and Elsing, A. (2002). Subversion of host defense mechanisms by adenoviruses. Curr. Top Microbiol. Immunol. *269*, 273-318.

Carfi, A., Smith, C. A., Smolak, P. J., McGrew, J., and Wiley, D. C. (1999). Structure of a soluble secreted chemokine inhibitor vCCI (p35) from cowpox virus. Proc. Natl. Acad. Sci. USA. *96*, 12379-12383.

Cheng, E. H., Nicholas, J., Bellows, D. S., Hayward, G. S., Guo, H. G., Reitz, M. S., and Hardwick, J. M. (1997). A Bcl-2 homolog encoded by Kaposi sarcoma-associated virus, human herpesvirus 8, inhibits apoptosis but does not heterodimerize with Bax or Bak. Proc. Natl. Acad. Sci. USA. *94*, 690-694.

Clem, R. J., and Miller, L. K. (1994). Control of programmed cell death by the baculovirus genes p35 and iap. Mol. Cell Biol. *14*, 5212-5222.

Colamonici, O. R., Domanski, P., Sweitzer, S. M., Larner, A., and Buller, R. M. L. (1995). Vaccinia virus B18R gene encodes a type I interferon-binding protein that blocks interferon alpha transmembrane signaling. J. Biol. Chem. *270*, 15974-15978.

Dairaghi, D. J., Fan, R. A., McMaster, B. E., Hanley, M. R., and Schall, T. J. (1999). HHV8-encoded vMIP-I selectively engages chemokine receptor CCR8. Agonist and antagonist profiles of viral chemokines. J. Biol. Chem. *274*, 21569-21574.

Damon, I., Murphy, P. M., and Moss, B. (1998). Broad spectrum chemokine antagonistic activity of a human poxvirus chemokine homolog. Proc. Natl. Acad. Sci. USA. *95*, 6403-6407.

Deane, D., McInnes, C. J., Percival, A., Wood, A., Thomson, J., Lear, A., Gilray, J., Fleming, S., Mercer, A., and Haig, D. (2000). Orf virus encodes a novel secreted protein inhibitor of granulocyte-macrophage colony-stimulating factor and interleukin-2. J. Virol. *74*, 1313-1320.

Didcock, L., Young, D. F., Goodbourn, S., and Randall, R. E. (1999). The V protein of simian virus 5 inhibits interferon signalling by targeting STAT1 for proteasome-mediated degradation. J. Virol. *73*, 9928-9933.

Endres, M. J., Garlisi, C. G., Xiao, H., Shan, L., and Hedrick, J. A. (1999). The Kaposi's sarcoma-related herpesvirus (KSHV)-encoded chemokine vMIP-I is a specific agonist for the CC chemokine receptor (CCR)8. J. Exp. Med. *189*, 1993-1998.

Everett, H., Barry, M., Sun, X., Lee, S. F., Frantz, C., Berthiaume, L. G., McFadden, G., and Bleackley, R. C. (2002). The myxoma poxvirus protein, M11L, prevents apoptosis by direct interaction with the mitochondrial permeability transition pore. J. Exp. Med. *196*, 1127-1139.

Everett, H., and McFadden, G. (1999). Apoptosis: an innate immune response to virus infection. Trends Microbiol. *7*, 160-165.

Farrell, H. E., Vally, H., Lynch, D. M., Fleming, P., Shellam, G. R., Scalzo, A. A., and Davis-Poynter, N. J. (1997). Inhibition of natural killer cells by a cytomegalovirus MHC class I homologue *in vivo*. Nature *386*, 510-514.

Farrell, P. J. (1998). Signal transduction from the Epstein-Barr virus LMP-1 transforming protein. Trends Microbiol. *6*, 175-177; discussion 177-178.

Favoreel, H. W., Van de Walle, G. R., Nauwynck, H. J., and Pensaert, M. B. (2003). Virus complement evasion strategies. J. Gen. Virol. *84*, 1-15.

Flint, S. J., Enquist, L. W., Krug, R. M., Racaniello, V. R., and Skalka, A. M. (2000). Principles of Virology (Washington, D.C., ASM Press).

Fruh, K., Gruhler, A., Krishna, R. M., and Schoenhals, G. J. (1999). A comparison of viral immune escape strategies targeting the MHC class I assembly pathway. Immunol. Rev. *168*, 157-166.

Gale, M., Jr., and Katze, M. G. (1998). Molecular mechanisms of interferon resistance mediated by viral-directed inhibition of PKR, the interferon-induced protein kinase. Pharmacol Ther *78*, 29-46.

Gangappa, S., van Dyk, L. F., Jewett, T. J., Speck, S. H., and Virgin, H. W. t. (2002). Identification of the *in vivo* role of a viral bcl-2. J. Exp. Med. *195*, 931-940.

Goodbourn, S., Didcock, L., and Randall, R. E. (2000). Interferons: cell signalling, immune modulation, antiviral responses and virus countermeasures. J. Gen. Virol. *81*, 2341-2364.

Graham, K. A., Lalani, A. S., Macen, J. L., Ness, T. L., Barry, M., Liu, L. Y., Lucas, A., Clark-Lewis, I., Moyer, R. W., and McFadden, G. (1997). The T1/35kDa family of poxvirus-secreted proteins bind chemokines and modulate leukocyte influx into virus-infected tissues. Virology *229*, 12-24.

Harte, M. T., Haga, I. R., Maloney, G., Gray, P., Reading, P. C., Bartlett, N. W., Smith, G. L., Bowie, A., and O'Neill, L. A. (2003). The poxvirus protein A52R targets Toll-like receptor signaling complexes to suppress host defense. J. Exp. Med. *197*, 343-351.

Henderson, S., Huen, D., Rowe, M., Dawson, C., Johnson, G., and Rickinson, A. (1993). Epstein-Barr virus-coded BHRF1 protein, a viral homologue of Bcl-2, protects human B cells from programmed cell death. Proc. Natl. Acad. Sci. USA. *90*, 8479-8483.

Hengel, H., Koopmann, J. O., Flohr, T., Muranyi, W., Goulmy, E., Hammerling, G. J., Koszinowski, U. H., and Momburg, F. (1997). A viral ER-resident glycoprotein inactivates the MHC-encoded peptide transporter. Immunity *6*, 623-632.

Hewitt, E. W., Duncan, L., Mufti, D., Baker, J., Stevenson, P. G., and Lehner, P. J. (2002). Ubiquitylation of MHC class I by the K3 viral protein signals internalization and TSG101-dependent degradation. Embo J. *21*, 2418-2429.

Hsu, D. H., de Waal Malefyt, R., Fiorentino, D. F., Dang, M. N., Vieira, P., de Vries, J., Spits, H., Mosmann, T. R., and Moore, K. W. (1990). Expression of interleukin-10 activity by Epstein-Barr virus protein BCRF1. Science *250*, 830-832.

Hu, F., Smith, C. A., and Pickup, D. J. (1994). Cowpox virus contains two copies of an early gene encoding a soluble secreted form of the type II TNF receptor. Virology *204*, 343-356.

Jefferies, W. A., and Burgert, H. G. (1990). E3/19K from adenovirus 2 is an immunosubversive protein that binds to a structural motif regulating the intracellular transport of major histocompatibility complex class I proteins. J. Exp. Med. *172*, 1653-1664.

Johnson, D. C., and Hill, A. B. (1998). Herpesvirus evasion of the immune system. Curr. Top Microbiol. Immunol. *232*, 149-177.

Katze, M. G., He, Y., and Gale, M., Jr. (2002). Viruses and interferon: a fight for supremacy. Nat. Rev. Immunol. *2*, 675-687.

Kledal, T. N., Rosenkilde, M. M., Coulin, F., Simmons, G., Johnsen, A. H., Alouani, S., Power, C. A., Luttichau, H. R., Gerstoft, J., Clapham, P. R., et al. (1997). A broad-spectrum chemokine antagonist encoded by Kaposi's sarcoma-associated herpesvirus. Science *277*, 1656-1659.

Kotwal, G. J., Isaacs, S. N., McKenzie, R., Frank, M. M., and Moss, B. (1990). Inhibition of the complement cascade by the major secretory protein of vaccinia virus. Science *250*, 827-830.

Krathwohl, M. D., Hromas, R., Brown, D. R., Broxmeyer, H. E., and Fife, K. H. (1997). Functional characterization of the C---C chemokine-like molecules encoded by molluscum contagiosum virus types 1 and 2. Proc. Natl. Acad. Sci. USA. *94*, 9875-9880.

Krmpotic, A., Busch, D. H., Bubic, I., Gebhardt, F., Hengel, H., Hasan, M., Scalzo, A. A., Koszinowski, U. H., and Jonjic, S. (2002). MCMV glycoprotein gp40 confers virus resistance to CD8+ T cells and NK cells *in vivo*. Nat. Immunol. *3*, 529-535.

Lalani, A. S., Barrett, J. W., and McFadden, G. (2000). Modulating chemokines: more lessons from viruses. Immunol. Today *21*, 100-106.

Lalani, A. S., Graham, K., Mossman, K., Rajarathnam, K., Clark-Lewis, I., Kelvin, D., and McFadden, G. (1997). The purified myxoma virus gamma interferon receptor homolog M-T7 interacts with the heparin-binding domains of chemokines. J. Virol. *71*, 4356-4363.

Lehner, P. J., Karttunen, J. T., Wilkinson, G. W., and Cresswell, P. (1997). The human cytomegalovirus US6 glycoprotein inhibits transporter associated with antigen processing-dependent peptide translocation. Proc. Natl. Acad. Sci. USA. *94*, 6904-6909.

Levitskaya, J., Coram, M., Levitsky, V., Imreh, S., Steigerwald-Mullen, P. M., Klein, G., Kurilla, M. G., and Masucci, M. G. (1995). Inhibition of antigen processing by the internal repeat region of the Epstein-Barr virus nuclear antigen-1. Nature *375*, 685-688.

Lilley, B. N., and Ploegh, H. L. (2004). A membrane protein required for dislocation of misfolded proteins from the ER. Nature *429*, 834-840.

Loparev, V. N., Parsons, J. M., Knight, J. C., Panus, J. F., Ray, C. A., Buller, R. M., Pickup, D. J., and Esposito, J. J. (1998). A third distinct tumor necrosis factor receptor of orthopoxviruses. Proc. Natl. Acad. Sci. USA. *95*, 3786-3791.

Lubinski, J., Wang, L., Mastellos, D., Sahu, A., Lambris, J. D., and Friedman, H. M. (1999). In vivo role of complement-interacting domains of herpes simplex virus type 1 glycoprotein gC. J. Exp. Med. *190*, 1637-1646.

Mahr, J. A., and Gooding, L. R. (1999). Immune evasion by adenoviruses. Immunol. Rev. *168*, 121-130.

McFadden, G., and Murphy, P. M. (2000). Host-related immunomodulators encoded by poxviruses and herpesviruses. Curr. Opin. Microbiol. *3*, 371-378.

McNees, A. L., and Gooding, L. R. (2002). Adenoviral inhibitors of apoptotic cell death. Virus Res. *88*, 87-101.

Miller, C. G., Shchelkunov, S. N., and Kotwal, G. J. (1997). The cowpox virus-encoded homolog of the vaccinia virus complement control protein is an inflammation modulatory protein. Virology *229*, 126-133.

Miskin, J. E., Abrams, C. C., Goatley, L. C., and Dixon, L. K. (1998). A viral mechanism for inhibition of the cellular phosphatase calcineurin. Science *281*, 562-565.

Miura, M., Friedlander, R. M., and Yuan, J. (1995). Tumor necrosis factor-induced apoptosis is mediated by a CrmA-sensitive cell death pathway. Proc. Natl. Acad. Sci. USA. *92*, 8318-8322.

Moore, P. S., Boshoff, C., Weiss, R. A., and Chang, Y. (1996). Molecular mimicry of human cytokine and cytokine response pathway genes by KSHV. Science *274*, 1739-1744.

Mossman, K., Upton, C., Buller, R. M. L., and McFadden, G. (1995). Species specificity of ectromelia virus and vaccinia virus interferon-γ binding proteins. Virology *208*, 762-769.

Moutaftsi, M., Mehl, A. M., Borysiewicz, L. K., and Tabi, Z. (2002). Human cytomegalovirus inhibits maturation and impairs function of monocyte-derived dendritic cells. Blood *99*, 2913-2921.

Murphy, P. M. (2001). Viral exploitation and subversion of the immune system through chemokine mimicry. Nat. Immunol. *2*, 116-122.

Neilan, J. G., Lu, Z., Afonso, C. L., Kutish, G. F., Sussman, M. D., and Rock, D. L. (1993). An African swine fever virus gene with similarity to the proto-oncogene bcl-2 and the Epstein-Barr virus gene BHRF1. J. Virol. *67*, 4391-4394.

Nishio, M., Garcin, D., Simonet, V., and Kolakofsky, D. (2002). The carboxyl segment of the mumps virus V protein associates with Stat proteins *in vitro* via a tryptophan-rich motif. Virology *300*, 92-99.

Orange, J. S., Fassett, M. S., Koopman, L. A., Boyson, J. E., and Strominger, J. L. (2002). Viral evasion of natural killer cells. Nat. Immunol. *3*, 1006-1012.

Palosaari, H., Parisien, J. P., Rodriguez, J. J., Ulane, C. M., and Horvath, C. M. (2003). STAT protein interference and suppression of cytokine signal transduction by measles virus V protein. J. Virol. *77*, 7635-7644.

Panus, J. F., Smith, C. A., Ray, C. A., Smith, T. D., Patel, D. D., and Pickup, D. J. (2002). Cowpox virus encodes a fifth member of the tumor necrosis factor receptor family: A soluble, secreted CD30 homologue. Proc. Natl. Acad. Sci. USA. *99*, 8348-8353.

Parisien, J. P., Lau, J. F., Rodriguez, J. J., Sullivan, B. M., Moscona, A., Parks, G. D., Lamb, R. A., and Horvath, C. M. (2001). The V protein of human parainfluenza virus 2 antagonizes type I interferon responses by destabilizing signal transducer and activator of transcription 2. Virology *283*, 230-239.

Parry, C. M., Simas, J. P., Smith, V. P., Stewart, C. A., Minson, A. C., Efstathiou, S., and Alcami, A. (2000). A broad spectrum secreted chemokine binding protein encoded by a herpesvirus. J. Exp. Med. *191*, 573-578.

Penfold, M. E., Dairaghi, D. J., Duke, G. M., Saederup, N., Mocarski, E. S., Kemble, G. W., and Schall, T. J. (1999). Cytomegalovirus encodes a potent alpha chemokine. Proc. Natl. Acad. Sci. USA. *96*, 9839-9844.

Puehler, F., Schwarz, H., Waidner, B., Kalinowski, J., Kaspers, B., Bereswill, S., and Staeheli, P. (2003). An interferon-gamma-binding protein of novel structure encoded by the fowlpox virus. J. Biol. Chem. *278*, 6905-6911.

Rathmell, J. C., and Thompson, C. B. (2002). Pathways of apoptosis in lymphocyte development, homeostasis, and disease. Cell *109 Suppl*, S97-107.

Ray, C. A., Black, R. A., Kronheim, S. R., Greenstreet, T. A., Sleath, P. R., Salvesen, G. S., and Pickup, D. J. (1992). Viral inhibition of inflammation: cowpox virus encodes an inhibitor of the interleukin-1 beta converting enzyme. Cell *69*, 597-604.

Reading, P. C., Symons, J. A., and Smith, G. L. (2003). A soluble chemokine-binding protein from vaccinia virus reduces virus virulence and the inflammatory response to infection. J. Immunol. *170*, 1435-1442.

Reusch, U., Muranyi, W., Lucin, P., Burgert, H. G., Hengel, H., and Koszinowski, U. H. (1999). A cytomegalovirus glycoprotein re-routes MHC class I complexes to lysosomes for degradation. Embo J. *18*, 1081-1091.

Rosengard, A. M., Liu, Y., Nie, Z., and Jimenez, R. (2002). Variola virus immune evasion design: expression of a highly efficient inhibitor of human complement. Proc. Natl. Acad. Sci. USA. *99*, 8808-8813.

Rosenkilde, M. M., Kledal, T. N., Brauner-Osborne, H., and Schwartz, T. W. (1999). Agonists and inverse agonists for the herpesvirus 8-encoded constitutively active seven-transmembrane oncogene product, ORF-74. J. Biol. Chem. *274*, 956-961.

Rother, R. P., Rollins, S. A., Fodor, W. L., Albrecht, J. C., Setter, E., Fleckenstein, B., and Squinto, S. P. (1994). Inhibition of complement-mediated cytolysis by the terminal complement inhibitor of herpesvirus saimiri. J. Virol. *68*, 730-737.

Saifuddin, M., Parker, C. J., Peeples, M. E., Gorny, M. K., Zolla-Pazner, S., Ghassemi, M., Rooney, I. A., Atkinson, J. P., and Spear, G. T. (1995). Role of virion-associated glyco sylphosphatidylinositol-linked proteins CD55 and CD59 in complement resistance of cell line-derived and primary isolates of HIV-1. J. Exp. Med. *182*, 501-509.

Salio, M., Cella, M., Suter, M., and Lanzavecchia, A. (1999). Inhibition of dendritic cell maturation by herpes simplex virus. Eur. J. Immunol. *29*, 3245-3253.

Saraiva, M., and Alcami, A. (2001). CrmE, a novel soluble tumor necrosis factor receptor encoded by poxviruses. J. Virol. *75*, 226-233.

Saraiva, M., Smith, P., Fallon, P. G., and Alcami, A. (2002). Inhibition of type 1 cytokine-mediated inflammation by a soluble CD30 homologue encoded by ectromelia (mousepox) virus. J. Exp. Med. *196*, 829-839.

Sarid, R., Sato, T., Bohenzky, R. A., Russo, J. J., and Chang, Y. (1997). Kaposi's sarcoma-associated herpesvirus encodes a functional bcl-2 homologue. Nat. Med. *3*, 293-298.

Scheffner, M., Werness, B. A., Huibregtse, J. M., Levine, A. J., and Howley, P. M. (1990). The E6 oncoprotein encoded by human papillomavirus types 16 and 18 promotes the degradation of p53. Cell *63*, 1129-1136.

Seet, B. T., Johnston, J. B., Brunetti, C. R., Barrett, J. W., Everett, H., Cameron, C., Sypula, J., Nazarian, S. H., Lucas, A., and McFadden, G. (2003). Poxviruses and immune evasion. Annu. Rev. Immunol. *21*, 377-423.

Seet, B. T., and McFadden, G. (2002). Viral chemokine-binding proteins. J. Leukoc Biol. *72*, 24-34.

Seet, B. T., Singh, R., Paavola, C., Lau, E. K., Handel, T. M., and McFadden, G. (2001). Molecular determinants for CC-chemokine recognition by a poxvirus CC- chemokine inhibitor. Proc. Natl. Acad. Sci. USA. *98*, 9008-9013.

Shisler, J. L., Senkevich, T. G., Berry, M. J., and Moss, B. (1998). Ultraviolet-induced cell death blocked by a selenoprotein from a human dermatotropic poxvirus. Science *279*, 102-105.

Smith, C. A., Davis, T., Wignall, J. M., Din, W. S., Farrah, T., Upton, C., McFadden, G., and Goodwin, R. G. (1991). T2 open reading frame from Shope fibroma virus encodes a soluble form of the TNF receptor. Biochem. Biophys Res. Commun. *176*, 335-342.

Smith, C. A., Hu, F. Q., Smith, T. D., Richards, C. L., Smolak, P., Goodwin, R. G., and Pickup, D. J. (1996). Cowpox virus genome encodes a second soluble homologue of cellular TNF receptors, distinct from CrmB, that binds TNF but not LT alpha. Virology *223*, 132-147.

Smith, C. A., Smith, T. D., Smolak, P. J., Friend, D., Hagen, H., Gerhart, M., Park, L., Pickup, D. J., Torrance, D., Mohler, K., et al. (1997). Poxvirus genomes encode a secreted, soluble protein that preferentially inhibits beta chemokine activity yet lacks sequence homology to known chemokine receptors. Virology *236*, 316-327.

Smith, V. P., Bryant, N. A., and Alcami, A. (2000). Ectromelia, vaccinia and cowpox viruses encode secreted interleukin-18-binding proteins. J. Gen. Virol. *81*, 1223-1230.

Smith, V. P., Bryant, N. A., and Alcami, A. (2003). Soluble chemokine binding proteins encoded by viruses. In Chemokines in viral infections, S. Mahalinham, ed. (Eurekah).

Sozzani, S., Luini, W., Bianchi, G., Allavena, P., Wells, T. N., Napolitano, M., Bernardini, G., Vecchi, A., D'Ambrosio, D., Mazzeo, D., et al. (1998). The viral chemokine macrophage inflammatory protein-II is a selective Th2 chemoattractant. Blood 92, 4036-4039.

Spear, G. T., Lurain, N. S., Parker, C. J., Ghassemi, M., Payne, G. H., and Saifuddin, M. (1995). Host cell-derived complement control proteins CD55 and CD59 are incorporated into the virions of two unrelated enveloped viruses. Human T cell leukemia/lymphoma virus type I (HTLV-I) and human cytomegalovirus (HCMV). J. Immunol. *155*, 4376-4381.

Spriggs, M. K., Hruby, D. E., Maliszewski, C. R., Pickup, D. J., Sims, J. E., Buller, R. M., and VanSlyke, J. (1992). Vaccinia and cowpox viruses encode a novel secreted interleukin-1-binding protein. Cell *71*, 145-152.

Steegenga, W. T., Riteco, N., Jochemsen, A. G., Fallaux, F. J., and Bos, J. L. (1998). The large E1B protein together with the E4orf6 protein target p53 for active degradation in adenovirus infected cells. Oncogene *16*, 349-357.

Stine, J. T., Wood, C., Hill, M., Epp, A., Raport, C. J., Schweickart, V. L., Endo, Y., Sasaki, T., Simmons, G., Boshoff, C., et al. (2000). KSHV-encoded CC chemokine vMIP-III is a CCR4 agonist, stimulates angiogenesis, and selectively chemoattracts TH2 cells. Blood *95*, 1151-1157.

Streblow, D. N., Soderberg-Naucler, C., Vieira, J., Smith, P., Wakabayashi, E., Ruchti, F., Mattison, K., Altschuler, Y., and Nelson, J. A. (1999). The human cytomegalovirus chemokine receptor US28 mediates vascular smooth muscle cell migration. Cell *99*, 511-520.

Symons, J. A., Alcami, A., and Smith, G. L. (1995). Vaccinia virus encodes a soluble type I interferon receptor of novel structure and broad species specificity. Cell *81*, 551-560.

Talley, A. K., Dewhurst, S., Perry, S. W., Dollard, S. C., Gummuluru, S., Fine, S. M., New, D., Epstein, L. G., Gendelman, H. E., and Gelbard, H. A. (1995). Tumor necrosis factor alpha-induced apoptosis in human neuronal cells: protection by the antioxidant N-acetylcysteine and the genes bcl-2 and crmA. Mol. Cell Biol. *15*, 2359-2366.

Tewari, M., Telford, W. G., Miller, R. A., and Dixit, V. M. (1995). CrmA, a poxvirus-encoded serpin, inhibits cytotoxic T-lymphocyte-mediated apoptosis. J. Biol. Chem. *270*, 22705-22708.

Thome, M., and Tschopp, J. (2001). Regulation of lymphocyte proliferation and death by FLIP. Nat. Rev. Immunol. *1*, 50-58.

Tomasec, P., Braud, V. M., Rickards, C., Powell, M. B., McSharry, B. P., Gadola, S., Cerundolo, V., Borysiewicz, L. K., McMichael, A. J., and Wilkinson, G. W. (2000). Surface expression of HLA-E, an inhibitor of natural killer cells, enhanced by human cytomegalovirus gpUL40. Science *287*, 1031.

Tomazin, R., Boname, J., Hegde, N. R., Lewinsohn, D. M., Altschuler, Y., Jones, T. R., Cresswell, P., Nelson, J. A., Riddell, S. R., and Johnson, D. C. (1999). Cytomegalovirus US2 destroys two components of the MHC class II pathway, preventing recognition by CD4+ T cells. Nat. Med. *5*, 1039-1043.

Tortorella, D., Gewurz, B. E., Furman, M. H., Schust, D. J., and Ploegh, H. L. (2000). Viral subversion of the immune system. Annu. Rev. Immunol. *18*, 861-926.

Ulane, C. M., Rodriguez, J. J., Parisien, J. P., and Horvath, C. M. (2003). STAT3 ubiquitylation and degradation by mumps virus suppress cytokine and oncogene signaling. J. Virol. *77*, 6385-6393.

Upton, C., Mossman, K., and McFadden, G. (1992). Encoding of a homolog of IFN-γ receptor by myxoma virus. Science *258*, 1369-1372.

van Berkel, V., Barrett, J., Tiffany, H. L., Fremont, D. H., Murphy, P. M., McFadden, G., Speck, S. H., and Virgin, H. I. (2000). Identification of a gammaherpesvirus selective chemokine binding protein that inhibits chemokine action. J. Virol. *74*, 6741-6747.

Vanderplasschen, A., Mathew, E., Hollinshead, M., Sim, R. B., and Smith, G. L. (1998). Extracellular enveloped vaccinia virus is resistant to complement because of incorporation of host complement control proteins into its envelope. Proc. Natl. Acad. Sci. USA. *95*, 7544-7549.

Wang, X. W., Gibson, M. K., Vermeulen, W., Yeh, H., Forrester, K., Sturzbecher, H. W., Hoeijmakers, J. H., and Harris, C. C. (1995). Abrogation of p53-induced apoptosis by the hepatitis B virus X gene. Cancer Res. *55*, 6012-6016.

Webb, L. M., Clark-Lewis, I., and Alcami, A. (2003). The gammaherpesvirus chemokine binding protein binds to the N. terminus of CXCL8. J. Virol. *77*, 8588-8592.

Werness, B. A., Levine, A. J., and Howley, P. M. (1990). Association of human papillomavirus types 16 and 18 E6 proteins with p53. Science *248*, 76-79.

Wiertz, E. J., Jones, T. R., Sun, L., Bogyo, M., Geuze, H. J., and Ploegh, H. L. (1996a). The human cytomegalovirus US11 gene product dislocates MHC class I heavy chains from the endoplasmic reticulum to the cytosol. Cell *84*, 769-779.

Wiertz, E. J., Tortorella, D., Bogyo, M., Yu, J., Mothes, W., Jones, T. R., Rapoport, T. A., and Ploegh, H. L. (1996b). Sec61-mediated transfer of a membrane protein from the endoplasmic reticulum to the proteasome for destruction. Nature *384*, 432-438.

Xiang, Y., and Moss, B. (1999). IL-18 binding and inhibition of interferon gamma induction by human poxvirus-encoded proteins. Proc. Natl. Acad. Sci. USA. *96*, 11537-11542.

Yang, T. Y., Chen, S. C., Leach, M. W., Manfra, D., Homey, B., Wiekowski, M., Sullivan, L., Jenh, C. H., Narula, S. K., Chensue, S. W., and Lira, S. A. (2000). Transgenic expression of the chemokine receptor encoded by human herpesvirus 8 induces an angioproliferative disease resembling Kaposi's sarcoma. J. Exp. Med. *191*, 445-454.

Yao, Z., Fanslow, W. C., Seldin, M. F., Rousseau, A. M., Painter, S. L., Comeau, M. R., Cohen, J. I., and Spriggs, M. K. (1995). Herpesvirus Saimiri encodes a new cytokine, IL-17, which binds to a novel cytokine receptor. Immunity *3*, 811-821.

Ye, Y., Shibata, Y., Yun, C., Ron, D., and Rapoport, T. A. (2004). A membrane protein complex mediates retro-translocation from the ER lumen into the cytosol. Nature *429*, 841-847.

Yewdell, J. W., and Hill, A. B. (2002). Viral interference with antigen presentation. Nat. Immunol. *3*, 1019-1025.

York, I. A., Roop, C., Andrews, D. W., Riddell, S. R., Graham, F. L., and Johnson, D. C. (1994). A cytosolic herpes simplex virus protein inhibits antigen presentation to CD8+ T lymphocytes. Cell *77*, 525-535.

Ziegler, H., Thale, R., Lucin, P., Muranyi, W., Flohr, T., Hengel, H., Farrell, H., Rawlinson, W., and Koszinowski, U. H. (1997). A mouse cytomegalovirus glycoprotein retains MHC class I complexes in the ERGIC/cis-Golgi compartments. Immunity *6*, 57-66.

Zimring, J. C., Goodbourn, S., and Offermann, M. K. (1998). Human herpesvirus 8 encodes an interferon regulatory factor (IRF) homolog that represses IRF-1-mediated transcription. J. Virol. *72*, 701-707.

Zou, P., Isegawa, Y., Nakano, K., Haque, M., Horiguchi, Y., and Yamanishi, K. (1999). Human herpesvirus 6 open reading frame U83 encodes a functional chemokine. J. Virol. *73*, 5926-5933.

5

Evasion Mechanisms of Bacteria

*Michael M. Frank**

Abstract

This chapter reviews some aspects of molecular mimicry used in evasion strategies of bacteria. For long term survival a bacterium must occupy a niche where it can survive, reproduce and spread. Pathogenic bacteria in invading man must deal with a wide assortment of innate and adaptive host defense responses and have themselves evolved a wide variety of bacterial protective mechanisms. In many cases this involves the use of the host's cellular or biochemical pathways to subvert the host attack. For example, a mammalian host has evolved sophisticated mechanisms to separate self from non-self, so that attack on host structures by innate and adaptive immune responses with resultant tissue damage is minimized. Some bacteria have evolved mechanisms to disguise themselves as self, thereby thwarting the host defense process. In similar ways bacteria use host structures for invasion of cells and tissues, and intracellular survival. Analysis of these mechanisms has often led to the finding of convergent evolution, the evolution of bacterial elements that have similar function to host elements but unique structure. In some cases there is actual incorporation of host genes by horizontal transfer. There is almost no aspect of the host defense process that some group of organisms has not subverted for pathogenic use.

1. Introduction

A bacterial species to survive must define an ecologic niche in which it can remain protected, viable, and multiply for prolonged periods of time. If the organism is extremely virulent, the host may succumb to infection before it can transfer to a new host and therefore it will have participated in its own demise. It is in the organism's interest to produce an indolent niche infection that allows for sufficient multiplication for it to thrive and infect others but to not be lethal. An interesting example that illustrates this point is that of *Listeria monocytogenes*, an intracellular Gram positive pathogen (Portnoy et al., 2002). The organism is taken up into

*For correspondence email Frank007@mc.duke.edu

phagocytic vacuoles, produces a pore forming protein, listeriolysin, that allows it to escape from the vacuole and then grow in the cytoplasm. Mutants of the organism that are more destructive to host cells in tissue culture, which suggests that they have the potential for greater damage, tend to be far less virulent *in vivo*. Most organisms are capable of a range of infections. For example the Streptococcus, one of the most studied bacterial pathogens of man, produces infections that range from pharyngitis to lethal sepsis or fasciitis. One would assume that the mild infections are the programmed life cycle preferences, but that the characteristics of the organism, the circumstances of the infection, and a variety of host defense factors, define the behavior of the individual infection.

Infection and invasion by microbes have always been major dangers of animal and plant life. Over the course of evolution sophisticated mechanisms have developed to control or destroy infectious agents. In turn the agents have over time developed highly sophisticated mechanisms for avoiding host attack. From the point of view of the infected host, it is essential to recognize, identify and localize an infectious agent. Often the host then attempts to rid the body of the infectious agent. Commensals, like the normal gut flora, are required for normal bodily function and must not be eliminated.

2. Mammalian host defense mechanisms

The mammalian host has an extensive and overlapping armamentarium of weapons to provide these various aspects of host defense. The integument and mucosa act as a physical barrier to invasion. If the organism penetrates the integument or a mucosal surface, it encounters an environment that contains plasma proteins like the host defense proteins of the complement system. These include lectins like mannose binding lectin, that recognizes and binds to the repeating polysaccharide structures common to many bacteria, and initiates attack. Fixed and circulating phagocytes have multiple receptors that facilitate host defense. A recent focus of attention has been the Toll-like receptors on phagocytes, a number of which also recognize repeating polysaccharide structures. Some Toll like receptors recognize lipoteichoic acids or other common chemical structures on microbial surfaces: β2 integrins also recognize a number of bacterial carbohydrates like mannan. The β2 integrins as well as other receptors that recognize complement activation products, mediate phagocytosis, killing and cytokine release. Immunoglobulin Fc region receptors on phagocytes also bind and phagocytose antibody coated bacteria. A variety of cells, like PMN and platelets, release polyspecific bacterial peptides that directly destroy microorganisms. Various cells on being damaged by an invading microorganism produce and release signaling molecules, often peptides that signal danger and damage, to alert defense systems in the host. These in turn up regulate and activate phagocytic cell receptors, cause fever which contributes to organism destruction, initiate an inflammatory process designed to wall off the invader and, ultimately, contribute to induction of a specific adaptive immune response. Adaptive immunity, a relatively late development in phylogeny, can recognize a specific pathogen and generate specific cells and inflammatory molecules that interact with and damage the pathogen in a milieu of thousands of proteins, complex carbohydrates and lipids. The sophistication of the system is extraordinary.

It is customary to divide the host defense response to pathogens into an innate immune response and an adaptive immune response. The innate immune response includes the specific defenses in place, coded for by germ line genes, designed to destroy an invader upon entrance to the body. Here there is only a limited measure of specificity. The adaptive immune response recognizes and confronts a specific pathogen in a world of millions of potential specific pathogens. Clearly in evolutionary terms the innate immune response did not have the power or specificity to deal effectively with many injurious pathogens. The adaptive immune response evolved to provide a heightened ability to recognize and confront an invader and also the capacity to recall a prior exposure to the pathogen and mount a specific response rapidly on reexposure. The adaptive immune response usually is not able to control a specific pathogen until a period of time has passed in which specific cells with the capacity to recognize the pathogen are generated or expanded in number. It is during this time that the body is highly reliant on the elements of innate immunity to provide protection. The innate immune system evolved first and is present in extremely primitive organisms. The adaptive immune system appeared first in fish and in its evolution incorporated elements of the innate immune system into its functional elements. Thus many elements of the innate immune system have a role to play in the function of the adaptive immune system as well.

3. Bacterial subversion of host defense processes

Confronted with a complex innate and adaptive immune system that has developed and evolved over eons, organisms, particularly bacteria for the purpose of this chapter, have their own defense mechanisms, which have allowed them to survive and escape these sophisticated host defense mechanisms. Some of these will be reviewed here. There is an additional element that comes about as a by-product of this potentially destructive interaction between parasite or invader and host. The host defense mechanisms, both innate and adaptive immunity, although breathtaking in their ability to deal with invaders, occasionally mediate host damage. Many microorganisms have evolved surface structures that (in terms of chemical specificity), overlap with or interact with structures in the host. As will be discussed below, in many cases the structures that mimic host elements do not have the same amino acid backbone as the host structures but have the same function. Thus the mimicry may not be of structure but of function. For example the contact surface formed by a bacterial protein with a host structure may be functionally similar to a host proteins in that it may have contact points permitting adherence that mimic the contact points on the host protein. Functionally the bacterial protein and the host protein are similar even though they may have quite a different overall structure. In most cases this represents an example of convergent evolution. In some cases genetic elements from the host have been incorporated into the microbe through horizontal transfer. At times, an immune response or elements of innate immunity, responding to the microbe, will recognize in part or in whole, tissue elements in the host and will produce unwanted local inflammation, an autoimmune response with resulting organ damage, or an abnormal metabolic response. The story of autoimmunity resulting from microbial infection with an immune response that cross-reacts with host tissues is rapidly expanding and is beyond the scope of this chapter.

To gain a niche, bacteria frequently have evolved or acquired adhesion molecules allowing them to attach to host structures. These structures may mimic host adhesion molecules and are discussed in greater detail below. Some bacteria produce toxins that mimic regulatory molecules in the host. For example toxins may relax smooth muscle, relax or constrict blood vessels or break down the vessel endothelial barrier, allowing cells circulating in the bloodstream to enter tissues. Individual organisms have evolved almost every method for host evasion or molecular mimicry that can be envisioned.

One of the most critical difficulties encountered by the host in developing systems to provide innate and adaptive immunity, is the difficulty recognizing self from nonself. As mentioned, it is essential to differentiate self from non-self with a high degree of fidelity if, one is to attack non-self with highly destructive processes. Therefore, both the innate and adaptive immune system mechanisms for recognizing host tissues and blood elements and not responding to them. Microorganisms have evolved ways of making themselves look like host elements and thus avoiding induction of an immune response with resulting danger,as well as for escaping the efferent or destructive elements of innate and adaptive immunity.

4. Evasion mechanisms used by *Streptococcus pyogenes*

For purposes of this chapter we will use *Streptococcus pyogenes* as our poster boy and introduce elements that pertain to the evasion of the host defense process as seems appropriate. As discussed above most pathogenic bacteria have a wide array of adhesins allowing them to attach to and damage tissues. Again the mimicry is usually functional rather than structural. There are many adhesion molecules on the surface of cells and in tissue matrix that allow for the evolution of a normal inflammatory response or normal repair mechanisms. Bacteria have learned to generate molecules that form similar anchors to tissue, thus facilitating the invasion process. In some cases they actually trigger the receptor mediated internalization mechanism with vacuole formation, and come to reside within the cell where they are safe from destructive processes. Internalization may facilitate the process of persistent carriage and also may facilitate penetration into deeper tissues. Obviously, once inside the cell, organisms must trigger processes that allow them to escape the vacuole, or alternatively, resist lysosomal and other destructive processes. Again we are dealing most often with functional rather than structural mimicry.

The M protein on *S. pyogenes* has long been known to bind fibrinogen and albumin present in plasma of the host (Whitnack et al., 1984). It is believed that surface bound plasma proteins disguise the organism and render it less likely to be recognized by immune cells or elements like Toll-like receptors or some complement proteins that recognize repeating polysaccharide structures on the pathogen. It is reported that the presence of fibrinogen also decreases complement binding to the bacterial surface (Chhatwal et al., 1985). Here the mimicry involves binding an array of host molecules on the bacterial surface to avoid attack. The streptococcal M protein also specifically binds a number of other factors from plasma that function to down regulate host defense including a number of complement regulatory molecules (Johnsson et al., 1998; Perez-Caballero et al., 2000; Blackmore et al., 1998; Sharma et al., 1997). Two molecules that act as cofactors for the cleavage of C3b, the first major opsonin to be deposited on the bacteria, are factor H and FHL-1 (factor H like). These molecules bind to C3b and when C3b is in the form of a complex with these cofactors, it can be cleaved by the circulating protein factor I to iC3b. It is in the interest of the bacterium to destroy the function of C3b. This molecule not only can interact with C5 and the later acting proteins of the complement cascade to form a complement lesion on the bacterial surface, but, more importantly, C3b acts as a nidus for the deposition of proteins of the alternative pathway including large amounts of additional C3b. C3b on an organism is recognized by CD35, present on all phagocytes, facilitating binding, phagocytosis and killing of the organisms. C3b is converted to iC3b by the action of factors H, FHL-1 and factor I. These two C3 fragments, C3b and iC3b, represent the most important opsonins of the complement system. Some organisms use deposited C3 fragments to facilitate entry into cells by phagocytosis, but in most cases, phagocytosis leads to destruction. The less C3b and iC3b on the organism surface, the more likely it is to avoid destruction. The fewer C3b deposited, the fewer iC3b can be generated. M protein binds factor H and FHL-1(Johnsson et al., 1998; Perez-Caballero et al., 2000; Blackmore et al., 1998; Sharma et al., 1997). SCR 7 of factor H is reported to contain the binding site for M protein, it is reported that there are two binding sites for factor H on M protein and there may be yet an additional binding site on factor H. Current thinking is that of the regulatory molecules FHL-1 is the C3b regulatory factor bound with the higher affinity. M protein also specifically binds C4 binding protein, C4bp, which regulates an early step in the complement cascade that involves the activity of C4 (Blom et al., 2000; Carlsson et al., 2003; Accardo et al., 1996). The binding site recognized on C4bp by protein M appears to be in the same region as the site that recognizes C4b. Since C4bp is a multichain macromolecule with multiple binding sites, although bound to the bacterium by one of these binding sites, it retains biological activity, and down regulates the classical complement pathway by inhibiting the action of C4. In the case of FHL-1 which facilitates the degradation of C3b and which also remains functionally active when bound, binding acts to down regulate the classical, lectin and alternative pathway. An additional protein on the streptococcus, Fba, appears to strengthen the affinity of Protein M for the C3b binding factors and may bind these regulatory factors in the absence of M protein (Pandiripally et al., 2002). It is also recognized that there are a number of M like proteins of *S. pyogenes*, coded for by separate genes, which can also bind complement regulatory factors. There is even reported to be a site on M protein that binds the C3 regulatory factor MCP (membrane cofactor protein, CD46). This C3 regulatory molecule is usually membrane bound on host cells and whether this binding activity represents an mechanism for cellular adhesion or the MCP molecule is shed and acquired from plasma is not known (Giannakis et al., 2002)

The hyaluronic acid capsule of *S. pyogenes* is strongly anti-phagocytic (Dale et al., 1996). Hyaluronic acid is a component of joint fluid and thus it acts as a tissue mimic. It covers deposited complement peptides deposited on the bacterial cell wall beneath the capsule, and prevents interaction with phagocytes with complement receptors. In many situations the capsule also acts as a partial barrier to the deposition of complement proteins such as C3. This is certainly true of the capsule of *S. aureus* and the hyaluronic acid capsule of Group B Streptococcus (Marques et al., 1992; Cunnion et al., 2001).

An interesting complement regulatory molecule termed Sic (streptococcal inhibitor of complement) is secreted by streptococci (Åkesson et al., 1996). This molecule, a highly polymorphic extracellular protein, is made by serotype M1 Group A Streptococcus strains and contributes to bacterial persistence in the mammalian upper respiratory tract. New variants of Sic arise frequently and in screening the blood of normal individuals one can often find antibody to the protein in blood and nasal washings (Hoe et al., 2000). The highly antigenic sites on the Sic protein tend to be the segments that are polymorphic rather than the constant regions of the protein (Hoe et al., 2001). From the point of view of the streptococcus, a highly polymorphic and rapidly mutating form of Sic is highly advantageous. Antibody made to the polymorphic segments of the protein only bind to those Sic molecules with that form, and may not bind to molecules that have mutated this segment. The protein is unlikely to be neutralized by preexisting antibody or by products of the immune response and thus can continue to function. Purified Sic protein has been found to bind two plasma proteins, clusterin and histidine-rich glycoprotein, both of which are known to bind to the complex of the late acting complement proteins, the membrane attack complex (Fernie-King et al., 2001). Experiments were performed to determine whether Sic binds to the membrane attack complex. It was found to bind poorly to the individual complement proteins C5, C6, C7 C8 and C9 that make up the complex as exist in plasma, but to bind well to the membrane attack complex, a complex of the above proteins. It specifically binds to the complement complex on erythrocytes consisting of C5b67 and C5b678 and inhibits the lysis by complement of antibody-sensitized erythrocytes. Of great interest is the finding that bacteria in which this activity has been deleted, colonize the mucosa of the upper airway of mice poorly (Lukomski et al., 2000). It is assumed that this property of Sic is unrelated to its complement regulatory property. Since, we do not believe that the membrane attack complex plays a role in the destruction of the gram-positive streptococci, it would appear that the importance of Sic in colonization of mucosa is the more important property.

Pathogenic streptococci also synthesize and secrete a peptidase that specifically cleaves C5a termed SPCA, Streptococcal C5a peptidase (Cleary et al., 1992). The gene is under the control of a regulator, the *virR* locus. Streptococci have at least 3 IgG binding proteins, Arp, Sir and protein H, all part of the M protein family (Jarva et al., 2003). Finally they secrete an endoglycosidase EndoS from M1 Group A Streptococci that removes Asn297-linked glycan of IgG, which is reported to inhibit C1q binding by IgG. (Collin et al., 2001). Streptococci also have a gene, *fcrA* that codes for an IgG Fc receptor. Fcγ receptors bind IgG coated particles like bacteria and facilitate phagocytosis (La Penta et al., 1994). *fcrA* encodes an IgG FcRIIa mimic that presumably inhibits opsonization. This group of genes (*Sic*, *SPCA* and *fcrA*), are all expressed in M (+) bacteria and are co-regulated (La Penta et al., 1994). The C5a peptidase is quite specific and does not recognize the intact protein C5. C5a, a cleavage product of C5, is the most important complement derived chemotactic peptide and the secreted C5a peptidase interacts with and destroys the activity of C5a. The biological consequences of this cleavage have been studied (Ji et al., 1996). Streptococci with the SPCA deleted, when injected into an air pouch on the back of an animal, attract far more neutrophils into the pouch. C5a generated by the host innate immune response is not destroyed and attracts neutrophils to the site of infection. The path of the bacteria after injection is also different in the presence and absence of the peptidase. In the presence of the protein, which destroys C5a, most bacteria that leave the pouch traffic to the spleen. In its absence a significantly greater number traffic to the lymph nodes. Like Sic, the C5a peptidase is reported to play a role in nasopharyngeal colonization of mice and antibody to the protein is reported to prevent nasopharyngeal colonization (Ji et al., 1997).

Some streptococci in log phase growth secrete a protein that has sequence homology with CD11b, one chain of the two-chain β2 integrin on phagocytes termed CR3 that binds iC3b (Lei et al., 2001). CD11b is the chain that contains the recognition site for iC3b. The bacterial protein blocks this site and also blocks the phagocyte IgG receptor FcγRIII (CD 16). Some CR3 appears to be directly linked to FcγRIII in the granulocyte membrane, which may explain this phenomenon. CR3 act as a signaling partner for a number of receptors and intracellular biochemical pathways and blocking of CR3 also blocks generation of reactive oxygen species. In some studies the bacterial protein blocks anti-CR3 mediated release of several cytokines. The bacterial protein was given the unfortunate designation Mac that is also used for the membrane attack complex.

The streptococcus is also reported to form a protein that mimics sialyl-Lewis(x) and acts as a ligand for selectin positive cells (Hirota et al., 1998). Since adherence to tissue is of great importance in pathogenesis, this protein assumes great importance. For a complete discussion of streptococcal adhesions and their receptors see Cunningham (2002). Finally this organism is reported have a surface protein, protein H, as well as other proteins that specifically bind and block the activity of C1q (Berge et al., 1997; Koroleza et al., 1998). Bound C1q is reported to increase the binding of fibronectin to the bacterium and thereby also increases binding to tissue sites. In another organism, the gonococcus, where the effect of the binding of C1q has been examined further, C1q binding is also reported to inhibit complement mediated lysis by blocking activation of the classical complement pathway. C1 and C1q binding structures have also been described on a variety of *E. coli* (van den Berg et al., 1996). Usually it is not in the interest of the bacterium to bind proteins that initiate the complement cascade, unless these proteins are bound in an inactive form. One of the *E. coli* structures that binds C1q has much of the activity of the true C1q receptor found on many cells. However, the protein differs from the true C1q receptor in that it binds both the globular heads of C1q and the collagen like arms of C1q. Thus it binds both ends of the molecules and is therefore uniquely designed to bind, in a functionally inactive form, the C1q (van den Berg et al., 1996).

5. Other well defined strategies used by bacteria

Many organisms produce factors that destroy C3. One of the most recently described and nicely studied is a factor produced and released by *Staphylococcus aureus* that binds and destroys fluid phase C3, cleaving it and preventing it from binding to the bacterial surface (Lee et al., 2004). Other bacteria have evolved related mimics of host tissues. As mentioned, the area of adhesion is one in which molecular mimics are common. *Neisseria gonorrhoeae* and *Neisseria meningiditis* have been found to synthesize lipopolysaccharide core polysaccharide that binds various glycosphingolipids as exist on the surface of many cells (Harvey et al., 2001). *Campylobacter jejuni* lipopolysaccharide binds GM1 and GM2, present on the surface of cells in the GI tract, and it has been shown that the presence of these binding structures facilitates invasion into cells (Guerry, et.al 2002). On the surface of mammalian cells, sialic acid residues often play a critical role in regulating activation of the alternative pathway by providing charged sites that increase the efficiency of factor H as a co-factor for factor I. Increasing the efficiency of factor I facilitates the destruction of C3b if deposited on host cells. Bacteria have incorporated sialic acid into various structures to facilitate destruction of C3b and a number of major pathogens, like the K1 strain of *E. coli*, have sialic acid as the major component of their capsules (Cunningham, 2002). Thus they escape opsonization. It is of interest that some strains of *E. coli* have the gene for a global regulator, nanR, that monitors the level of sialic acid in the environment and controls the expression of sialic acid on the organism surface (Vimr et.al. 2004). Presumably in the presence of host cells with a high sialic acid content, the bacterium up regulates its own sialic acid.

The capsule of the encapsulated bacterium activates the alternative complement poorly and acts as a partial barrier to the deposition of complement on the bacterial surface. However, the bacterium must bring molecules of various sizes into the cell for growth and the capsule must therefore allow passage of macromolecules. The capsule forms a loose structure that complement can penetrate. If the alternative pathway is activated at the surface of the bacterium, it can deposit complement proteins on the surface of the bacterium beneath the capsule (Joiner et al., 1984). However opsonic peptides, deposited deep to the capsule interact poorly with phagocyte complement receptors; the capsule is a barrier to binding and is therefore strongly anti-phagocytic. It is for this reason that anti-capsular antibodies are so important in protection against bacterial infection and that the goal of many antibacterial vaccines is the efficient production of anticapsular antibodies.

The presence of anti-capsular antibody tends to direct complement deposition, particularly the deposition of C3, the major opsonin, to the surface of the capsule rather than to the cell wall (Joiner, et al., 1984). Complement proteins can form ester or amide bonds with many surface constituents of the bacterium including components of the capsule. A major proportion of the deposited C3 actually binds to the capsule bound antibody, particularly IgG bound to the capsule, and both the capsule bound C3 and the antibody bound C3 is available for binding to phagocyte complement receptors.

Other examples of structural or functional mimicry are common. *Bordetella pertussis* synthesizes a major fimbrial subunit that binds to sulfated sugars, which are ubiquitous in the human respiratory tract (Geuijen et al., 1996). These peptides from the pertussis mimic fibronectin, which binds to the same sugars. Interesting examples of structural mimicry have been reported for Salmonella species (Stebbins et.al. 2001). These organisms, using a type III secretion system, that is the injection of bacterial proteins into host cells, inject two proteins SopE and SptP that have profound effects on the function of host cells. SopE, an exchange factor for Rac and Cdc42 induces rearrangement of the actin cytoskeleton and membrane ruffling by activating Rho family GTPases. This facilitates the internalization of the Salmonella. StpP acts after internalization of the organisms to down regulate these same signals, promoting bacterial survival. SptP interacts with an intracellular protein, Rac-1 to form a complex that has been shown structurally to resemble the host transition state complex GAP (GTPase activating protein) bound to Rho GTPase. In addition GTP kinases of the Ras/Rho family require a critical arginine to be added to complete a catalytic site for GTP hydrolysis. Arg 209 on SptP serves this function in a way which differs completely from the normal host protein. It extends from an α helix in SptP rather than being inserted as part of a loop which is usually the case.

Yersinia mediates internalization by structurally mimicking host integrin ligands (Stebbins et al., 2001). The bacterium, using a process of convergent evolution, forms a structure with contact points similar to fibronectin that binds to the host cell integrin with high affinity and sets off the process of internalization. Although not a true example of molecular mimicry, the formation of superantigens is of particular interest (Baker et al., 2003). Antigen presenting cells present antigen to T lymphocytes as peptides in the antigen-binding pocket of appropriate Class1 or Class 2 Major Histocompatibility Complex proteins. The lymphocytes that are triggered to respond, have an appropriate corresponding configuration in their T cell antigen receptor and only those T cells that recognize the antigen in question will respond. Microbes like streptococci and staphylococci often produce proteins that act as superantigens. Many pathogenic, secreted exotoxins have this property. These proteins are capable of linking the T cell receptor to the the antigen presenting molecules on the antigen presenting cell, at a site remote from the antigen binding groove, and can trigger the T cell without regard to antigenic specificity. This in turn leads to the release of immune response activating of cytokines disrupting the normal host response process. Here the mimicry represents a modification of the normal antigen presentation process. From the cell's point of view, the mimicry is of normal antigen presentation, so there is functional mimicry, but the process differs from true antigen presentation.

It should be clear from this chapter, which cannot present an exhaustive catalogue of all molecular mimicry used by bacterial pathogens to facilitate a productive life cycle, that mimicry is essential for that life cycle as well as for the pathologic process. Humans and bacteria have lived together throughout human evolution and adaptations have taken place over time that allow these organisms to continue to infect man and cause disease. For critical elements of protection ,multiple overlapping mechanisms of defense have evolved. These multiple mechanisms, mounted by the microbes for their own defense, in major part explain why curing infection is not easily accomplished.

References

Accardo P., Sanchez-Corral P., Criado O., Garcia E., and Rodriguez de Cordoba S. (1996). Binding of human complement component C4b-binding protein (C4BP) to Streptococcus pyogenes involves the C4b-binding site. J. Immunol. *157*, 4935-9.

Åkesson P., Sjöholm A.G., and Björck L. (1996). Protein SIC, a novel extracellular protein of *Streptococcus pyogenes* interfering with complement function. J. Biol. Chem. *271*, 1081-1088.

Baker M.D., and Acharya K.R. (2003) Superantigens. Structure, function, and diversity. [Review] [101 refs] Methods in Molec. Biol. *214, 1-31.*

Berge A., Kihlberg B.M., Sjöholm A.G., and Björck L. (1997). Streptococcal protein H forms soluble complement-activating complexes with IgG, but inhibits complement activation by IgG-coated targets. J. Biol. Chem. *272*, 20774-20781

Blackmore T.K., Fischetti V.A., Sadlon T.A., Ward H.M., and Gordon D.L. (1998). M protein of the group A Streptococcus binds to the seventh short consensus repeat of human complement factor H. Infection and Immunity. *66*, 1427-31.

Blom A.M., Berggard K., Webb J.H., Lindahl G., Villoutreix B.O., and Dahlback B. (2000). Human C4b-binding protein has overlapping, but not identical, binding sites for C4b as streptococcal M proteins. J. Immunol. *164*, 5328-36.

Carlsson F., Berggard K., Stalhammar-Carlemalm M., and Lindahl G. (2003). Evasion of phagocytosis through cooperation between two ligand-binding regions in *Streptococcus pyogenes* M protein. J. Exp. Med. *198*, 1057-68.

Chhatwal G.S., Dutra I.S., and Blobel H. (1985). Fibrinogen binding inhibits the fixation of the third component of human complement on surface of groups A, B, C, and G streptococci. Microbiol. and Immunol. *29*, 973-80.

Cleary P.P., Prahbu U., Dale J.B., Wexler D.E., and Handley J. (1992) Streptococcal C5a peptidase is a highly specific endopeptidase. Infection and Immunity. *60*, 5219-23.

Collin M., and Olsen A., (2001) EndoS, a novel secreted protein from *Streptococcus pyogenes* with endoglycosidase activity on human IgG. EMBO J. *20*, 3046-55.

Cunningham M.W., (2002) Pathogenesis of Group A Streptococcal infections. Microbiol. Rev. *13*, 470-511.

Cunnion, K.M., Lee, J.C. and Frank, M.M. (2001) Capsule production and growth phase influence binding of complement to *Staphylococcus aureus*. Infect. Immunity 69(11), 6796-6803.

Dale J.B., Washburn R.G., Marques M.B., and Wessels M.R. (1996). Hyaluronate capsule and surface M protein in resistance to opsonization of group A Streptococci. Infection and Immunity. *64*,1495-501.

Fernie-King B.A., Seilly D.J., Willers C., Würzner R., Davies A., and Lachmann P.J. (2001). Streptococcal inhibitor of complement (SIC) inhibits the membrane attack complex by preventing uptake of C567 onto cell membranes. Immunol. *103*, 390-398.

Geuijen C.A., Willems R.J., and Mooi F.R. (1996). The major fimbrial subunit of *Bordetella pertussis* binds to sulfated sugars. Infection and Immunity. *64*, 2657-65.

Giannakis E., Jokiranta T.S., Ormbsy R.J., Duthy T.G., Male D.A., Christiansen D., Fischetti V.A., Bagley C., Loveland B.E., and Gordon D.L. (2002) Identification of a streptococcal binding site on membrane cofactor protein (CD46). J. Immunol. *168*, 4585-4592.

Guerry P., Szymanski C.M., Prendergast M.M., Hickey T.E., Ewing C.P., Pattarini D.L., and Moran A.P. (2002). Phase variation of *Campylobacter jejuni* 81-176 lipooligosaccharide affects ganglioside mimicry and invasiveness *in vitro*. Infection and Immunity. *70*, 787-93.

Harvey H.A., Swords W.E., and Apicella M.A. (2001). The mimicry of human glycolipids and glycosphingolipids by the lipooligosaccharides of pathogenic Neisseria and Haemophilus. J. Autoimmunity. *16*, 257-62.

Hirota K., Murakami K., Nemoto K., Ono T., Matsuo T., Kumon H., and Miyake Y. (1998). Fosfomycin reduces CD15s-related antigen expression of *Streptococcus pyogenes*. Antimicrobial Agents and Chemotherapy. *42*, 1083-7.

Hoe N.P., Kordari P., Cole R., Liu M., Palzkill T., Huang W., Mclellan D., Adama G.J., Hu M., Vuopio-Varkila J., Cate T.R., Pichichero M.E., Edwards K.M., Eskola J., Low D.E., and Musser J.M. (2000). Human immune response to streptococcal inhibitor of complement, a serotype M1 group A Steptococcus extracellular protein involved in epidemics. J. Infectious Diseases. *182*,1425-36.

Hoe N.P., Vuopio-Varkila J., Vaara M., Grigsby D., DeLorenzo D., Fu Y.X. Dou S.J., Pan X., Nakashima K., and Musser J.M. (2001). Distribution of Streptococcal inhibitor of complement variants in pharyngitis invasive isolates in an epidemic of serotype M1 Group A Streptococcus infection. J. Infectious Diseases. *183*, 633.

Jarva H., Jokiranta T.S., Würzner R., and Meri S. (2003). Complement resistance mechanisms of streptococci Molecular Immunol. *40*, 95-107.

Ji Y., Carlson B., Kondagunta A., and Cleary P.P. (1997) Intranasal immunization with C5a peptidase prevents nasopharyngeal colonization of mice by the Group A Streptococcus. Infect and Immun *65*, 2080-87.

Ji Y., McLandsborough L., Kondagunta A., and Cleary P.P. (1996). C5a peptidase alters clearance and trafficking of group A streptococci by infected mice. Infection and Immunity. *64*, 503-10.

Johnsson E., Berggard K., Kotarsky H., Hellwage J., Zipfel PF., Sjobring U., and Lindahl G. (1998). Role of the hypervariable region in streptococcal M proteins, binding of a human complement inhibitor. J. Immunol. *161*, 4894-901.

Joiner, K.A., Brown, E.J., and Frank, M.M. (1984) Complement and Bacteria, Chemistry and Biology in Host Defense. Annual Review of Immunology 2, 461-492.

Koroleza I.V., Sjöholm A.G., and Schalén C. (1998). Binding of complement subcomponent C1q to streptococcus pyogenes, evidence for interactions with the M5 and FcRA76 proteins. FEMS Immunol. Med. Microbiol. *20*, 11-20.

Kotarsky H., Hellwage J., Johnsson E., Skerka C., Svensson H.G., Lindahl G., Sjobring U., and Zipfel P.F. (1998). Identification of a domain in human factor H and factor H-like protein-1 required for the interaction with streptococcal M proteins. J. Immunol. *160*, 3349-54.

La Penta D., Zhang X.P., and Cleary P.P. (1994). *Streptococcus pyogenes* type IIa IgG Fc receptor expression is co-ordinately regulated with M protein and streptococcal C5a peptidase. Molecular Microbiol. *12*, 873-9.

Lee LY. Hook M. Haviland D. Wetsel RA. Yonter EO. Syribeys P. Vernachio J. and Brown EL.(2004) Inhibition of complement activation by a secreted *Staphylococcus aureus* protein. J. Infect. Dis. *190(3)*, 571-9.

Lei B., DeLeo F.R., Hoe N.P., Graham M.R., Mackie S.M., Cole R.L., Liu M., Hill H.R., Low D.E., Federle M.J., Scott J.R., and Musser J.M. (2001). Evasion of human innate and acquired immunity by a bacterial homolog of CD11b that inhibits opsonophagocytosis. Nature Medicine. *7*, 1-8.

Lukomski S., Hoe N.P., Abdi I., Rurangirwa J., Kordari P., Liu M., Dou S.J., Adams G.G., and Musser J.M. (2000). Nonpolar inactivation of the hypervariable streptococcal inhibitor of complement gene (sic) in serotype M1 Streptococcus pyogenes significantly decreases mouse mucosal colonization. Infect. and Immun. *68*,535-42.

Marques M.B., Kasper D.L., Pangburn M.K. and Wessels M.R. (1992) Prevention of C3 deposition by capsular polysaccharide is a virulence mechanism of Type III Group B Streptococci. Infect. and Immun. *60*, 3986-*93*,

Pandiripally V., Gregory E., and Cue D.(2002). Acquisition of regulators of complement activation by *Streptococcus pyogenes* serotype M 1. Infect. Immun. *70*, 6206-14.

Perez-Caballero D., Alberti S., Vivanco F., Sanchez-Corral P., and Rodriguez de Cordoba, S. (2000). Assessment of the interaction of human complement regulatory proteins with group A Streptococcus. Identification of a high-affinity group A Streptococcus binding site in FHL-1. Euro. J. Immunol. *30*, 1243-53.

Portnoy D.A., Auerbach V, and Glomski I.J. (2002) The cell biology of Listeria monocytogenes infection: the intersection of bacterial pathogenesis and cell mediated immunity. J. Cell Biol *158*, 409-414

Sharma A.K., and Pangburn M.K. (1997). Localization by site-directed mutagenesis of the site in human complement factor H that binds to *Streptococcus pyogenes* M protein. Infection and Immunity. *65*,484-7.

Stebbins C.E., and Galan J.E. (2001) Structural mimicry in bacterial virulence. Nature *412*, 701-705.

van den Berg R.H., Faber-Krol M.C., van de Klundert J.A., van Es L.A., Daha M.R. (1996). Inhibition of the hemolytic activity of the first component of complement C1 by an Escherichia coli C1q binding protein. J. Immunol. *156*, 4466-73

Vimr E.R., Kalivoda K.A., Deszo E.L., and Steenbergen S.M. (2004). Diversity of microbial sialic acid metabolism. Microbiol. Mol. Biol. Rev. *68*, 132-53.

Whitnack E., Dale J.B., and Beachey E.H. (1984). Common protective antigens of group A streptococcal M proteins masked by fibrinogen. J. Exp. Med.*159*, 1201-12.

6

Subversion of the Immune Response by Helminths

*Michael J. Doenhoff**

Abstract

Many years ago it was observed that helminth parasites were immunosuppressive and this property was deemed to be an important element in the ability of these organisms to evade host immune responsiveness and thus establish enduring chronic infections. Subsequent studies on anti-helminth immunity played a significant part in increasing our knowledge about the complexity and range of immune effector mechanisms, particularly by helping to distinguish between two sup-populations of T helper cells. Recent research has taken an unexpected turn that stems from observations that these parasites can exert profound modulatory (rather than merely suppressive) influences on host immune responsiveness. Possibilities of using infections of, and factors from these parasites to ameliorate human immunological disorders such as inflammatory bowel disease and asthma, and as sources of pharmacologically active agents, are now being actively explored.

1. Introduction

Organisms that infect vertebrates are classified as either microparasites or macroparasites (Anderson and May, 1979). Microparasites comprise viruses, bacteria and protozoa: they are generally capable of direct replication in their hosts, often at high rates and their infections tend to be short-lived and induce immune responses that protect against re-infection. Microparasites also generally have short generation times and it is clear that genetic diversity and/or antigenic variation are two characteristics of these organisms that contribute significantly to the ability of many microparasites to persist in the face of adaptive immune responses generated by their hosts (See Chapter 4, Alcami).

Helminths (Table 1) are macroparasites and they generally do not replicate directly in their vertebrate hosts; instead by sexual reproduction they generate stages with the attributes necessary for onward transmission. In contrast to the

*For correspondence email m.doenhoff@bangor.ac.uk

Table 1. An outline classification of parasitic helminth genera (in italics) mentioned in this chapter

- Phylum: Nematoda
    - Class: Enoplea
        - Subclass: Enoplia
            - Order: Tichurida
                - Genera: *Trichinella, Trichuris*
    - Class: Rhabditea
        - Subclass: Rhabditia
            - Order: Strongylida
                - Genera: *Strongyloides, Necator, Ancylostyma*
            - Order: Ascaridida
                - Genera: *Ascaris, Toxocara*
            - Order: Spirurida
                - Genera: *Wuchereria, Brugia, Loa, Onchocerca*
- Phylum: Platyhelminthes.
    - Class: Trematoda
        - Subclass: Digenea
            - Genera: *Schistosoma, Fasciola, Paragonimus*
    - Class: Cestoidea
        - Subclass: Cestodaria
            - Genera: *Taenia, Echinococcus*

(Adapted from Schmidt and Roberts, 2000)

high rates of mortality caused by some microparasites (*e.g.*, malaria), deaths of humans as a result of overwhelming worm burdens are rare and some processes must operate to account for this fact. The adaptive immune response has obvious candidature for such a process, but at least two other factors are likely to limit the size of worm burdens: namely, transmission rates which allow infection loads to build up only slowly and genetic or otherwise determined predispositions that prevent most host animals becoming too heavily infected.

2. To what extent do adaptive immune responses control helminth infections?

If adaptive immune responses exerted a limiting effect on parasites it would be expected that the latter would be more infective, live longer, and/or become more fecund in hosts that were immune-deficient. Very few helminths 'opportunistically' infect immunosuppressed humans, however, suggesting that specific immune-responsiveness may play only a small host-protective role, at least in this host species. The only well-established exception to this generalization is *Strongyloides stercoralis* (Potter et al., 2003; Keiser and Nutman, 2004), hyperinfections of which are common in those being treated with immunosuppressive drugs or infected with human T-cell lymphocytrophic virus (HTLV)-1, though surprisingly, and for unknown reasons, *S. stercoralis* appears not to be opportunistic in human immunodeficiency virus (HIV)-infected subjects (Viney et al., 2004). A recent study has found that *Schistosoma mansoni* as well as *S. stercoralis* infections are more prevalent in HTLV-1 carriers (Porto et al., 2005).

Once established many helminth infections of vertebrates are chronic and although reiterated cycles of parasite death and reinfection contribute to this chronicity, evidence from epidemiological studies, self-inflicted infections or emigration of infected individuals to non-endemic areas indicates that individual helminth parasites can survive for a long time. For example, field studies indicated *Schistosoma mansoni* worms had a life-span of 7.5 years (Fulford et al., 1995) and some schistosome infections lasting 30 years or more have been reported (Wallerstein, 1949; Berberian et al., 1953; Harris et al., 1984; Chabasse et al., 1985). The life-spans of filarial worms are similarly typically measured in years (Subramanian et al., 2004; Vanamail et al., 1996) and as a broad generalization these tissue-dwelling nematodes tend to survive longer than the worms of species that live most of their lives in the gastrointestinal tract (*e.g.*, an estimate of 12 months for *Ascaris suum*; Elkins et al., 1986), which in turn tend to be longer-lived than free living nematodes (Gems, 2000). More detailed evidence on the longevity of nematode infections was given by Behnke (1987) and Behnke et al. (1992). The ability of many helminth parasites to survive for long periods indicates that the immune responses they induce, if any, must be relatively ineffective, at least with respect to their ability to remove the inducing organisms.

Some *in vivo* evidence is, however, consistent with immune responses having a host-protective anti-parasite role in humans, including a decrease in infection loads that is often observed with increasing host age, a predisposition to harbour infections of differing intensity, and evidence of 'concomitant immunity'.

2.1. Infection loads decrease with increasing host age

Epidemiological observations indicating that individuals become immune as they get older because infection intensities reach a plateau and level off or begin to decline in early adulthood, have been recorded for schistosomes (Fulford et al., 1992; Wilkins et al., 1987; Woolhouse et al., 1991; Kabatereine et al., 1999), intestinal helminths (Bundy, 1988) and filarial worms (Day et al., 1991a; 1991b). The long-standing question whether the levelling-off or decline in infection intensities with age is due to host adaptive immune responsiveness or to other factors (*e.g.*, in the case of schistosomiasis, reduced water contact – Warren, 1973) cannot be easily resolved. Peak infection intensity levels occur in different age-classes in different endemic areas, with higher intensity infections being found in younger age groups when infection transmission rates are high and low transmission rates giving lower peaks in older people. Acquisition of protective immune responsiveness is consistent with this 'peak shift' epidemiological pattern (Woolhouse, 1998; Dunne and Mountford, 2001). It has however been observed that age-related differences in intensity of *S. mansoni* infection in respective communities in Senegal (Gryseels, 1994) and Kenya (Ouma et al., 1998) which had only very recent experience of the disease were similar to those found in populations which had experienced endemic schistosomiasis at the same (high) levels for many years; *i.e.*, an indication that changes which occur with age *per se* may be as important as immunological experience gained over a long period of infection.

2.2. Aggregation of parasite infection loads and host predisposition to infection

In helminth-infected communities the pattern of distribution of worm loads is generally best described as negatively binomial; *i.e.*, the majority of hosts have no or only small infections, while a minority harbour the majority of worms (Anderson and May, 1985). This has been shown for a variety of helminth species infecting humans, including hookworm (Croll and Ghadirian, 1981; Schad and Anderson, 1985), *Ascaris* (Haswell-Elkins et al., 1989), *Trichuris trichiura* (Bundy, 1986) and schistosomes (Polderman, 1979; Lehman et al., 1976), and also those infecting other animal species in natural conditions (Shaw et al., 1998).

Studies showing a positive correlation between pre- and post-treatment infection levels provide evidence for individuals being 'pre-disposed' to carry a particular level of infection (Haswell-Elkins et al. 1987; Chan et al., 1994; Tingley et al., 1988; Behnke et al., 2000).

Patterns of aggregation may be largely a reflection of heritable differences between individuals (Quinnell, 2003), including the genes enabling effective immune responses to be generated against the infecting agent, but other factors such as behaviour or nutrition of the host, and environmental factors, acting singly or in combination with host genetic predisposition are quite likely to be involved (Bundy and Medley, 1992; Bundy, 1988; Maizels et al., 1993).

Results from modelling indicate that heterogeneity in hosts' predisposition to exposure to infection plays an important role in generating aggregation (Bundy and Medley, 1992; Chan and Isham, 1998; Galvani, 2003), but modelling studies also indicate aggregation is likely to decrease as immunity to infection increases (Galvani, 2003). The latter observation is consistent with observations that the degree of aggregation often tends to be lower in adults than in the young, for example, with respect to *Trichuris trichiura* and *Ascaris lumbricoides* (Bundy et al., 1988) and *Wuchereria bancrofti* (Das et al., 1990) in humans, but apparently not *H. polygyrus* in wood mice (Quinnell, 1992). Also in contrast to humans, results of experiments with *H. polygyrus* in mice indicated that acquired resistance was the factor that most consistently led to variability and aggregation of worm numbers (Tanguay and Scott, 1992).

2.3. Concomitant immunity

The term 'concomitant immunity' was coined by Gershon et al. (1967) to describe the phenomenon in which animals bearing one tumour are resistant to another graft of the same tumour, although the first continues to grow. The concept was first used in a parasitological context to describe schistosome-induced resistance to reinfection (Smithers and Terry, 1969). Evidence for concomitant immunity was also found in experimental infections of cestodes (Mitchell, 1979) and nematodes (Miller, 1984), but the evidence for its occurence in human helminth infections is difficult to delineate because of the confounding factors of naturally occurring and continuous turnover of an infected individual's parasite populations (Maizels et al., 1993).

Factors that individually or collectively may affect the dynamics of an infection include age, sex and nutrition. Host genes other than those required for immune-responsiveness may be involved in control of the infection and the environment in which the host lives and the transmission rate of the infecting agent may individually or collectively also affect the dynamics of an infection. All may act in concert with, independently of, or against adaptive immunity.

Such complexity means it is often difficult to conclude what role immune responses play in defining an observed pattern of infection. The existence of many mechanisms which appear to allow helminths to subvert and/or evade immunity does however suggest that immune responses play some role in controlling these infections and that the parasites have adapted accordingly.

Variations in the strategies seemingly employed by different helminths to subvert and evade host immune responses are to some extent related to differences in their life-cycles. Gastrointestinal (GI) and filarial nematodes will therefore be considered separately, followed by trematodes (mainly schistosomes) and cestodes.

3. Gastrointestinal (GI) nematodes

3.1. Passive exploitation of host-dependent factors

The lumen, mucosal surfaces and tissues of the gastrointestinal tract are rendered potentially suitable habitats for parasitization because of their relative inaccessibility to some immunological effector mechanisms. Many GI helminths additionally exploit naturally occurring weaknesses in host immunocompetence, such as occur during lactation and in neonatal life. Thus, lactating mice and rats were less capable of expelling infections of, respectively, *Nippostrongylus brasiliensis* (Houdijk et al., 2003) and *Trichuris muris* (Selby and Wakelin, 1975) than non-lactating controls.

Hypobiosis and the periparturient rise in faecal egg counts is characteristic of many GI nematodes: having large numbers of infective forms produced coincidentally with the presence of susceptible neonate hosts helps ensure transmission and is of considerable agricultural importance (Gibbs,1986). The less than fully functional immune responses of neonates are in turn potentially helpful to parasite survival.

Poor nutrition could be one of the factors contributing to the temporary loss of resistance during the preparturient period (Donaldson et al., 1998) and malnourishment alone tends to compromise immune-responsiveness, and render hosts more susceptible to GI helminth infection (Koski and Scott, 2001). This may in part account for the coincidence in distribution of some human GI helminth infections (*e.g.*, *Ascaris* and hookworm) with areas in which human malnourishment is widespread (Crompton, 1984). Stress and old age are other host-dependent factors from which nematode infections are known to benefit with higher worm burdens or longer survival rates (see Behnke, 1987; 1992 for reviews of the literature.)

The best opportunity for parasite survival is provided by the minority of individuals that carry the majority of worms as a result of aggregation of infection loads (see above), but the factors differentiating between heavily infected and lightly infected or uninfected hosts have not been readily identified. Selective breeding of mice with different susceptibilities to *T. muris* (Wakelin, 1975) or *Heligmosomoides polygyrus* (formerly *Nematospiroides dubius*; Brindley and

Dobson, 1981) confirmed that host susceptibility to primary nematode infections is genetically controlled, and results from mouse strains and different breeds of other animals indicated that the genetic basis predisposing to heavy or light primary infections is complex and can seldom be related to a specific adaptive immune response. The application of recently-developed technologies for genomic and proteomic analysis are, however, enabling progress to be made in identification of relevant genes. Thus, for example, linkage mapping techniques based on microsatelite markers are being used to identify loci associated with resistance traits to *H. polygyrus* in mice (Behnke et al., 2003) and a proteomic and genomic search for the factor responsible for the difference in susceptibility to *Trichinella spiralis* in BALB/c (resistant) and C57BL/10 (susceptible) mice indicated that the latter did not have a gene for a lectin produced by enteric epithelial (goblet) cells (Pemberton et al., 2004).

Low intensity infections of some species of GI nematode persist longer than heavier infections (*e.g.*, *T. muris* in mice – Behnke and Wakelin, 1973; *Ostertagia ostertagi* in lambs – Michel et al., 1978). Explanations for this are analogous to those for the 'peak shift' phenomenon mentioned above: lighter infections are likely to be less effective than heavier infections at eliciting protective immune responses. However the rule is not invariable as Wakelin and Lloyd (1976) found that light and heavy infections of *T. spiralis* were eliminated with similar kinetics.

The worms causing a GI infection tend not to be evenly distributed along the length of the infected intestine (*e.g.*, *H. polygyrus* in the mouse; Lewis and Bryant, 1976), and aggregation of worm loads thus occurs in an individual animal as well as in the host population as a whole. As Behnke (1987) has argued, if immunomodulatory factors with only a local effect mediate the parasite's survival, collective activity will be required to prevent worm expulsion because the efforts of only a single worm may be insufficient. Immune reactions against a parasite in the intestine may in turn be focused in a limited area and regular change of location would thus be an easy way for a parasite to avoid them (Miller, 1984). As an example, the dog hookworm *Ancylostoma caninum* was found to change its site of attachment regularly every few hours (Kalkofen, 1970).

GI helminth species can be sub-divided into those in which the egg/juvenile stages are ingested orally and spend their whole lives in the alimentary tract and species with stages that are found in other tissues at some time. Of the latter, species of *Trichinella* are somewhat unusual amongst helminths in having larvae able to invade muscle cells, a niche that seems immunologically 'silent' and in which they can survive for many months. The *Trichinella* larva transforms the muscle cell into a 'nurse cell', a change believed to be effected by the parasite secreting proteins that reprogramme host genomic expression (Despommier, 1998). Boonmars et al. (2004) have found evidence that the muscle cells parasitized by *Trichinella spiralis* larvae undergo apoptosis.

Adult *Trichinella* worms are also partially intracellular in as much as they are found threaded through syncytium-like remnants of cells of the columnar epithelium of the intestine (Despommier, 1993). *T. muris* worms are similarly located in intracellular channels (Lee and Wright, 1978) and this characteristic may provide an anchorage that delays immune response-dependent expulsion of adult worms, which does nevertheless occur eventually.

The time of expulsion of the adults of some species is worm dose-dependent, indicating intensity of infection is critical in eliciting an immune response that is active against the primary infection. In addition to dose, the frequency of infection of the host, as well as differences between strains of host, can be important determinants of the course of infection. Most natural GI helminth infections are likely to result from exposure to several successive small doses of parasite ('trickle infections'), which are likely to engender different, perhaps less host-protective immune responses when compared with those induced by single large-dose infections (the latter being more usual in experimental work). Natural infections of *T. muris* have been found mostly to be low-level and laboratory experiments confirmed that low intensity infections survived longer than heavier ones (Behnke and Wakelin, 1973). The worms derived from trickle infections of *N. brasiliensis* administered to young rats also survived longer than those from single larger infections (Jenkins and Phillipson, 1971). In contrast, but perhaps understandable in the context of immune evasion, heavy primary *H. polygyrus* infections were found to survive longer than low-level infections in BALB/c mice (which have intermediate responder status). One interpretation is that only the larger worm burden had an immunosuppressive effect (see below) that was sufficiently large to prolong the parasite's survival.

3.2. Stage-specific restriction of antigenicity, antigen shedding and antigen excretion

By expressing a different set of antigens each stage of an infection could evade the immune responses induced by preceding stages. Indeed, a corollary of the concept of concomitant immunity is that the stage(s) of an infection that induce protective immune responses differ in their antigenicity, immunogenicity, and/or susceptibility to immune attack from the stages that are the targets of those immune responses. The earlier larval stages of GI helminth infections tend to be more immunogenic and thus potentially more capable of inducing protective immunity, and/or more susceptible to immune attack than the adults. *H. polygyrus* is an example - chemical abbreviation of low doses very early (*i.e.*, at the larval stage) induced immune responses that protected against subsequent infections, while delaying the chemotherapy until the parasites matured gave poorer host-protection (Wahid and Behnke, 1996).

After infection nematode larvae undergo several moults before reaching adulthood and antigenic differences between the stages of the infection have been reported, for example, for *Toxocara canis* (Page et al., 1992), *Haemonchus contortus* (Raleigh et al., 1996) and *Oesophagostomum dentatum* (Joachim et al., 1999). Hypothetically each larval moult provides an opportunity for enhancing survival of the parasite through removal of antigens that are the targets of protective immunity and/or a means to slough off potentially damaging surface-adherent antibodies.

The sloughing off of surface components from mature adults would also serve an immune-subversive purpose. Very early research on the cuticular surfaces of nematodes questioned whether they were antigenic at all, but the application of immunofluorescent techniques (Sadun, 1963) made it clear they were. Surface labelling techniques and *in vitro* culture confirmed that nematode parasites can shed surface components that were antigenic (*e.g.*, Philipp et al., 1980; Philipp and Rumjaneck, 1984) and Blaxter et al. (1992) saw a 'disposable surface coat' as an important mechanism for immune evasion, with *T. canis* being a good example of how coat-turnover provides a means of removing surface-adherent antibodies (Smith et al., 1981) and eosinophils (Badley et al., 1987). Behnke et al. (1992) however argued that shedding of surface antigens by some species of lumen-dwelling nematode (*e.g. N. brasiliensis* or *H. polygyrus*) might be irrelevant to a parasite's survival as surface-specific antibodies had not been shown pose a threat to their survival. An exception is *T. spiralis* where the onset of expulsion was found to be correlated with a strong serum IgA response (Almond and Parkhouse, 1986) and rapid expulsion was induced in infected rats by transfer of IgE antibodies and thoracic duct lymphocytes from immune donors (Ahmad et al., 1991). The composition of the surface of *T. spiralis* is however different from that of other nematodes (Blaxter et al., 1992).

Early work on *N. brasiliensis* showed that the worms derived from rats given repeated small daily infections or a second large reinfection after a primary infection had been rejected showed 'adaptation' to an immune environment. The adapted worms were stunted, less susceptible to immune attack and of relevance in the present context, less antigenic (Jenkins and Phillipson, 1972; Ogilvie, 1974).

Parasite-derived material shed from the surface or that is excreted/secreted by the parasite can be potently immunogenic and antigenic (Lightowlers and Rickard, 1988). These factors may be parasite-protective, for example by inducing the host to generate immune responses that have no effect on parasite survival and/or by mopping up antibodies that can compromise parasite survival. Perhaps not surprisingly GI helminth antigens have however seldom been found in free or immune complex form in the circulation of their hosts, though Ey (1988) suggested that the survival of *H. polygyrus* larvae over several days in the murine host's intestinal tissues (*i.e.*, before they entered the gut lumen as adults) was due to secretion of antigens that mopped up anti-parasite antibody.

3.3 Acquisition and mimicry of host antigens

Hypotheses that parasites had prolonged survival times because they and their hosts had antigens in common were first formalized about 4 decades ago (Sprent, 1959; Dineen, 1963; Damian, 1964). As reported by Behnke et al. (1992), however, evidence for the presence of host blood group A antigen on or in *T. spiralis*, *A. suum*, *N. americanus* and *T. canis* was published much earlier (Oliver-González and Torregrossa, 1944; Oliver-González, 1949), but because extracts of whole parasites were used in these studies there was no evidence that the host constituent was on the parasites' surface. There has only been an occasional follow-up of these early studies; for example, on *T. canis* by Smith et al. (1983) and *A. suum* by Kennedy and Qureshi (1986). The concept of 'molecular mimicry' (Damian, 1987) as an immune evasion strategy has thus received more evidence-based support from studies on other helminths, particularly schistosomes (see below).

3.4. Intra-specific antigenic variation

Genetic differences between individuals of a parasite species are likely to be a prerequisite for adaptation to, and survival in the environments provided by hosts that are themselves genetically (and phenotypically) variable. As stated by

Maizels and Kurniawan-Atmadja (2002), in the evolutionary race between hosts and parasites monomorphic species of the latter would be driven to extinction. However, the long lifespans that are typical of many individual adult helminth parasites, and apart from a few exceptions, their inability to replicate themselves in vertebrate hosts predisposes against their being able to evade or subvert host immunity by genetically-driven variation to the same extent as microparasites such as the African trypanosomes.

Genotypic variation within species of GI helminths has now been documented by many studies (for review, see Maizels and Kurniawan-Atmadja, 2002; Wakelin et al., 2002), but evidence for antigenic variation relevant to immune evasion is scarce. Studies on antisera and parasites derived from infected humans however showed that individual worms of *A. lumbricoides* (Fraser and Kennedy, 1991) and *T. trichiura* (Currie et al., 1998) were antigenically heterogeneous. Laboratory studies on different strains of *T. spiralis* (Wassom et al., 1988; Goyal and Wakelin, 1993), *T. muris* (Koyama and Ito, 1996; Bellaby et al., 1996) and *H. polygyrus* (Tang et al., 1995) directly demonstrated or implied the existence of intraspecific antigenic differences.

Some work has been done to characterize the intraspecific variability of acetylcholinesterase (ACH) in *N. brasiliensis*. As mentioned above, worms of this species have the apparently unique ability among GI helminth parasites to adapt and survive in immune (rat) hosts. Adaptation is accompanied by a diminution in immunogenicity of the parasite (Jenkins, 1972) and a change in the electrophoretic isozyme pattern of its acetylcholinesterase (Edwards et al., 1971). *N. brasiliensis* secretes different forms of ACH during the course of infection, probably due to post-translational processing (Blackburn and Selkirk, 1992). Some intestinal nematodes secrete copious amounts of ACH (*e.g., Nematodirus battus*; Douch et al., 1988), but the immunological significance of this, if any, is not yet clear - inhibition of mucus secretion by the intestine (Philipp, 1984) and immunomodulatory or anti-inflammation roles (Lee, 1996) have however been suggested.

3.5. Suppression, modulation and regulation of the immune response

Suppression of adaptive immune responsiveness would be expected to be one survival strategy commonly adopted by parasites, with the caveat that the ability of the host to survive should not be impaired to the extent that survival of the immunosuppressive parasite is also compromized. Experimental studies have shown that immune responses against heterologous antigens were suppressed during GI nematode infections in mice infected with *T. muris* (Lubiniecki and Cypess, 1975), *T. spiralis* (Jones et al., 1984; Alkarmi et al., 1995), *N. brasiliensis* (McElroy and Befus, 1987; Liwski et al., 2000) and *H. polygyrus* (Ali and Behnke, 1984).

H. polygyrus is seemingly one of the most overtly immunosuppressive of all GI nematode parasites. It can inhibit the immunological expulsion of other GI nematode infections such as *T. muris* (Jenkins and Behnke, 1977; Behnke et al., 1984), *T. spiralis* (Behnke et al., 1993) and *N. brasiliensis* (Colwell and Wescott, 1973). It is one of the few examples where host immunosuppression has been shown to have a parasite-enhancing effect as the presence of relatively few adult

H. polygyrus worms prevented the induction of host-protective immunity that could be generated by administration of irradiated homologous larvae (Behnke et al., 1983). *H. polygyrus* worms were, however, susceptible to the inflammatory changes induced by *T. spiralis* when the latter were at their most intense phase (Behnke et al., 1992).

Mast cells have been identified as an effector mechanism in immune expulsion of some helminths (Else and Finkelman, 1998; Marshall, 2004) and *H. polygyrus*-induced down-regulation of the murine host's ability to generate a mastocytosis may account for the defective immune reactivity against itself as well as other concurrently infecting nematodes such as *T. spiralis* (Dehlawi and Wakelin, 1988; Behnke et al., 1993).

Much of recent work on the immunology of metazoan parasite infections has been framed in the context of differences in immune responses driven by, respectively, T helper 1 and T helper 2 (Th1 and Th2) thymus-derived lymphocytes (Sher and Coffman, 1992; Abbas et al., 1996): indeed, analysis of the differences between the immune responses induced, respectively, by the protozoan parasite *Leishmania major* and *N. brasiliensis* contributed significantly to the emergence of the Th1/Th2 paradigm (Mosmann and Coffman, 1989). Th1 and Th2 cells are distinguished by the cytokines they produce when they have been activated. Activation of Th1 cells is promoted by interleukin (IL)-12 (Langrish et al., 2004) that is produced by monocytes, macrophages and dendritic cells: the cytokine they mainly produced was gamma interferon (IFN-γ) and they promote production of antibodies of the IgG2a isotype. Th1-driven responses are classically considered to be more 'pro-inflammatory' than Th2 responses and they contribute to host protection against a wide range of viral, bacteria and unicellular protozoan parasites.

Th2 cell activity is associated with production of IL-4, IL-5, IL-9, IL-10 and/or IL-13, promotion of antibodies of the IgE and IgG1 isotypes and mobilization of mast cells, basophils and eosinophils. Th2-type responsiveness is a general characteristic of helminth infections. Thus, predominanly Th2 cytokines were produced *in vitro* by lymphocytes from a well-studied helminth-infected population of Ethiopean immigrants to Israel (Bentwich et al., 1996). Th2-related cytokines and/or antibody isotype profiles were likewise found during infections of *A. lumbricoides* (Cooper et al., 2000; Turner et al., 2003) and *T. trichiura* (Faulkner et al., 2002) and were associated with protective immunity against these two parasite species (Jackson et al., 2004).

In experimental animals GI helminths also induce Th2-type responses, which in turn are responsible for the expulsion of the worms. Examples include infections of *N. brasiliensis* (McKenzie et al., 1998; Liu et al., 2002); *T. spiralis* (Grencis et al., 1991) and *T. muris* (Faulkner et al., 1998: see Finkelman et al., 2004; Hayes et al., 2004, for more extensive reviews). While Th2 cells are a principal source of IL-4, a *N. brasiliensis* infection induced production of this cytokine also in eosinophils (Shinkai et al., 2002) and basophils (Min et al., 2004).

Strains of mice that on the other hand were unable to self-cure infections of *T. muris* rapidly produced Th1-polarized immune responses (Else and Grencis, 1991; Else et al., 1993). In the same vein, *in vivo* depletion of the Th1 cytokine IFN-γ and the blocking of IL-4 promoted survival of *T. muris* in mice (Else et al., 1994) while administration of IL-12, which stimulates Th1 cells and IFN-γ production, promoted chronicity of *T. muris* (Bancroft et al., 1997). Administration of IL-12 did not however cause an on-going *T. muris*-induced Th2 response to switch to a Th1 profile (Bancroft et al., 2001).

GI helminth infections thus induce a pattern of host responses that is polarized by Th2 cells and which may include the production of IL-4, IL-5, IL-9 and IL-13, and increased IgE and IgG1 antibody synthesis. Pathological changes resulting from Th2 activity include mastocytosis and eosinophilia, and inflammatory enteropathy at the site of infection, as well as goblet cell hyperplasia, neuromuscular disfunction, and changes in smooth muscle contractility and epithelial cell function. Although it is generally accepted that these reactions are in turn responsible for the expulsion of the parasites, it has not been possible to identify a single mechanism of immune-mediated expulsion that is effective against all GI nematodes; rather, the removal of each species of GI nematode appears to involve a different subset of Th2-type responses (Maizels and Holland, 1998; Else and Finkelman, 1998; Lawrence, 2003; Finkelman et al. 2004, Hayes et al., 2004; Maizels et al., 2004).

Modulation of host immune-responsiveness such that it becomes less Th2-like would in theory thus provide a means by which GI nematodes could prolong their survival. Mice with chronic infections of the potently immunosuppressive *H. polygyrus* did indeed show poor Th2 (and Th1) responsiveness (Ben-Smith et al., 2003) and Th1-type responses were similarly associated with *T. muris* infections that had been established by low infection doses (Bancroft et al., 1994) or trickle infections (Bancroft et al. 2001), both of which modes resulted in prolongation of the infections. Antigen dose has an effect on whether an immune response will become mainly Th1 or Th2 (Boonstra et al., 2003) and it may be a general characteristic of host reactions to parasites that introduction of large doses induces predominantly Th2 responses, while low dose stimulation activates Th1 cells (Constant and Bottomly, 1997).

Th1 and Th2 cells cross-regulate their respective activities and cytokines produced by Th2 cells have been considered to be 'anti-inflamatory' (Opal and De Palo, 2000). IL-10 is particularly important in this respect through its capacity to inhibit cytokine production by Th1 cells and thus dampen down Th1-driven inflammatory responses (Mosmann and Moore, 1991; Pestka et al., 2004). As might be expected therefore, abrogation of host IL-10 production by gene-knockout or administration of anti-IL-10 antibodies has resulted in prolongation of GI helminth infections, for example *T. muris* (Schopf et al., 2002) and *T. spiralis* (Helmby and Grencis, 2003). However, deletion of the capacity to produce IL-10 also resulted in an overall increase in enteric pathology and morbidity and curtailed host survival (Schopf et al., 2002), which may help explain why a capacity to completely neutralize the effects of one or more Th2 cell cytokines appears not to have evolved as an immune-subversion strategy by helminths.

3.6. Characterization of immunosuppressive, immunomodulatory and evasion-related factors produced by GI helminths

Demonstrations that infections of GI helminths were potently immunosuppressive were naturally followed by some attempts to replicate the immunosuppressive effect with excretory/secretory (E/S) products and other non-living extracts (Lightowlers and Rickard, 1988). Extracts from *H. polygyrus* were successful in suppressing homologous immunity *in vitro* (Pritchard et al., 1984; Pritchard and Behnke, 1985) and the *in vitro* antibody response to a heterologous antigen (Crawford et al., 1989). So far only basic characteristics of the immunosuppressive factor(s) of this parasite in terms of heat sensitivity and molecular size have been described (Pritchard et al., 1994; Telford et al., 1998). Proteinaceous constituents of extracts of *A. suum* (Faquim-Mauro and Macedo, 1998; Boitelle et al., 2003; Oshiro et al., 2004) and *T. spiralis* (Gerencer et al., 1992) have also been found to be immunosuppressive in mice *in vivo*, and an extract from 4th stage larvae of *Ostertagia ostertagi* inhibited *in vitro* responsiveness of bovine lymphocytes (Gomez-Munoz et al., 2004).

Injections of the E/S products of adult *N. brasiliensis* intraperitoneally into mice generated an adherent peritoneal exudate cell population that suppressed antigen-induced lymphocyte proliferation (Allen and MacDonald, 1998). Proteins secreted by *N. brasiliensis* acted as adjuvants for Th2-type responses in mice challenged with the heterologous antigen hen egg lysozyme (Holland et al., 2000) and dendritic cells pulsed with E/S products of this parasite primed mice for Th2 responsiveness (Balic et al., 2004). Similarly, the pseudocoelomic body fluid of *A. suum* could skew the immune response to ovalbumin in a Th2 direction, and this activity was apparently due to factors other than the ABA-1 allergen known to be present in this parasite (Paterson et al., 2002).

The active moieties in most of the above-listed studies have yet to be better characterized at the molecular level. Studies involving better-characterized immunoregulatory factors produced by helminths have indicated that *T. muris* and *Trichinella* spp., have the capacity to synthesize molecules that are homologous with mammalian macrophage migration inhibitory factor (MIF; Wu et al., 2003). MIF is involved in regulation of innate and acquired immunity (Calandra and Roger, 2003; Lolis and Bucala, 2003), but its production by parasites is paradoxical as it is a pro-inflammatory cytokine. Synthesis of a homologue of the Th1 cytokine IFN-γ by *T. muris* has been reported by Grencis and Entwistle (1997), an observation consistent with a Th1-driven immune milieu being conducive to survival of GI nematodes. Similarly, Riffkin et al. (2000) reported results indicating that *Trichostrongylus colubriformis* can produce tumour necrosis factor (TNF)-α or its homologue.

Granulocytes such as eosinophils neutrophils, platelets and macrophages have the potential to react against intestinal helminths in an antibody-dependent or –independent manner and such reactivity could in turn result in the release of reactive oxygen radicals. These parasites are known to produce relatively large amounts of anti-oxidant enzymes such as superoxide dismutase (SOD), glutathione peroxidase, glutathione S-transferase, catalase, and the more recently discovered peroxidoxins which are postulated to have a parasite-protective role (Callahan et al., 1988; Bruschi and Lucchi, 2001). Consistent with this hypothesis, extracts of adapted *N. brasiliensis* were found to have higher concentrations of superoxide dismutase than those from normal worms of this species (Knox and Jones, 1992), and *H. polygyrus* extracts were considerably richer in several oxygen-scavenging enzymes than those of *N. brasiliensis* (Smith and Bryant, 1986).

The surfaces and excretions of nematodes are rich in glycoconjugates (Dell et al., 1999), as are the surfaces of host cells and tissues. The last decade has seen a great increase in our understanding how pathogens interact with the innate immune system, the latter recognizing pathogen-associated molecular patterns via pattern recognition Toll-like receptors (TLR; Janeway and Medzhitov, 2002). Much of this knowledge derives from work on microbial pathogens and the implications therefrom are now beginning to influence the direction of research on parasitic helminths (*e.g.*, Gruden-Movsesijan et al., 2003). Interactions between glycanic residues and lectin-like molecules modulate many of the interactions between granulocytes, lymphocytes and other host cells involved in expression of innate and acquired immunity, thus affecting cell trafficking and activatory signalling at the cell surface and glycanic molecules may thus play a considerable role in the Th2-bias of immune responsiveness to helminth infections (Thomas and Harn, 2004). Interestingly, immunization with antigens from the free-living nematode *Caenorrhabditis elegans* induced an antigen-specific Th2-skewed response, as did immunization with antigens from the filarial parasite *Brugia malayi*, and the Th2-inducing property of both nematodes was ablated by prior destruction of the glycanic residues of their antigens by treatment with sodium metaperiodate (Tawill et al., 2004).

Several intestinal nematode species have been shown to produce or to have genes coding for lectins (Loukas and Maizels, 2000) and these parasite-derived carbohydrate-binding molecules may have immunomodulatory properties (Loukas et al., 1999; Young and Meeusen, 2002; Van Die et al., 2004).

The lumen of the intestine is of course naturally rich in proteases for food digestion, but proteases are also one of the main products of degranulating mast cells and mice rendered deficient in mast cell protease-1 by gene-knockout showed delayed expulsion of *T. spiralis* (Lawrence et al., 2004). Biochemical characterization of nematode extracts and the mining of genomic data indicate that nematodes synthesise cysteine protease inhibitors (cystatins) and serine- and small serine protease inhibitors (serpins and smapins, respectively) which are likely to have parasite-protective roles including immune evasion and immunoregulation (Zang and Maizels, 2001). It has been suggested that a cystatin produced by *N. brasiliensis* modulates antigen processing in host APCs (Dainichi et al., 2001). A metalloprotease in the E/S products of the hookworm *N. americanus* illustrates the other side of this coin – it cleaved and inactivated the chemoattractant eotaxin, and thus may prevent accumulation of eosinophils and other granulocytes to the site of infection (Culley et al., 2000). Secretions of the hookworm *N. americanus* have also been shown to cause apoptosis of T-cells (Chow et al., 2000), which might be a strategy to reduce inflammation caused by infiltrating leucocytes and thereby create a site of 'immune privilege'.

4. Filarial and other tissue-dwelling nematodes

Filarial nematodes cause the major human tropical diseases lymphatic filariasis (*Brugia malayi* and *Wuchereria bancrofti*) and onchocerciasis (*Onchocerca volvulus*). Of these, only *B. malayi* is readily amenable for studies in experimental animals. Other animal models which have been studied include *Acanthocheilonema viteae* in jirds, *Dirofilaria immitis* in dogs, and *Brugia pahangi* in jirds and cats. Studies with *Litomosoides sigmoidentis* in mice are now proving particularly valuable in unravelling the dynamic relationships between immune effector mechanisms and stages of the filarial life cycle (Maizels et al., 2004).

Adult filarial worms are 'tissue-dwelling' and thus adapted to living in a very different set of habitats from GI nematodes. Filarial nematodes also differ in another important respect in so far as the former have no stages that are free-living. Behnke et al. (1992) argued that the absence of a reservoir with infective potential in the form of free living parasites may have been compensated for by longer survival of adult filarial worms through the evolution of more efficient immune-evasion strategies and indeed, 'the filariae are renowned for the chronicity of the infections they cause and for the success of their survival strategies in the face of host resistance'.

Filarial species display a variety of life-cycle adaptations. Adults of *O. volvulus* reside under the skin of their human hosts where they become encapsulated in nodules formed by host reactivity, but in which they can survive for long periods. The microfilariae of *O. volvulus* also remain mainly in the skin. *Loa loa* adults are also found subcutaneously, but can migrate and their micofilariae are blood-borne. The adults of *B. malayi*, *B. pahangi* and *W. bancrofti* live in the lymphatics and their microfilariae enter the blood stream via the thoracic duct. Adults of the heartworm *D. immitis* (infecting mainly dogs) live in the right side of the heart and pulmonary artery.

Survival of filarial nematodes depends on transmission of the microfilarial stage of the infection to the respective species of invertebrate that serves as the intermediate host, rendering the long-term persistence of microfilariae particularly important for the success of these parasitic life-cycles.

As for GI helminth infections, a Th2-type immune responsive state is typically associated with human filarial infections; namely, high IgE and IgG4 antibody levels, a Th2 > Th1 cytokine profile, and increased deployment of eosinophils, basophils and mast cells. T-cell populations secreting IL-4 were thus found in the blood of both microfilaraemic and elephantiasis patients (Yazdanbakhsh et al., 1993) and Mitre et al. (2004) showed that a major source of IL-4 in human *B. malayi* infections might be filarial antigen-driven basophils.

The immunology of filarial infections in experimental animals has many parallels with human infections (Lawrence and Devaney, 2001) and animal models allow the roles of each of the different life cycle stages to be tested separately. Thus, studies with *B. pahangi* in mice indicate that both adult worms (Lawrence et al., 1994) and L3 larvae (Osborne and Devaney, 1998) induce IL-4 synthesis, with NK1.1 T cells being the source of the cytokine in the latter study. Moieties that are glycan-rich (Tawill et al., 2004), as well as other types of molecule (such as the 'polyproteins' found in the E/S products of *D. immitis* - Tezuka et al., 2003), may be important in inducing Th2-type immune responsiveness.

It has however been realized that patterns of cytokine production and cellular interactions during filarial infection are complex. Th1 cells may be involved in the early human response to infection as stimulation of peripheral blood mononuclear cells (PBMC) from unexposed donors with *B. malayi* L3 larvae resulted in an increase in frequency of cells expressing pro-inflammatory Th1 cell-type cytokines, but not Th2 cytokines (Babu and Nutman, 2003). Similarly, proliferating PBMC from subjects putatively immune to *O. volvulus* (Elson et al., 1995) and *B. malayi* (Dimock et al., 1996) patients produced Th1-type cytokines. Consistent with these observations Hogarth et al. (1995) found that clearance of *O. volvulus* microfilariae was accelerated in IL-4 knockout mice which had normal eosinophilia, but lacked an IgE response.

In humans resistance to reinfection is age-related and has characteristics of concomitant immunity (Vanamail et al., 1989; Day et al., 1991a; MacDonald et al., 2002). Antibodies specific for cuticle antigen are acquired in an age-dependent manner (Day et al., 1991b), but this antigen appears not yet to have been characterized. In experimental animals L3 larvae can both induce and be the targets of mmune responses that protect against infection and results of experiments involving the use of irradiated L3 larvae as a resistance-inducing vaccine have indicated that Th2 responses are important (Devaney and Osborne, 2000). The mechanisms by which L3 larvae are killed are unclear, but eosinophils and IgE have been proposed as the mediators (Abraham et al., 2004). Paradoxically, in *L. sigmoidontis* infections in mice a Th2 response involving IL-4 is required to prevent development of the parasite in C57BL/6 mice that are normally resistant, but BALB/c mice do not develop resistance despite the induction of a Th2 response (Le Goff et al., 2002). IL-5 was shown to have greater impact than IL-4 on parasite containment, particularly of adult worms (Volkmann et al., 2003) and Saeftel et al. (2001;2003) showed that both IFN-γ and IL-5 were needed to protect mice against *L. sigmoidontis* infection and that neutrophils may be involved in controlling adult worms.

4.1. Stage-specific restriction of antigenicity and antigen shedding and excretion

The capacity to shed surface antigens which are the targets of potentially damaging antibody has been found in the microfilariae of some filarial species, including *A. viteae* (Apfel and Meyer, 1990) and the horse parasite *Onchocerca cervicalis* (Edwards et al., 1990). Gatrill et al. (1991) found differences between the IgM antibodies from *A. viteae*-infected BALB/c and C57BL/10 mice with regard to strength of binding and promotion of leukocyte adherence to the surface of microfilaria, and the rate of shedding of surface-bound antibodies, and suggested such differences may help account for the microfilaraemias being more chronic in BALB/c than C57BL/10 mice. Changes in the antigenicity of the different maturational stages of filarial larvae, and that may be involved in immune-evasion, have also been observed in *D. immitis* (Abraham et al., 1988; Grieve, 1990), *O. volvulus* (Miller et al., 1991), *B. malayi* (Carlow et al., 1987), *B. pahangi* (Marshall and Howells, 1986; Devaney et al., 1996), *W. bancrofti* (Maizels et al. 1986) and *A. viteae* (Apfel et al., 1992).

Similarities between the surface of the tissue-dwelling larvae of *T. canis* and the filarial species *D. immitis* that were of potential relevance to immune evasion were noted by Grieve (1990). *T. canis* larvae have a marked capacity for turnover of a mucin-like protein that has been partially characterized (Page et al., 1992; Loukas et al., 2000) and proposed to be an immune-evasion device (Gems and Maizels, 1996). Experimental evidence suggests that the surfaces of adult filarial worms are generally less changeable than those of their larvae with regard to turnover rate and antigenic variation (see Behnke et al., 1992 for review). As also argued by these authors, this is perhaps because adult worms have alternative mechanisms to survive host immunity, particularly the capacity for immunomodulation (see below).

Circulating antigens are found in patients with lymphatic filariasis, with levels of anti-microfilarial sheath antibodies and circulating filarial antigens being inversely related (Ravindran et al., 2000). Circulating antigens and immune complexes containing antigens derived from filarial worms have also been found in human infections of *O. volvulus* (Thambiah et al. 1992; Chandrashekar et al., 1993). Antibody and Fc receptors were found by Gray and Lawrence (2002) to play a role in clearance of *B. pahangi* microfilariae in mice and antigens released into the circulation may be a means of inactivating anti-parasite antibody. Furthermore, peripheral blood lymphocytes from subjects with circulating immune complexes were less responsive to filarial antigens (Dasgupta et al., 1987).

With regard to characterization of filarial circulating antigens, a 200kDa phosphorylcholine-containing glycoprotein was prominent in the circulating *W. bancrofti* immune-complexes (Lunde et al., 1988) and acetylcholinesterase from *O. voluvlus* and *B. pahangi* was found in serum from, respectively, infected human and *Mastomys natalensis* donors (Rathaur et al., 1992). Detection of a circulating antigen from *W. bancrofti* using monoclonal antibody Og4C3 (More and Copeman, 1990) has been widely used in the diagnosis of human bancroftian filariasis infection, but the significance of this or other antigenaemias in the host parasite relationship and in the context of immune subversion, if any, has not been defined.

4.2. Acquisition of host antigens and intra-specific antigenic variation

Serum albumin is the host-derived antigen most often found associated with filarial parasites. It has been found on the cuticular surfaces of microfilariae of *W. bancrofti* (Maizels et al., 1984; Kar *et al*, 1993), *B. malayi* (Mania and Kar, 1994) and *B. pahangi* (Maizels et al., 1985). A similar dichotomy occurs in *Onchocerca* spp, with albumin having been found on the microfilariae of *O. gibsoni* (Forsyth et al., 1984), but not *O. volvulus* (Taylor et al., 1986). Serum albumin was the principal protein present in detergent extracts of surface radio-labelled microfilariae of *L. carinii*, and it was also present in extracts of adult male and female worms (Philipp et al., 1984).

The problems inherent in seeking evidence for antigenic variation in helminth parasites have been outlined by Maizels and Kurniawan-Atmadja (2002) and include a lack of clonal populations, which in turn means a lack of sufficient study material that is needed to make comparisons between individual worms, and/or a reliance on samples from laboratory-reared parasites which may have reduced variability compared with their naturally occurring counterparts. The evidence for filarial worms having adopted antigenic variation as an immune-evasion device is therefore insubstantial. However, Ravindran et al. (1994) did show that sera from five *W. bancrofti*-infected individuals had antibody reactivity that reacted against the surface of microfilariae obtained from other patients, but not against their own microfilariae. This result is consistent with the observation of a significant correlation between amicrofilaraemia and the presence of antibody reacting with the sheath of *B. malayi* microfilariae (McGreevy et al., 1980), though a later similar study on patients with bancroftian filariasis failed to confirm this inverse relationship (Satapathy et al., 2001). As suggested by Maizels and Lawrence

(1991), the widespread adoption of tolerance-induction as an immune-evasion mechanism by filariae (see below) predicates against antigenic variation being also of value in this respect, and antigens that might be expected to be exposed to immune recognition (for tolerance induction) seem to be highly conserved within and between filarial species.

4.3. Suppression, modulation and regulation of immune responses

Human infections of *B. malayi* and *W. bancrofti*, species in which the adults inhabit human lymphatic vessels, progress through a series of phases which, as illustrated by Maizels and Lawrence (1991), followed a sequence of parasite establishment and maturation in the lymphatics; appearance of microfilariae in blood; parasite-specific anergy associated with microfilaraemia; breakdown of tolerance and a pathogenic response to incoming parasites. The last stage corresponds in time with disappearance of microfilariae from the blood and the onset of elephantiasis in some subjects, though others remain both asymptomatic and amicrofilaraemic.

Opinions differ as to whether the sequence of parasitological and pathogenic manifestations in infected subjects represents an inevitable progression from infection to inflammatory disease (Bundy et al., 1991) or a high degree of plasticity in host responsiveness (Ottesen, 1992). The results of many studies have however been consistent in showing that lymphocytes in peripheral blood of subjects with microfilaraemia were generally hyporesponsive to *in vitro* stimulation with parasite-derived antigens, while blood lymphocytes from amicrofilaraemics, including those with symptoms of chronic obstructive pathology of the lymphatics – elephantiasis – tended to respond better to antigenic stimulation (Ottesen et al., 1977; Piessens et al., 1980a; Nutman et al., 1987a; Yazdanbakhsh et al., 1993; see also O'Connor et al., 2003 for a recent review). Lymphocytes of children who were free from infection, but born 17-19 years previously to mothers with Bancroftian filariasis were also hyporesponsive (Steel et al., 1994), perhaps as a result of the children having been sensitized to filarial antigens prenatally (Weil et al., 1983). Impairment of lymphocyte responsiveness in patients with microfilaraemia was with respect to stimulation by antigens from microfilariae, but not adult worms (Mahanty et al., 1996a). Yazdanbakhsh et al. (1993) found that some amicrofilaraemics and elephantiasis patients had lymphocytes that were not responsive to *in vitro* stimulation with filarial antigens, and there was a negative correlation between lymphocyte proliferation and IgG4 antibody levels. Elevated specific IgG4 levels were deemed to be an indicator of active infection and the authors suggested that active infection thus resulted in lymphocyte hyporesponsiveness irrespective of a patient's clinical status (Yazdanbakhsh et al. 1993).

Prolongation of the survival of microfilariae is critical for filarial nematode survival as it helps increase the chances of transmission of the parasite to the invertebrate intermediate host. Suppression of the *in vitro* responsiveness of blood lymphocytes has been found to be most profound and most difficult to reverse during the microfilaraemic stage of lymphatic filariasis infection (Sartono et al., 1995a), though partial recovery of lymphocyte responsiveness occurred after microfilarial densities were reduced either by chemotherapy (Lammie et al., 1988; Sartono et al., 1995b) or as a result of seasonally-related changes in disease transmission (Sartono et al., 1999). Hence, the *in vitro* responsiveness of blood-borne B lymphocytes from microfilaraemic donors was also defective when compared with cells from patients with elephantiasis in so far as the former produced little antibody in response to stimulation by either filarial antigen or pokeweed mitogen (Nutman et al., 1987a).

The unresponsiveness of lymphocytes from microfilaraemics is considered to be a form of antigen-specific immunological tolerance that is more likely due to anergy rather than clonal deletion (Nutman et al., 1987b; Maizels and Lawrence, 1991).

Antibody of the IgG4 isotype is prominent in patients with Bancroftian filariasis (Ottesen et al., 1985) and infected, but microfilaraemic subjects were found to have much higher ratios of IgG4:IgE antibody specific for adult worm antigens in their sera than patients with elephantiasis (Kurniawan et al., 1993) or chronic disease (Atmadja et al., 1995). The IgG4 antibody that predominates in microfilaraemic patients blocked the allergic reactivity of IgE *in vitro*, which may explain why microfilaraemic patients rarely suffer allergic reactions despite the abundance of antigen and IgE-sensitized mast cells and basophils (Hussain et al., 1992).

In summary therefore, hyporesponsiveness of blood lymphocytes to filarial antigens and high IgG4:IgE ratios are characteristic of microfilaraemic patients, while infected subjects who are amicrofilaraemic and asymptomatic or suffering from elephantiasis have more immunoresponsive blood lymphocytes and a low ratio of IgG4:IgE antibody.

In onchocerciasis the stages of disease pathogenesis are not as starkly demarcated as for the lymphatic filariases. Unresponsiveness to specific antigen stimulation *in vitro* has nevertheless been noted in blood samples from human subjects infected with *O. volvulus* (Greene et al., 1983; Ward et al., 1988; Gallin et al., 1988), though enhanced parasite-specific cellular immune responsiveness was found in early infections with this species (Cooper et al., 2001). As for lymphatic filariasis, *in vitro* lymphocyte responsiveness was restored after chemotherapy of *O. volvulus* (Soboslay et al., 1994; Akuffo et al., 1996).

4.4. Mechanisms of induction of lymphocyte hyporesposiveness in lymphatic filariasis

Results of early searches for factors that would account for *in vitro* hyporesponsiveness of lymphocytes from microfilaraemic patients in lymphatic filariasis implicated a population of 'suppressor' cells as these were present in greater numbers in blood from patients with microfilaraemia than with elephantiasis (Piessens et al., 1980a; 1980b; 1982). The cell population in question lost its suppressive activity after overnight culture in the absence of antigen (Piessens et al., 1980b) and responsiveness of the cultured lymphocytes to parasite antigens increased if the suppressor population, identified as T cells, was removed (Piessens et al., 1982). King et al. (1992) subsequently suggested that an absence of parasite-specific lymphocyte precursor cells was the reason for low responses of blood-derived T and B cells to *in vitro* stimulation with filarial antigens, but this was in turn followed by observations indicating that in patients with microfilaraemia there was nevertheless a subset of Th2 cells secreting IL-4 and IL-5 that was preferentially induced relative to Th1 cells that secreted IFN-γ (King et al., 1993; Mahanty et al., 1993).

Several studies indicate the filarial infection-induced inhibition of lymphocyte proliferation may be mediated indirectly by antigen-presenting cells (APCs). Thus, at the onset of microfilaraemia *B. pahangi*-infected jirds were found to have adherent cells that inhibited lymphocyte proliferation (Prier and Lammie, 1988). Similarly, APCs from *B. malayi*-infected mice suppressed the proliferation of, but not cytokine production by antigen-stimulated T cells (Allen et al., 1996). Osborne and Devaney (1999) showed that IL-10 was involved in APC-induced suppression of proliferation of mitogen-driven Th1 cells in *B. pahangi*-infected mice, but MacDonald et al. (1998) showed that *B. malayi* L3 larvae implanted into the peritoneal cavity of mice recruited a population of IL-4-dependent macrophages that blocked the proliferation of T cells independently of the presence of IL-10. IL-4-dependent alternatively-activated macrophages (Gordon, 2003; Noël et al., 2004) may thus be primarily responsible for suppression of lymphocyte responsiveness (Loke et al., 2000; Maizels et al., 2004). Two genes, one for an eosinophil chemotactic factor (Ym1) and the other a cysteine-rich protein (Fizz1), were particularly highly expressed in such macrophages in *B. malayi*-infected mice (Loke et al., 2002), and expression of the same two macrophage genes is also characteristic of murine *L. sigmodontis* and *N. brasiliensis* infections, indicating this may be a general feature of APCs (including B cells) during nematode infections (Nair et al., 2005).

Semnani et al. (2004) showed that human Langerhans cell activity was downregulated by L3 larvae of *B. malayi*, leading to their having a diminished capacity to activate CD4$^+$ cells. Earlier these workers had shown that live *B. malayi* microfilariae, and to a lesser extent microfilarial extracts, induced apoptosis in human dendritic cells, and inhibited production of IL-12 and IL-10 by the human cells (Semnani et al., 2003).

More direct effects of the parasites on lymphocyte populations have also been shown. Thus, apoptosis of CD4$^+$ spleen cells was found in association with low antigen-induced proliferation from mice infected with *B. pahangi* microfilariae (Jenson et al., 2002). The lymphocytes could be rescued from apoptotic death and the proliferative defect reversed by incorporation of aminoguanidine and it was suggested that the apoptosis was due to the induction of nitric oxide production by IFN-γ (O'Connor et al., 2000; Jenson et al., 2002).

The production of IL-10 is pivotal in the down-regulation of lymphocyte activity during chronic human lymphatic filarial infections in which there is a failure to clear microfilariae (Mahanty and Nutman, 1995; Mahanty et al., 1996b; 1997; Cooper et al., 2001). Variants of the sequence of the promoter section of the IL-10 gene have recently been found to be associated with differences in the responsiveness of lymphocytes to *in vitro* stimulation with *O. volulus* antigen (Timmann et al., 2004). In murine infections also IL-10 has been implicated in prolonging survival of microfilariae of *L. sigmodontis* (Hoffmann et al., 2001) and of antagonizing Th2 responses capable of controlling this parasite species (Specht et al., 2004).

Together with IL-10, transforming growth factor (TGF)-β was found to be involved in the unresponsiveness of human lymphocytes to microfilariae (King et al., 1993; Doetze et al., 2000; Satoguina et al., 2002). Current opinion is that these two cytokines are produced by a subset of CD4$^+$CD25$^+$ regulatory T cells (Tregs) which regulates adaptive immune responses to many exogenous antigens including helminth parasites (Maizels et al., 2004). Steel and Nutman (2003) found that CD4⁺CD25⁺ T cells in blood from microfilaraemic subjects expressed high levels of the cytotoxic T lymphocyte antigen (CTLA)-4, a co-stimulatory surface receptor molecule that is involved in regulation of T-cell differentiation and activity (Alegre et al., 2001). IL-10 and TGF-β have also been shown to suppress IgE and promote IgG4 production (Akdis et al., 2004), which implicates Treg cells in generating the high ratios of IgG4:IgE found in microfilaraemic individuals (Kurniawan et al., 1993). Antibody-mediated killing of Treg cells in filarial worm-infected mice reverses immune hyporesponsiveness and also results in clearance of the parasite (Taylor et al., 2005).

4.5. Identification and characterization of immunomodulatory factors from filariae

Crude extracts of *B. pahangi* worms and microfilariae inhibited lymphocyte mitogenic activity when added to *in vitro* cell cultures (Leiva and Lammie, 1989). Similarly the E/S products of adult *B. phangi* worms induced peritoneal exudates cells that blocked *in ivtro* lymphocyte proliferation (Allen and MacDonald, 1998). One immunoregulatory factor that was discovered through experiments of this nature is ES-62, a tetrameric phosphoryl choline (PC)-bearing 62kDA protein with aminopeptidase activity (Harnett et al., 1989). It is the major constituent of the E/S products of *A. viteae* (Harnett et al., 1989) and has been shown to have marked immunomodulatory properties (Harnett et al., 2004). It influenced the maturation of dendritic cells in a Th2 direction (Goodridge et al., 2004) and exposure to ES-62 down-regulated that capacity of macrophages to produce the pro-inflammatory cytokines IL-12 and tumour necrosis factor (TNF)-α via Toll-like receptor (TLR)4 (Goodridge et al., 2005a). The molecule suppressed activation of B and T lymphocytes by disrupting signalling pathways involved in cell proliferation (Goodridge et al., 2005b), though cells of the B1 subset found mainly in mouse pleural and peritoneal cavities were stimulated to proliferate and produce IL-10 (Wilson et al., 2003).

The recent rapid progress in elucidating the genetic constitution of the filarial nematodes (Maizels et al., 2001a) has provided powerful novel means of identifying parasite molecules that may be involved in immune evasion. As pointed out by Maizels et al. (2001b), recent advances in filarial genomics have enhanced conventional biochemical and immunological approaches of characterizing parasite products and also allowed identification of abundantly-expressed parasite genes which may be immunomodulatory, and other genes that encode homologues of host proteins that regulate immunity. Thus, the possession by filarial nematodes of the capacity to produce homologues of mammalian cytokines such as transforming growth factor (TGF)-β (Maizels et al., 2001b) and MIF (Pastrana et al., 1998; Zang et al., 2002) has obvious connotations for immunoregulation.

Genes coding for cysteine and serine protease inhibitors (cystatins and serpins, respectively) and for an aspartyl protease inhibitor have also been identified in filarial nematodes (Maizels et al., 2001b). A recombinant derivative of a cystatin secreted by *O. volvulus* upregulated production of IL-10 and suppressed polyclonally stimulated and antigen-driven proliferation in cultures of PBMC, a Th2-orienting effect (Schonemeyer et al., 2001), but up-regulated nitric oxide production by macrophages treated with IFN-γ, a Th1 proinflammatory effect (Hartmann and Lucius, 2003). Comparison of cystatin homologues from *O. volvulus* and *C. elegans* indicated that the capacity to induce IL-10 was an adaptation to the parasitic life-style (Schierack et al., 2003). A cystatin from *L. sigmodontis* has been shown to suppress lymphocyte proliferation (Pfaff et al., 2002) and one from *B. malayi* inhibited both cysteine protease activities in the endosome/lysosome compartment of a human lymphocyte line and class II MHC antigen processing and presentation by APCs (Manoury et al., 2001). The BM-CPI-2 cystatin of *B. malayi* differs from its equivalents in *C. elegans* at a specific site which is considered to be responsible for inhibition of antigen processing enzymes (Murray et al., 2005). A serpin from *B. malayi* was claimed to inhibit neutrophil serine proteases (Zang et al., 1999), but a subsequent study has contradicted this observation (Stanley and Stein, 2003).

The surfaces of adult worms and microfilariae will be sites of initial antibody and/or immune cell-mediated attrition, including antibody-dependent cell mediated cytotoxicity (ADCC). For possible protection against toxic oxygen derivatives that may be generated in these reactions, *B. malayi* worm surfaces have been found to be rich in a 29kDa glutathione peroxidase (Cookson et al., 1992), and superoxide dismutase has been shown to be present as a less abundant constituent (Tang et al., 1994). Thioredoxin peroxidases have been found in both *B. malayi* (Ghosh et al., 1998) and *O. volvulus* (Chandrashekar et al., 1989; Lu et al., 1989). The L3 larvae of several filarial species produce a set of 'abundant larval transcript' (ALT) proteins that are immunogenic and have potential as a vaccine, but that have been included in a review of filarial immune evasion genes (Maizels et al., 2001b). Finally, lipopolysaccharides from *O. volvulus* (or from the *Wolbachia* bacteria that are endosymbiotic in this and other filarial nematodes - Taylor, 2002) induced the production of TNF-α and IL-10 by monocytes (Brattig et al., 2000). A variety of other molecules which filarial nematodes produce or have the genes for, and which could be involved in evasion and/or regulation of host immune responses, have been documented by Maizels et al. (2001a; 2001b).

5. Schistosomes and other trematodes

Trematode parasites have indirect life-cycles that require a snail as an intermediate host for asexual replication and a vertebrate definitive host for sexual reproduction. While worms of other trematode species are hermaphrodite, schistosome adults are dioecious.

Schistosoma mansoni has been the most-studied species of this class of parasite, mainly because it causes a serious human disease in the tropics and its life-cycle is relatively easy to maintain in the laboratory. Several other schistosome species also infect humans, *S. haematobium*, *S. japonicum* and *S. mekongi* being the most prevalent, but laboratory life-cycles of these are more difficult to maintain than those of *S. mansoni* and their immunology has thus been much less studied. *Fasciola hepatica* and *Paragonimus westermani* have been the only other trematode species on which experiments relevant to evasion and regulation of immunity have been conducted.

Schistosome infections of humans and other vertebrates commence with penetration of parasite cercariae through the skin. The cercariae transform into larvae, (schistosomula) and these are carried around the body in the blood stream for several days, with relatively large numbers being found in the lungs during this time because they are temporarily trapped in a tissue particularly rich in blood capillaries. By 7-10 days after percutaneous infection (of mice) young worms begin to accumulate in the hepatic portal vein, having become trapped between liver sinusoids and mesenteric capillaries because of their increased size. The worms grow and differentiate into adult males and females which pair up. *S. mansoni* and *S. japonicum* worms migrate to the mesenteric capillaries, while *S. haematobium* worms migrate to blood vessels of the bladder and other urino-genital tissues. In mice female schistosomes begin producing eggs after 4-5 weeks. Some eggs become extravasated and pass through host tissues into the lumen of the intestine (*S. mansoni* and *S. japonicum*) or ureter and bladder (*S. haematobium*), to be excreted. On contact with fresh water the miracidia hatch and continue the life cycle if they infect an appropriate snail host. However, many schistosome eggs remain embedded in host tissues where they become surrounded by large inflammatory lesions – schistosome egg granulomas - that are the main cause of the symptoms of urinary (*S. haematobium*) or intestinal schistosomiasis (*S. mansoni* and *S. japonicum*).

Warren (1964) showed that the pathogenesis of *S. mansoni* in mice was satisfactorily representative of human infections, though there are some subtle differences between the two hosts (Fallon, 2000; Cheever et al., 2000). Nevertheless, the mouse model of schistosomiasis mansoni has been used extensively to investigate the roles of cells and cytokines in schistosome egg-induced immunopathogenesis in particular, and local immune responses in general (Sandor et al., 2003).

5.1. Resistance to schistosome infection

Studies of schistosome infections in rhesus monkeys provided the first example of concomitant immunity in experimental parasitology, whereby a host is protected against reinfection by an existing infection (Smithers and Terry, 1969). The decline in intensity of schistosome infections in humans is consistent with concomitant immunity operating also in this host, but the factors responsible are not known for certain, and may in any case not have any connection with antigen-specific immune responses that are acquired as a result of experience of infection (Gryseels, 1994). However, the distribution of human schistosome infections is aggregated, as it is for other parasitic infections, and onward transmission is thus reliant mainly on a few people who carry large infection loads. A gene denoted as SM1 that putatively influences intensity of infection in humans has been located on chromosome 5q31-q33 (Marquet et al., 1996), a region containing some other genes that encode immune response-related molecules.

Resistance to reinfection may be mediated by different mechanisms in different host species. Results from the *Schistosoma mansoni* / rhesus monkey model were consistent with induction of anti-larval immunity by adult worms (Smithers and Terry, 1967), but experiments investigating the basis of concomitant immunity in mice indicated that resistance to reinfection was not due to specific anti-larval or anti-worm immune responses (Harrison et al., 1982): rather, it seems that secondarily incoming schistosome larvae failed to survive in mice with patent infections because egg-associated pathology generated by the primary infection caused 'leakiness' in hepatic and other vasculature, which in turn resulted in a failure of the incoming larvae to become trapped in the liver where they would begin developing into adults (Wilson et al., 1983). It is not clear to what extent, if any, egg-induced immunopathology is responsible for protection against infection in humans, though in this host species Th2-type immune responsiveness and/or IgE antibodies have generally been implicated in manifestation of resistance to infection (Dunne et al., 1992; Rihet et al., 1991; Medhat et al., 1998; Demeure et al., 1993; Webster et al., 1996; Rodrigues et al., 1999).

In experimental animals host-protective immunity against infection can be induced by administering irradiated larvae (Taylor and Bickle, 1986) or by drug-treating early infections (Bickle et al., 1990; Mountford et al., 1989; Bergquist et al., 2004), but not by administration of equivalent numbers of non-irradiated larvae that give rise to single-sex worm-alone infections (Harrison et al., 1982). The failure of non-attenuated larvae to induce protective immune responses may be due to their causing the host to produce more of the anti-inflammatory IL-10 and less pro-inflammatory IL-12 than irradiation-attenuated larvae (Hogg et al., 2003 Mountford and Trottein, 2004). Furthermore, IL-10 deficient mice gave higher antibody titres, greater lymphocyte proliferative responses, increased frequency of both IFN-γ- and IL-4-secreting cells, and heightened *in vivo* and *in vitro* schisosomulicidal immune responses than wild type mice (Hoffmann et al., 1999).

5.2. Intra-specific antigenic variation in schistosomes

Ealier analysis of isoenzyme polymorphism (Fletcher et al., 1981; LoVerde et al., 1985a) and the application of more modern genetic methods (Curtis et al., 2002) has confirmed that schistosomes are genetically heterogeneous, but there is so far little evidence that this inherent potential for variation is exploited as a means of immune evasion or subversion. This could in part be due to mice having been extensively used for maintenance of *S. mansoni* life cycles in the laboratory and successive passages in this host resulting in decreased genetic variability (LoVerde et al., 1985b). However, lack of genetic plasticity with regard to evasion of host-protective immunity was also indicated by the failure of an attempt to selectively breed a sub-population of *S. mansoni* that was insusceptible to resistance induced by vaccination with irradiated cercariae (Lewis et al., 1985).

5.3. The schistosome surface and modulation of surface antigenicity

Schistosomes in their vertebrate hosts are not protected by an outer chitinous-like cuticle as found on nematode worms: instead they have a tegument consisting of a syncitium of cytoplasm covered by a plasma membrane that provides essential physiological and absorptive functions (Asch and Read, 1975; Rogers and Bueding, 1975). The surfaces of young schistosome larvae (schistosomula) are initially bounded by a conventional trilaminate plasma membrane, but a few hours after skin penetration this is replaced by a membrane consisting of two closely apposed lipid bilayers (Hockley and McLaren, 1973).

The heptalaminate membrane is an apparent adaptation to an intravascular habitat as it is also a trait of other blood flukes (McLaren and Hockley, 1977). It may also be an important device to allow adult worms to survive for many years in an environment that is ostensibly immunologically hostile; for example, by rendering the mature parasite insusceptible to immune sera which are toxic to larvae (Dean, 1977; Abath and Werkhauser, 1996) and immature worms (Jwo and LoVerde, 1989).

Indirect immunofluorescence tests have shown that schistosomula lost much of their surface antigenicity within 24 hours of infection, apparently due to antigen shedding (Pearce et al., 1986) and thereafter the heptalaminate surface membrane seems in some way to shield antigens near the worm surface that may be susceptible to immune attack, particularly by antibodies. Expression of different surface proteins is developmentally-regulated during maturation of the parasite (Simpson, 1990). Staining of carbohydrates on larval surfaces using labelled lectins was markedly less on lung schistosomula than on newly transformed schistosomula (Simpson et al., 1983) and proteins were harder to label on the surface of adult worms than on schistosomula (Snary et al., 1980). Polyspecific rabbit antisera raised against worm extracts stained the surface of intact worms only weakly and diffusely, but reacted much more intensely and focally after the outer membrane, particularly that on male worm dorsal tubercles, had been damaged by anti-schistosome drug treatment (Doenhoff et al., 1988; Fallon et al., 1992). In related studies sera from *S. mansoni*-infected mice which could act synergistically with a schistosomicidal drug in killing adult worms had no cidal effect on worms in the absence of drug (Doenhoff and Bain, 1978), thus supporting the notion that the adult worm possesses mechanisms for withstanding attack from antibodies produced in response to the infection.

Monitoring of host-derived antigens (see below) and other labels on the surface of adult worms indicated the outer membrane was being turned over with an estimated *in vivo* half-life of between 5 and 6 days (Wilson and Barnes, 1977; Saunders et al., 1987). Shed membrane is most likely replaced with new material from membranous bodies that are numerous in the tegument (Hockley and McLaren, 1973).

5.4. Acquisition and mimicry of host antigens

Following the suggestion by Sprent (1963) that parasites were likely to have evolved surface antigens similar to those of the host in order to avoid detection by the immune response, host-derived or host-like antigens were detected in extracts of schistosome worms and the work of Smithers and his colleagues (Smithers et al., 1969) demonstrating that schistosome worms incorporated host-derived antigens has become a text-book example of a mechanism of avoiding immune recognition. Thus *S. mansoni* worms that had been grown in mice and surgically transferred to rhesus monkeys were killed if the monkeys had been previously immunized against mouse erythrocytes, but survived after suffering only a temporary setback in unimmunized monkeys. Host antigens were acquired by larvae after skin penetration and before the lung stage of migration, and protection against antibodies present in rhesus monkeys that were resistant to infection correlated with acquisition of the host red cell antigens (Clegg, 1974). Evidence for the acquisition of host antigens was also obtained using some other paired combinations of hosts, for example by transferring worms from mice to normal and anti-mouse immunized hamsters (Cioli, 1976; Boyer et al., 1977).

By growing schistosomula in culture with human blood of known specificity it was shown that the acquired host antigens were A, B, H or Lewis antigens, but not rhesus, M, N, S or Duffy antigens, the parasites being able also to take up A and B antigens from serum, probably in the form of glycolipids (Goldring et al., 1977). It has since been shown that the parasite can synthesize glycoconjugates containing the Lewis-X epitope (Srivatsan et al., 1992).

Evidence that schistosomes took up other host molecules followed, these including major histocompatibility complex (MHC) antigens (Sher et al., 1978; Simpson et al., 1983) and the protease inhibitors alpha-2 macroglobulin (Damian et al., 1973; Kemp et al., 1976) and contrapsin (Modha et al., 1988). It is not clear, however, exactly what function the uptake of protease inhibitors serves, nor what their relationship is with protease inhibitors that schistosomes themselves have the capacity to synthesize (Blanton et al., 1994; Ghendler et al., 1994). A scenario whereby parasite proteases might be shielded from immune recognition by interaction with host-derived protease inhibitors such as contrapsin was proposed by Modha and Doenhoff (1994).

The heterospecific immunoglobulins (Ig) that have been found on adult worms (Kemp et al., 1977) may provide a means of protection against specific immune detection. Worms were shown to have Fc receptors at or near the surface (Torpier et al., 1979; Tarleton and Kemp, 1981), one molecule with such properties being paramyosin (Loukas et al., 2001). Worm surfaces with attached immunoglobulin would be prone to activate complement (Skelly, 2004), which may however be prevented by paramyosin also having anti-complementary activity similar to mammalian CD59 (Marikovsky et al., 1990; Parizade et al., 1994; Deng et al., 2003). Complement-inhibitory decay accelerating factor (DAF) is also present on worms (Pearce et al.,1990) and schistosomula were shown to be able to acquire DAF from normal human erythrocytes and thus be protected from complement-mediated killing (Fatima et al., 1991).

5.5. Blocking antibodies

In animals with normal unattenuated infections the egg stage has been found to be a more potent inducer of anti-parasite antibody production than either the larval or worm stages (Ambroise-Thomas and Andrews, 1976; Omer Ali et al., 1988) and most of the antibody in *S. mansoni*-infected mice that was reactive against the schistosomulum surface was due to epitopes shared between the parasite's larvae and eggs (Omer Ali et al., 1988).

Schistosome survival may be prolonged by 'blocking antibodies' specific for carbohydrate epitopes induced by egg antigens. Thus, Khalife et al. (1986) found that removal of IgM from sera of infected subjects resulted in enhancement of the cytotoxic potential of those sera in ADCC reactions *in vitro*. Results of studies on *S. mansoni*-infected children in Kenya led to suggestions that egg antigens induced antibodies that blocked immune effector mechanisms directed against schistosomula (Butterworth et al., 1987) and that the blocking antibodies were anti-carbohydrate (Butterworth et al., 1988). Monoclonal antibodies with specificity for carbohydrate epitopes could also block *in vitro* killing of schistosomula in complement-dependent (Yi et al., 1986) or ADCC (Dunne et al., 1987) reactions. A corollary of the blocking antibody hypothesis is that immunity to schistosomes is effected by antibodies with specificity for peptide (rather than carbohydrate) epitopes and Omer Ali et al. (1989) did indeed find that human chronic infection serum was less intensely reactive against carbohydrate epitopes on the schistosomulum than acute infection serum, and that mice in which immunity had been induced by vaccination with irradiated cercariae lacked high levels of anti-carbohydrate antibody. Eberl et al. (2001) obtained a similar result in chimpanzees. In humans there is a negative correlation between levels of IgG2 antibody and resistance to infection (Demeure et al., 1993), an observation that is consistent with the blocking antibody hypothesis in so far as IgG2 antibodies are mainly reactive against carbohydrate epitopes (Langley et al., 1994).

5.6. Modulation of acquired immune-responsiveness by larvae and worms

As already mentioned, only irradiated or otherwise attenuated larvae of *S. mansoni* induce host-protective immune responses in mice. The failure of normal larvae to immune-prime the host and the survival of this stage of the infection may in part be due to its secreting the eicosanoids PGE_2 and PGD_2 which inhibit inflammatory responses in the skin by inducing the production of IL-10 (Ramaswamy et al., 2000). Through activation of prostanoid receptors on Langerhans cells, the PGD_2 produced by the larvae may also prevent these cells migrating from the skin to draining lymph nodes (Angeli et al., 2001; Hervé et al., 2003). Intriguingly, the larval enzyme responsible for synthesis of PGD_2 has been identified as the *S. mansoni* 28kDa glutathione S-transferase (Hervé et al., 2003) that otherwise has been shown to have potential as a vaccine for human schistosomiasis (Capron et al., 2001). Additional strategies that may protect schistosomula are their ability to abrogate the transcription of many genes implicated in normal dendritic cell function (Trottein et al., 2004) and to induce an anti-inflammatory phenotype in endothelium cells (Trottein et al., 1999b).

During transformation from cercariae to schistosomula *S. mansoni* larvae secrete protease activity that facilitates host penetration by degrading skin proteins (McKerrow and Salter, 2002). The secreted proteases may also promote larval survival by generating inflammation at the site of skin penetration (Fallon et al., 1966) and by rendering the larvae more resistant to complement by mediating the the release of the complement-activating glycocalyx from the surface of cercariae (Marikovsky et al., 1988). Further work by this research group indicated that following release of the glycocalyx the larvae expressed a complementary regulatory protein (Marikovsky et al., 1990), which was subsequently identified as a surface-exposed form of paramyosin (Deng et al., 2003). Larval protease activity has been implicated in regulation of host IgE production (Verwaerde et al., 1986) and direct protection from antibody attack may be provided by larval proteases through their ability to degrade immunoglobulins (Darani et al., 1997; Pleass et al., 2000).

A cathepsin-S like protease from the sparganum *Spirometra mansoni* has also been found to cleave IgG (Kong et al., 1994), while a cysteine protease present in metacercariae of *P. westermani* has had a wide range of immunosuppressive properties ascribed to it (Hamajima et al., 1994). In the same vein, a cathepsin L protease isolated from juvenile and adult *F. hepatica* prevented antibody-mediated eosinophil attachment to newly excysted juveniles (Carmona et al., 1993).

Other proteins secreted by *S. mansoni* larvae with immunomodulatory potential include a molecule of 16.8kDa that exerts anti-inflammatory activity by inducing production of IL-1 receptor antagonist (IL-Ra) by keratinocytes (Ramaswamy et al., 1996) and a ~23kDa protein that causes T cells to apoptose (Chen et al., 2002). Larval E/S products also induced mast cell degranulation and the production of IL-4 that would favour Th2-type responsiveness (Machado et al., 1996).

Schistosomes possess anti-oxidant enzymes (superoxide dismutase, glutathione peroxidase and glutathione S-transferase) that may protect them against immune cell-mediated attack (LoVerde, 1998). Adult schistosome worms release material into the blood stream (*i.e.*, circulating antigens) in sufficient quantities for methods for their detection to have become useful in diagnosis of human infection (reviewed in Doenhoff et al., 2004). Two antigens, circulating anodic antigen and circulating cathodic antigen (respectively CAA and CCA) are excreted by larvae and worms (Van Dam et al., 1996a) and there has been progress in elucidation of their respective structures (Bergwerff et al., 1994; Van Dam et al., 1994). It is not clear what benefit the parasite gains from the considerable energy expenditure required for synthesis of these proteoglycans, though Capron et al. (1980) found an inverse relationship between circulating antigens and complement-dependent cytotoxic antibodies, and CAA has been shown to be a receptor for complement C1q (Van Dam et al., 1993), and may thus protect the parasite gut (where both CAA and CCA are synthesized) against complement-mediated attack. CAA shares Lewis-X epitopes with human granulocyte surface antigen and anti-Lewis-X antibodies thus evoked may help protect the parasite by mediating complement-dependent lysis of granulocytes (Van Dam et al., 1996b). Circulating immune complexes down-regulated the *in vitro* activity of PBMC, this activity being associated with a decrease in TNF-α production and increase in IL-10 production (Rezende et al., 1997).

Experimental hosts have been shown to be immunosuppressed during the pre-patent stages of *S. mansoni* infections (Mota-Santos et al., 1977; Trizio et al., 1980). Activation of lymphocytes by mitogens or antigens *in vitro* was inhibited by peptides or other factors in parasite extracts (Dessaint et al., 1977; Mazingue et al., 1986) and a tripeptide fragment derived from immunoglobulins degraded by worm proteinases inhibited macrophage activities (Auriault et al., 1985) and granuloma formation around *S. mansoni* eggs (Himmi et al., 1985). As in other instances of down-regulation of lymphocyte responsiveness, IL-10 has been implicated in the anergy of PBMC from schistosome-infected subjects to parasite antigen stimulation *in vitro* (King et al., 1996; Velupillai et al., 2000). Subsequently Smith et al. (2004) showed that suppression of T cell activation during the early stages of *S. mansoni* infection was effected by macrophages that had upregulated programmed death ligand-1 (PD-L1).

5.7. Modulation and regulation of immunopathological responses to schistosome eggs

Schistosome eggs present a considerably more potent immunogenic challenge to the host than larvae or adult worms. As well as inducing antibody responses, eggs induce granulomatous inflammatory reactivity that surrounds eggs retained in tissues of infected hosts. Trend-setting work by Ken Warren and his colleagues in the 1960s showed firstly that the granulomas induced by *S. mansoni* eggs (rather than the eggs alone) were responsible for the pathogenesis of schistosomiasis; and secondly, *S. mansoni* egg-induced granulomatous inflammation was a T cell-mediated immunological response with many of the characteristics of a Type IV hypersensitivity reaction. Schistosomiasis mansoni was thus an early example of a parasitic disease with an immune aetiology (Warren, 1975).

Warren and others having shown that pathogenesis in schistosomiasis had an immune aetiology, and that disease symptoms ameliorated with the reduction in granuloma size that occurs over time after infection (see below), it was a natural progression to test whether immunosuppression might be therapeutic. Granuloma formation was ablated in *S. mansoni*-infected athymic nude mice (Byram and Von Lichtenberg, 1977) and infected T-cell deprived or otherwise immunosuppressed mice (Byram et al., 1979; Lucas et al., 1980), confirming the T cell-dependence of the egg-induced lesions. However, immunosuppression also resulted in the mouse hosts suffering from a severe hepatoxicity reaction and earlier death, results that were consistent with the notion that granulomas were needed to protect host tissues against damage from toxic egg products (Von Lichtenberg, 1964).

Much work has since been done to characterize the aetiology and regulation of the immunopathology due to schistosome eggs, mainly *S. mansoni*. Th1-type reactivity is generally found prior to patency of *S. mansoni* infections, and perhaps very early after the onset of patency (*i.e.*, acute schistosomiasis - Pearce et al., 1991; Montenegro et al., 1999; de Jesus et al., 2002), but after patency immune-responsiveness becomes polarized towards Th2-ness (see reviews by Pearce and MacDonald, 2002; Hoffman et al., 2002; Sandor et al., 2003; Stavitsky, 2004; Wynn et al., 2004). Production of IL-10 appears to be important in the egg-induced switch to Th2-ness (Sher et al., 1991), this cytokine being produced by Th2 cells (Sher et al., 1991), B cells (Velupillai and Harn, 1994), and /or $CD4^+CD25^+$ regulatory T cells (Hesse et al., 2004) which may at the same time also suppress production of the Th1-promoting cytokine IL-12 (McKee and Pearce, 2004). Macrophages (Atochina et al., 2001) and dendritic cells (MacDonald et al., 2001; Faveeuw et al., 2002; De Jong et al., 2002; Cervi et al., 2004) have also been shown to be potential mediators in egg antigen-induced immunomodulation towards Th2-ness, though one recent study has shown that eggs can activate dendritic cells via a Toll-like receptor to produce IFN-γ (Trottein et al., 2004). This research group recently showed that dendritic cells are also activated via TLR3 by double-stranded RNA from *S. mansoni* eggs (Aksoy et al., 2005).

Experiments in mice that had had their cytokine and/or cytokine-receptor genes mutated or deleted did however indicate that optimum health and survival occurred in hosts with balanced immune responses. Exacerbated morbidity and early death, often with evidence of severe toxicity to hepatocytes occurred after experimentally-induced abnormal skewing towards either Th1- or Th2-directed responsiveness (Brunet et al. 1997; 1999; Fallon and Dunne, 1999; Hoffman et al., 2000; Patton et al., 2001; Rutitzky et al., 2001; Herbert et al., 2004). The excessive morbidity associated with hepatocellular damage that occurs in *S. mansoni*-infected mice with profoundly-skewed cytokine profiles has been ascribed to 'cytokine-induced shock' (Wynne et al., 2004 Imm reviews). However, eggs of this schistosome species have been found to produce a 31kDa glycoprotein antigen, omega-1, that has putative hepatotoxic properties in immunosuppressed mice (Dunne et al., 1981; 1991) and that could be neutralized by small volumes of passively-transferred specific antibody (Doenhoff et al., 1981; Dunne et al., 1981). The possibility that the hepatotoxicity associated with 'cytokine shock' can be prevented by anti-omega-1 antibody appears not yet to have been tested. In any case doubts have been expressed whether the *S. mansoni*-induced hepatotoxicity reactions observed in mice bear any relevance to human infections (Cheever et al., 2000).

Andrade and Warren (1964) and Cheever (1965) showed that as *S. mansoni* infections became more chronic granulomas formed in tissues around freshly-produced eggs became smaller, a process termed 'endogenous desensitization' (Domingo and Warren, 1968) or 'spontaneous modulation' (Boros et al., 1975), concomitant with which there was a marked amelioration of the symptoms of hepatosplenic disease (Warren, 1966). Although the aetiology of the granulomas around *S. japonicum* eggs in mice seems somewhat different from that against *S. mansoni* eggs - early studies indicated the former were more like foreign body reactions than T-cell-mediated hypersensitivity responses (Warren and Domingo, 1970) - modulation of both granuloma formation and hepatosplenic disease were also observed during experimental *S. japonicum* infections (Warren and Berry, 1972).

A reduction in the size of granulomas generated by eggs in acutely-infected animals could be achieved by transfer of T lymphocytes from chronically-infected donors, but not by transfer of their serum (Colley, 1976; Colley et al., 1979). However, some studies have indicated that B cells can also mediate in the down-modulation of granulomas (Cheever et al., 1985; Jankovic et al., 1998), perhaps indirectly because of an effect on T cells (Wynne et al., 2004). There are contradictory results regarding the involvement of IL-10 in down-modulation of granuloma size, with Wynn et al. (1998) finding no requirement for this cytokine, while Flores-Villaneuva et al. (1996) found it was needed. Th1 cytokines such as IL-12, IFN-γ and TNF-α may be involved in endogenous desensitization, as experimentally inducing their production during early infection inhibited granuloma formation (Sher et al., 1996; Hoffmann et al., 1998; Boros and Whitfield, 1999). Moreover, high pathology strains of mice that suffered more extensive and prolonged granulomatous reactivity had $CD4^+$ T cells with a stronger Th1 bias than low-pathology mouse strains, which in turn was due to the former having dendritic cells that produced IL-12 (Stadecker et al., 2004). A higher rate of apoptosis in the $CD4^+$ T cells of the low pathology mice may however have contributed to the reduced severity of immunopathology in these animals (Rutitzky et al., 2003).

Advanced and serious cases of human schistosomiasis are often characterized by Symmer's periportal or 'clay pipestem' fibrosis and although fibrotic deposition of collagen does occur in livers of *S. mansoni*-infected mice, there may be no experimental model fully representative of the periportal fibrotic lesions found in humans. A sub-set of CBA/J mice have, however, been found to develop fibrotic histopathology which is morphologically quite similar to that seen in humans (Henderson et al., 1993) and the mice that developed fibrosis did not produce regulatory anti-egg idiotype antibodies that were found in counterparts without fibrosis (Montesano et al., 1997; 2002). In the mice an association between hepatosplenic disease and high production of TNF-α (Adewusi et al., 1996) and a deficit in IL-10 production (Bosshardt et al., 1997) was observed.

The relationships between human host cellular activity and cytokine production patterns, severe hepatosplenic pathology and periportal fibrosis are not yet entirely clear. Results of an early study showed hepatosplenic disease was related to unbalanced T cell populations and absence of functional adherent suppressor cells (Tweardy et al., 1987). Antigen-stimulated lymphocytes from a human population in Kenya that had severe hepatosplenic disease, but little or no periportal fibrosis, produced high levels of the Th1 cytokines IFN-γ and TNF and low levels of Th2-biasing IL-5 (Mwatha et al., 1998). However, in a Sudanese population production of IFN-γ by blood cells was related to apparent protection against periportal fibrosis, while production of TNF-α was related to aggravated disease (Henri et al., 2002; Dessein et al., 2004). Consistent with this is the observation that periportal fibrosis is controlled by a major locus on human chromosome 6q22-q23 which is closely linked to the gene for the alpha chain of IFN-γ receptor, and that two polymorphisms in the IFN-γ gene are associated with periportal fibrosis (Chevillard et al., 2003).

Some recent experiments in mice have indicated that IL-13 is instrumental in development of fibrosis and blockade of the action of this cytokine prolonged the life of infected animals (Chiaramonte et al., 1999; 2001; 2003; Fallon et al., 2000; Mentink-Kane et al. 2004; Wynne et al., 2004).

5.8. Sensitizing and immunomodulatory molecules of schistosome eggs

Concurrently with work aimed at understanding how schistosome egg-induced granulomatous inflammation, hepatosplenic disease and fibrosis are induced and regulated, attempts were made to identify the particular egg-derived factors responsible for these phenomena.

Initial work suggested that sensitization of mice to give enhanced granulomatous hypersensitivity against *S. mansoni* eggs was relatively species-specific (Warren and Domingo, 1970a) and egg-stage specific (Warren and Domingo, 1970b). The veracity of these initial conclusions about species- and stage-specificity has been challenged by more recent results showing that sensitization for enhanced granulomatous reactivity against *S. mansoni* eggs can be induced by *S. haematobium* infections (Jacobs et al., 1999a), unisexual *S. mansoni* infections (Cheever et al., 1997) and circulating anodic and cathodic antigens (CAA and CCA respectively) which are proteoglycans secreted by worms (Jacobs et al., 1999b). Sensitization can also be achieved with keyhole limpet haemocyanin (De Vijver et al., 2004), which shares carbohydrate epitopes with *S. mansoni* eggs (Wuhrer et al., 2000), implying that glycans could be important in sensitizing for granuloma formation. Peptides derived from some egg antigens have however been shown to have the potential to sensitize T-cells, for example phosphoenolpyruvate carboxykinase, thioredoxin peroxidase-1 and Sm-p40, the latter being a member of the small heat shock protein family, (Asahi and Stadecker, 2003; Stadecker et al., 2004).

The Th2-polarization that occurs after *S. mansoni* infections become patent has been ascribed to the parasite's eggs and their antigenic constituents (Vella and Pearce, 1992; Sabin and Pearce, 1995), the activity having been shown to reside in the glycanic residues thereof (Okano et al., 1999). Egg-derived lacto-N-fucopentaose III (LNFPIII) that contains the Lewis X trisaccharide has been shown to have Th2-inducing properties through its action on dendritic cells via TLR4 (Thomas and Harn, 2004). LNFPIII can also be used as a Th2 adjuvant in responses induced against non-schistosome antigens (Okano et al., 2001). An alternative or additional route of Th2 response induction may be via a ~40kDa glycoprotein from *S. mansoni* eggs that binds IgE non-specifically and activates human basophils to release IL-4 (Haisch et al., 2001), an activity which the microbially-expressed recombinant form of the molecule also has (Schramm et al., 2003) and which is therefore independent of glycanic residues.

An ability to induce key IL-10-dependent immunoregulatory functions also resides in the lipid fraction of *S. mansoni* eggs. Thus, glycolipids stimulated the production of IL-10, as well as IL-6 and TNF-α, from peripheral blood monocytes (Van der Kleij et al., 2002) and lysophosphatidyl serine extracted from both *S. mansoni* eggs and worms modulated the activity of mature dendritic cells via TLR2 such that they induced the development of IL-10 producing regulatory T cells (Van der Kleij et al., 2002). A recent report indicates that responses to ligands of Toll-like receptors such as schistosome glycolipids and endotoxin are modulated in *S. haematobium*-infected children (Van der Kleij et al., 2004).

5.9. A subversive role for schistosome egg-induced immunopathogenesis

To what extent does granulomatous inflammation protect the host against parasite-inflicted damage? As indicated above, immunosuppressed *S. mansoni*-infected mice suffered from a severe hepatoxicity reaction and earlier death (Byram and Von Lichtenberg, 1977; Doenhoff et al., 1979; Byram et al., 1979; Lucas et al., 1980), results that were consistent with the notion that granulomas were needed to protect host tissues against damage from toxic egg products (Von Lichtenberg, 1964).

Much more recently it was found that mice deficient in IL-4/IL13 alternatively-activated macrophages had egg-induced granulomas that were in fact larger than those found in wild type controls, yet they suffered from increased hepatocyte damage and endotoxaemia (Herbert et al., 2004). In contrast to the hepatotoxicity observed in *S. mansoni*-infected, T-cell deprived mice, similarly immunosuppressed mice infected with *S. bovis* showed no evidence of hepatotoxicity and survived longer than comparably-infected immunologically-intact controls (Murare et al., 1987). Furthermore, the results of experiments in which serum or lymphocytes from *S. mansoni*-infected donors were transferred to immunosuppressed mice with homologous infections suggested that antibodies were better than granulomas at neutralizing egg-induced hepatotoxicity (Hassounah and Doenhoff, 1993). These results cast some doubt on the hypothesis that the principal role of granulomas is to protect tissues of infected hosts from the toxicity of the eggs within.

Experiments to test the effects of immunosuppression on schistosome infections in mice fortuitously revealed that immunosuppressed animals excreted markedly fewer parasite eggs than intact controls, a characteristic which applied to both *S. mansoni* (Doenhoff et al., 1978) and *S. bovis* (Murare and Doenhoff, 1987). The apparent immune-dependence of *S. mansoni* egg excretion was confirmed in mice with severe combined immunodeficiency (SCID) in which granulomatous inflammation was virtually absent (Cheever et al., 1999) and in mice that had been tolerized to egg antigens (Fallon and Dunne, 1999). Reduced rates of *S. mansoni* egg excretion were also found in human subjects that had low $CD4^+$ T cell counts as a result of co-infection with HIV (Karanja et al., 1997).

Interestingly, in the context of the possible difference in the aetiology of the granulomatous inflammation induced by *S. japonicum* eggs which was mentioned above, granuloma formation around eggs of this schistosome species was not completely ablated in SCID mice and the capacity to excrete eggs was also less deleteriously affected (Cheever et al., 1999).

The rate of egg excretion in *S. mansoni*-infected immunosuppressed mice could be partially restored by passive transfer of immune sera (Doenhoff et al., 1978, 1981; Dunne et al., 1983), and a more substantial enhancement of egg excretion rates was obtained by adoptive transfer of immune lymphocytes, with a numerical correlation being found in individual animals between the number of eggs in faeces and the mean diameter of granulomas around eggs in their livers (Doenhoff et al., 1985; 1986). The potential that lymphocytes from normal infected donors had to enhance egg excretion in schistosome-infected immunosuppressed hosts was at a maximum at the time that granulomas around eggs in the donors were at their largest, *i.e.*, approximately 8 weeks after infection, after which mean granuloma size and egg excretion-enhancing potential waned (Doenhoff, 1997) as expected because of 'endogenous desensitization'. Infected mice however continued to produce egg excretion-enhancing antibodies through chronic infection (Doenhoff, 1997).

The results outlined above indicate that egg-induced granulomatous inflammation promotes the onward transmission of infection by facilitating egg excretion. Damian (1987) reached a similar conclusion about the role of the granuloma from histological examination of *S. mansoni*-infected baboon tissue. The transmission of tuberculosis, another disease with pathogenesis due to granulomatous hypersensitivity, may similarly depend on immune response-facilitated release of the infectious agent from the host (Doenhoff, 1997; 1998).

This conclusion about the role of the granuloma in schistosome egg excretion may be too simplistic, however, as *S. mansoni*-infected mice lacking deficient in IL-4/IL13 alternatively activated macrophages excreted few eggs despite having larger granulomas than wild-type controls (Herbert et al., 2004). The process by which schistosome eggs become extravasated and pass through host tissues to reach the lumena of the intestine or bladder is likely to depend on many factors not yet fully investigated, though plateletets, another agent with potential for defending the host against infection, have also been found to be necessary for egg excretion (Ngaiza and Doenhoff, 1990) and several studies have shown that eggs modulate the activities of endothelial cells (Freedman and Ottesen, 1988; Ngaiza et al., 1993; File, 1995; Trottein et al., 1999b; Loeffler et al., 2002).

5.10. Immune-enhancement of schistosomes

Other observations indicate that schistosomes appear to have evolved a comprehensive set of strategies allowing them to benefit from the immune-response potential of their vertebrate hosts. These include immune responses generated against the larval glycocalyx which enhance worm burdens (Harn et al., 1989), though admittedly the host is not likely to see much glycocalyx-derived material as a result of natural infections. Worm burdens are decreased in mice with suppressed IgE responses (Amiri et al., 1994) and a requirement has been shown for IL-7 produced by keratinocytes (Wolowczuk et al., 1999; Roye et al., 2001).

The maturation of *S. mansoni* worms was retarded and their fecundity somewhat depressed in T-cell deprived or otherwise immunosuppressed mice compared with immunologically-intact controls (Coker, 1957; Weinman and Hunter, 1960; Harrison and Doenhoff, 1983) and Davies et al. (2001) formally demonstrated that $CD4^+$ T cells were necessary for normal worm maturation and fecundity. Schistosome worm viability (Davies et al., 2004) and fecundity (Amiri et al., 1992) are dependent on TNF-α, male worms playing the central role in transducing the host-derived signals for appropriate development and in turn facilitating female worm maturation (Hernandez et al., 2004). Stavitsky et al. (2003) suggested that the requirement of MIF for fecundity of *S. japonicum* was because it up-regulated expression of TNF-α.

6. Cestodes

Humans serve as definitive hosts for adults of three species of cestode: *Diphyllobothrium latum*, *Taenia solium* and *T. saginata* – the fish, pork and beef tapeworms respectively. Adult worms of these and other tapeworm species have a relatively benign presence, but larvae (cysts) infecting intermediate hosts can be the cause of serious pathology. Thus, humans are intermediate (as well as definitive) hosts for *T. solium*, the cysts of which cause the disease cysticercosis, and cysts of *Echinococcus granulosus* and *E. multilocularis* cause human unilocular and multilocular hydatid disease respectively.

Cestodes have been the subject of less experimental immunology than nematodes and schistosomes, probably mainly because of their lesser importance (in terms of numbers of people infected and lower economic cost in domestic animals), but also because, with the exception of *E. multilocularis* (Vuitton, 2003), convenient laboratory animal models are scarce. Nevertheless, sufficient information is available to indicate that cestodes share some characteristics with the other main groups of helminths with regard to evasion and subversion of the immune response.

Thus, prolongation of the survival of *E. granulosus* may be gained by antigen shedding (Holcman and Heath, 1997) and/or overt immunosuppression, the latter mediated by induction of high levels of nitric oxide from peritoneal exudate cells (Dai et al., 2003). The E/S products of *T. cracisseps* larvae taken early in infection were found to suppress *in vitro* lymphocyte responsiveness (Spolski et al., 2000). Control or inhibition of complement are strategies that may be possessed by *T. solium* (White et al., 1997) and *E. granulosus* (Ferreira et al., 2000).

Th2-oriented activity has been associated with successful establishment and prolonged survival of cestodes (Terrazas et al., 1998), while Th1-type responses mediated rejection (Godot et al., 2003; Rodriguez-Sosa et al., 2003). IL-10 has been implicated in regulation of immunopathology (Godot et al., 2000).

Carbohydrate-rich fractions from *E. granulosus* protoscoleces caused mouse spleen cells to produce IL-10 (Dematteis et al., 2001), while those from

Taenia crassiceps have been shown to be responsible for Th2 bias by inducing IL-6 production in macrophages via Toll-like receptors (Dissanayake et al., 2004). Purified 'antigen B', a lipoprotein secreted by *E. granulosus* protoscoleces induced patients' PBMC to produce the Th2 cytokines IL-4 and IL-13 (Rigano et al., 2001).

7. Consequences of helminth subversion of immune responses: A human perspective

A considerable volume of work on experimental helminth infections indicates they are often immunosuppressive and the majority of immune responses they induce against themselves, or that are produced against other antigens in helminth-infected hosts, have a Th2-bias. The consequences of this should not be oversimplified or excessively ramified in the context of the Th1/Th2 paradigm (Allen and Maizels, 1997), but it is legitimate to ask how they impact on the health of people that are infected with worms?

Helminth infections in humans rarely, if ever, cause immunosuppression to a life-threatening degree. However, similarities between helminths and HIV have been noted with regard to their capacity for 'immune-activation', with the resulting anergy possibly affecting the ability of hosts to cope with other infections and to respond to vaccinations (Borkow et al., 2001; Borkow and Bentwich, 2004). Decreased HIV viral loads were observed after successful treatment of intestinal worms (Wolday et al., 2002), though there was no evidence of faster progression of HIV disease in Ugandans carrying a variety of helminths (Brown et al., 2004) and from their long-term studies on HIV and *S. mansoni* co-infected individuals in Western Kenya Secor et al. (2004) concluded that the effects of HIV-1 on schistosomiasis were 'more profound' than *vice versa*.

Th2-inducing helminth infections would be expected to have an impact on infections that induce host-protective Th1 responses, for example mycobacterial and malaria infections, and the incidence of lepromatous leprosy in onchocerciasis hyperendemic areas was indeed double that in areas which had little or no onchocerciasis, but the same overall prevalence of leprosy (Prost et al., 1979). In Thailand a higher incidence of malaria was associated with *A. lumbroides* infections (Nacher et al., 2002b), but the helminth protected against severe cerebral malaria (Nacher et al., 2002a), results that are interpretable in terms of a helminth-induced Th2 shift decreasing the Th1-dependence responsible for both anti-sporozoite immunity and cerebral symptoms. In Senegal severe malaria attack and Ascaris infection were however associated (Le Hesran et al., 2004). The possible adverse effects of some helminth infections on the clinical outcome of malaria has recently been reviewed (Druillhe et al., 2005).

Helminth infections may impact on public health as a result of an effect on responsiveness to vaccines as reduced immune reactivity has been observed to vaccination with tetanus toxoid (Cooper et al., 1998; Nookala et al., 2004) and mycobacterial antigens (Stewart et al., 1999) in patients with *O. volvulus*, and to tetanus toxoid in humans infected with *S. mansoni* (Sabin et al., 1996). Production of IFN-γ in response to *in vitro* BCG stimulation was impaired in T cells from new-born children who had been sensitised *in utero* to filarial and schistosome antigens (Malhotra et al., 1999).

8. Concluding remarks: Helminth infections may be good for you!

The longevity of many helminth infections is attributable to two main factors - the evolution of a diverse array of mechanisms that enable host immune responses to be evaded or neutralized, and the (eventual) establishment of a balance between the parasite with pathogenic potential and the host with the capacity to protect itself by immune reactions.

Research on some mechanisms of immune evasion, such as the shedding of antigens, modulation of antigenicity and the acquisition of host antigens, is now not much in vogue and we remain still quite ignorant about how they work: for example, through what receptors do albumin and protease inhibitors bind to, respectively, micofilariae and schistosomes; what is the mechanism of incorporation and the distribution of glycolipid blood group antigens in the schistosome worm surface; how does the outer doubled lipid bilayer of schistosomes operate as a parasite-protective surface? Furthermore, the nature of the interaction between helminths and some aspects of host defence have not been fully investigated: for example microfilariae and schistosome worms, are in contact with blood plateletsfor extended periods, but as yet little is known about how these parasite stages avoid thrombotic engulfment.

Much of the research into mechanisms of immune evasion and subversion has been driven by a belief that a better understanding thereof would help the development of effective anti-parasite countermeasures, particularly through vaccination. Unfortunately, despite a cornucopia of possible immune-evasion strategies having been uncovered, anti-helminth vaccines remain rare entities. The newly emergent '-omic' technologies and methods for transfecting and inhibiting individual genes may, however, hasten the identification of molecules which have a key role in helminth immune evasion and subversion.

Paradoxically, recent work, particularly that investigating the capacity of helminths to regulate immune responsiveness, has led to the counterintuitive notion that infections of helminth 'parasites' may in fact be beneficial to human health.

The contextual frame for this development is the 'hygiene hypothesis', a term coined originally by Strachan (1989), which associates the rapid increase in the incidence of immunologically-related inflammatory disorders such as allergies and diabetes with improved standards of health care and cleanliness. Earlier derivations of this hypothesis (*e.g.*, Rook and Stanford, 1998) held that pathologies such as allergies, which are mainly Th2-driven, were due to immune systems not having encountered microbial infections which stimulated Th1 cell responses and which thus somehow controlled the allergically-inclined reactivity of Th2 cells. These ideas did not however account satisfactorily for an apparent relative absence of atopic allergy in areas where Th2-inducing helminth infections were common (Van den Bigelaar et al., 2001; Cooper, 2004) and a rise in Type-1 autoimmune diseases such as diabetes (Cooke et al., 2004), as well as Type-2 disorders. The view being promoted by recent more inclusive versions of the hypothesis is that helminth infections activate networks of regulatory cells and cytokines, with IL-10 having a pivotal role (Wills-Karp et al. 2001; Yazdanbakhsh et al., 2002; 2004; Wilson and Maizels, 2004).

Experimental evidence that helminth infection may prevent diseases in part or wholly attributable to an immunological derangement has been obtained with regard to *S. mansoni* in the prevention of diabetes (Maron et al., 1998; Zaccone et al., 2003) and atherosclerosis (Doenhoff et al., 2002) and *H. polygyrus* in the suppression of established colitis (Elliott et al., 2004). A positive result was obtained in a clinical trial in which *Trichuris suis* eggs were administered to prevent Crohn's disease (Summers et al., 2004). Future work is likely to take the concept of procuring 'benefit from parasites' further by isolating individual helminth-derived molecules with pharmacological properties and developing them into drugs (Imai and Fujita, 2004), such as might be done with the immunomodulatory ES-62 from *A. vitiae* (Harnett et al., 2004), IPSE from *S. mansoni* eggs (Schramm et al., 2003) and helminth glycolipids (Van der Kleij and Yazdanbakhsh, 2003).

It has sometimes been suggested that host-parasite relationships evolve to become less virulent, even symbiotic, but studies on evolution of the myxomatosis virus in Australian rabbits indicated that establishment of a balanced relationship between pathogen and host may require the former to retain a relatively high degree of virulence (Fenner, 1983). The apparent dependence of schistosome egg excretion on host granulomatous inflammatory responsiveness (Doenhoff, 1998) also suggests that loss of this parasite's ability to induce a pathological reaction would impair its survivability. The recent work indicating that helminth infections could have beneficial effects allows the nature of host-parasite relationships to be considered from a new perspective, namely that there is a trade-off between the latter's beneficial and harmful effects. Davies et al. (1980) suggested that 'far from being a rejection mechanism….(the immune response) should be regarded as a mechanism so structured as to facilitate symbiosis – a welcome mat'. The accomplished ability of helminths to evade and subvert host immune responsiveness may be an integral part of the welcoming process.

References

Abath, F. G. C., and Werkhauser, R. C. (1996). The tegument of *Schistosoma mansoni*: Functional and immunological features. Parasite Immunol. *18*, 15-20.

Abbas, A. K., Murphy, K. M., and Sher, A. (1996). Functional diversity of helper T lymphocytes. Nature *383*, 787-793.

Abraham, D., Grieve, R. B., and Mikagrieve, M. (1988). *Dirofilaria immitis*: surface properties of 3rd-stage and 4th-stage larvae. Exp. Parasitol. *65*, 157-167.

Abraham, D., Leon, O., Schnyder-Candrian, S., Wang, C. C., Galioto, A. M., Kerepesi, L. A., Lee, J. J., and Lustigman, S. (2004). Immunoglobulin E and eosinophil-dependent protective immunity to larval *Onchocerca volvulus* in mice immunized with irradiated larvae. Inf. Imm. *72*, 810-817.

Adewusi, O. I., Nix, N. A., Lu, X. T., Colley, D. G., and Secor, W. E. (1996). *Schistosoma mansoni*: Relationship of tumor necrosis factor- alpha to morbidity and collagen deposition in chronic experimental infection. Exp. Parasit. *84*, 115-123.

Ahmad, A., Wang, C. H., and Bell, R. G. (1991). A role for IgE in intestinal immunity - expression of rapid expulsion of *Trichinella spiralis* in rats transfused with Ige and thoracic-duct lymphocytes. J. Imm. *146*, 3563-3570.

Akdis, C. A., Blaser, K., and Akdis, M. (2004). Genes of tolerance. Allergy *59*, 897-913.

Aksoy, E., Zouain, C.S., Vanhoutte, F., Fontaine, J., Pavelka, N., Thieblemont, N., Willems, F., Ricciardi-Castagnoli, P., Goldman, M., Capron, M., Ryffel, B., and Trottein, F. (2005). Double-stranded RNAs from the helminth parasite Schistosoma activate TLR3 in dendritic cells. J. Biol. Chem. *280*, 277-283.

Akuffo, H., Maasho, K., Lavebratt, C., Engstrom, K., and Britton, S. (1996). Ivermectin-induced immunopotentiation in onchocerciasis: Recognition of selected antigens following a single dose of ivermectin. Clin. Exp. Imm. *103*, 244-252.

Alegre, M. L., Frauwirth, K.A., and Thompson, C.B. (2001). T-cell regulation by CD28 and CTLA-4. Nature Rev. Imm. *1*, 220-228.

Ali, N. M. H., and Behnke, J. M. (1984). Non-specific immunodepression by larval and adult *Nematospiroides dubius*. Parasitology *88*, 153-162.

Alkarmi, T., Ijaz, M. K., Dar, F. K., Abdou, S., Alharbi, S., Frossard, P., and Naiem, M. (1995). Suppression of transplant immunity in experimental trichinosis. Comp. Imm. Microbiol. Inf. Dis. *18*, 171-177.

Allen, J. E., Lawrence, R. A., and Maizels, R. M. (1996). APC from mice harbouring the filarial nematode, *Brugia malayi*, prevent cellular proliferation but not cytokine production. Int. Imm. *8*, 143-151.

Allen, J. E., and Macdonald, A. S. (1998). Profound suppression of cellular proliferation mediated by the secretions of nematodes. Parasite Imm. *20*, 241-247.

Allen, J. E., and Maizels, R. M. (1997). Th1-Th2: reliable paradigm or dangerous dogma? Imm. Today *18*, 387-392.

Almond, N. M., and Parkhouse, R. M. E. (1986). Immunoglobulin class specific responses to biochemically defined antigens of *Trichinella spiralis*. Parasite Imm. *8*, 391-406.

Ambroise-Thomas, P., and Andrews P. (1976). Development of fluorescent antibodies directed against larval stages, eggs, and adults of *Schistosoma mansoni* in mice harbouring unisexual or bisexual infections. Tropenmed. Parasitol. *27*, 483-488.

Amiri, P., Haakfrendscho, M., Robbins, K., McKerrow, J. H., Stewart, T., and Jardieu, P. (1994). Anti-immunoglobulin-E treatment decreases worm burden and egg production in *Schistosoma mansoni*-infected normal and interferon-gamma knockout mice. J. Exp. Med. *180*, 43-51.

Amiri, P., Locksley, R. M., Parslow, T. G., Sadick, M., Rector, E., Ritter, D., and McKerrow, J. H. (1992). Tumor necrosis factor-alpha restores granulomas and induces parasite egg-laying in schistosome-infected SCID mice. Nature *356*, 604-607.

Anderson, R. M., and May, R.M. (1979). Population biology of infectious diseases: Part 1. Nature *280*, 361-367.

Anderson, R. M., and May, R. M. (1985). Helminth infections of humans: mathematical models, population-dynamics, and control. Adv. Parasitol. *24*, 1-101.

Andrade, Z. A., and Warren, K.S. (1964). Mild prolonged schistosomiasis in mice: alterations on host response with time and the development of portal fibrosis. Trans. Roy. Soc. Trop. Med.Hyg. *58*, 53-57.

Angeli, V., Faveeuw, C., Roye, O., Fontaine, J., Teissier, E., Capron, A., Wolowczuk, I., Capron, M., and Trottein, F. (2001). Role of the parasite-derived prostaglandin D-2 in the inhibition of epidermal Langerhans cell migration during schistosomiasis infection. J. Exp. Med. *193*, 1135-1147.

Apfel, H., Eisenbeiss, W. F., and Meyer, T. F. (1992). Changes in the surface composition after transmission of *Acanthocheilonema viteae* 3rd stage larvae into the jird. Mol. Biochem. Parasitol. *52*, 63-74.

Apfel, H., and Meyer, T. F. (1990). Active release of surface proteins - a mechanism associated with the immune escape of *Acanthocheilonema viteae* microfilariae. Mol. Biochem. Parasitol. *43*, 199-210.

Asahi, H., and Stadecker, M. J. (2003). Analysis of egg antigens inducing hepatic lesions in schistosome infection. Parasitol. Int. *52*, 361-367.

Asch, H. L., and Read, C.P. (1975). Membrane transport in *Schistosoma mansoni*: transport of amino acids by adult males. Exp. Parasitol. *38*, 123-135.

Atmadja, A. K., Atkinson, R., Sartono, E., Partono, F., Yazdanbakhsh, M., and Maizels, R. M. (1995). Differential decline in filaria-specific IgG1, IgG4, and IgE antibodies in

Brugia malayi-infected patients after diethylcarbamazine chemotherapy. J. Inf. Dis. *172*, 1567-1572.

Atochina, O., Daly-Engel, T., Piskorska, D., McGuire, E., and Harn, D.A. (2001). A schistosome-expressed immunomodulatory glycoconjugate expands peritoneal Gr1(+) macrophages that suppress naive CD4(+) T cell proliferation via an IFN-gamma and nitric oxide-dependent mechanism. J. Immunol. *167*, 4293-4302.

Auriault, C., Joseph, M., Tartar, A., Bout, D., Tonnel, A. B., and Capron, A. (1985). Regulatory role of a tripeptide (TKP) from the 2nd constant domain of immunoglobulin-G .1. Inhibition of rat and human macrophage activities. Int. J. Immunopharm. *7*, 73-79.

Babu, S., and Nutman, T. B. (2003). Proinflammatory cytokines dominate the early immune response to filarial parasites. J. Immunol. *171*, 6723-6732.

Balic, A., Harcus, Y., Holland, M. J., and Maizels, R. M. (2004). Selective maturation of dendritic cells by *Nippostrongylus brasiliensis*-secreted proteins drives Th2 immune responses. Eur. J. Immunol. *34*, 3047-3059.

Bancroft, A. J., Else, K. J., and Grencis, R. K. (1994). Low level infection with *Trichuris muris* significantly affects the polarization of the CD4 response. Eur. J. Immunol. *24*, 3113-3118.

Bancroft, A. J., Else, K. J., Humphreys, N. E., and Grencis, R. K. (2001). The effect of challenge and trickle *Trichuris muris* infections on the polarisation of the immune response. Int. J. for Parasitol. *31*, 1627-1637.

Bancroft, A. J., Else, K. J., Sypek, J. P., and Grencis, R. K. (1997). Interleukin-12 promotes a chronic intestinal nematode infection. European J. Immunol. *27*, 866-870.

Behnke, J. M., and Wakelin, D. (1973). The survival of *Trichuris muris* in wild populations of its natural hosts. Parasitology *67*, 157-164.

Behnke, J. M. (1987). Evasion of immunity by nematode parasites causing chronic infections. Adv. Parasitol. *26*, 1-71.

Behnke, J. M., Barnard, C. J., and Wakelin, D. (1992). Understanding chronic nematode infections: evolutionary considerations, current hypotheses and the way forward. Int. J. Parasitol. *22*, 861-907.

Behnke, J. M., Iraqi, F., Menge, D., Baker, R. L., Gibson, J., and Wakelin, D. (2003). Chasing the genes that control resistance to gastrointestinal nematodes. J. Helm. *77*, 99-109.

Behnke, J. M., Ali, N. M. H., and Jenkins, S. N. (1984). Survival to patency of low-level infections with *Trichuris muris* in mice concurrently infected with *Nematospiroides dubius*. Ann. Trop. Med. Parasitol. *78*, 509-517.

Behnke, J. M., Cabaj, W., and Wakelin, D. (1992). Susceptibility of adult *Heligmosomoides polygyrus* to intestinal inflammatory responses induced by heterologous infection. Int. J. Parasitol. *22*, 75-86.

Behnke, J. M., De Clercq, D., Sacko, M., Gilbert, F. S., Ouattara, D. B., and Vercruysse, J. (2000). The epidemiology of human hookworm infections in the southern region of Mali. Trop. Med. Int. Hlth. *5*, 343-354.

Behnke, J. M., Hannah, J., and Pritchard, D. I. (1983). *Nematospiroides dubius* in the mouse: evidence that adult worms depress the expression of homologous immunity. Par. Immunol. *5*, 397-408.

Behnke, J. M., Wahid, F. N., Grencis, R. K., Else, K. J., Bensmith, A. W., and Goyal, P. K. (1993). Immunological rlationships during primary infection with *Heligmosomoides polygyrus* (*Nematospiroides dubius*): down-regulation of specific cytokine secretion (IL-9 and IL-10) correlates with poor mastocytosis and chronic survival of adult worms. Par. Immunol. *15*, 415-421.

Bellaby, T., Robinson, K., and Wakelin, D. (1996). Induction of differential T-helper-cell responses in mice infected with variants of the parasitic nematode *Trichuris muris*. Inf. Imm. *64*, 791-795.

Ben-Smith, A., Lammas, D. A., and Behnke, J. M. (2003). The relative involvement of Th1 and Th2 associated immune responses in the expulsion of a primary infection of *Heligmosomoides polygyrus* in mice of differing response phenotype. J. Helm. *77*, 133-146.

Bentwich, Z., Weisman, Z., Moroz, C., BarYehuda, S., and Kalinkovich, A. (1996). Immune dysregulation in Ethiopian immigrants in Israel: Relevance to helminth infections? Clin. Exp. Immunol. *103*, 239-243.

Berbarian, D. A., Paquin, H.O., Fantauzzi, A. (1953). Longevity of *Schistosoma haematobium* and *Schistosoma mansoni*, observations based on a case. J. Parasitol. *39*, 517-519.

Bergquist, R., Utzinger, J., Chollet, J., Shu-Hua, X., Weiss, N. A., and Tanner, M. (2004). Triggering of high-level resistance against *Schistosoma mansoni* reinfection by artemether in the mouse model. Am. J. Trop. Med. Hyg. *71*, 774-777.

Bergwerff, A. A., Vandam, G. J., Rotmans, J. P., Deelder, A. M., Kamerling, J. P., and Vliegenthart, J. F. G. (1994). The immunologically reactive part of immunopurified circulating anodic antigen from *Schistosoma mansoni* Is a threonine-linked polysaccharide consisting of -]6)-(Beta-D-Glcpa-(1-]3))-Beta-D- Galpnac-(1-] repeating units. J. Biol. Chem. *269*, 31510-31517.

Bickle, Q. D., Sacko, M., and Vignali, D. A. A. (1990). Induction of Immunity against *Schistosoma mansoni* by drug (Ro11-3128)-terminated infections: analysis of surface antigen recognition. Par. Immunol. *12*, 569-586.

Blackburn, C. C., and Selkirk, M. E. (1992). Characterization of the secretory acetylcholinesterases from adult *Nippostrongylus brasiliensis*. Mol. Biochem. Parasitol. *53*, 79-88.

Blanton, R. E., Licate, L. S., and Aman, R. A. (1994). Characterization of a native and recombinant *Schistosoma haematobium* serine protease inhibitor gene product. Mol. Biochem. Parasitol. *63*, 1-11.

Blaxter, M. L., Page, A. P., Rudin, W., and Maizels, R. M. (1992). Nematode surface coats: actively evading immunity. Parasitol. Today *8*, 243-247.

Boitelle, A., Scales, H. E., Di Lorenzo, C., Devaney, E., Kennedy, M. W., Garside, P., and Lawrence, C. E. (2003). Investigating the impact of helminth products on immune responsiveness using a TCR transgenic adoptive transfer system. J. Immunol. *171*, 447-454.

Boonmars, T., Wu, Z., Nagano, I., and Takahashi, Y. (2004). Expression of apoptosis-related factors in muscles infected with *Trichinella spiralis*. Parasitology *128*, 323-332.

Boonstra, A., Asselin-Paturel, C., Gilliet, M., Crain, C., Trinchieri, G., Liu, Y. J., and O'Garra, A. (2003). Flexibility of mouse classical and plasmacytoid-derived dendritic cells in directing T helper type 1 and 2 cell development: Dependency on antigen dose and differential toll- like receptor ligation. J. Exp. Med. *197*, 101-109.

Borkow, G., and Bentwich, Z. (2004). Chronic immune activation associated with chronic helminthic and human immunodeficiency virus infections: Role of hyporesponsiveness and anergy. Clin. Microbiol. Rev. *17*, 1012-1030.

Borkow, G., Weisman, Z., Leng, Q. B., Stein, M., Kalinkovich, A., Wolday, D., and Bentwich, Z. (2001). Helminths, human immunodeficiency virus and tuberculosis. Scand. J. Inf. Dis. *33*, 568-571.

Boros, D. L., Pelley, R.P., and Warren, K.S. (1975). Spontaneous modulation of granulomatous hypersensitivity in schistosomiasis mansoni. J. Immunol. *114*, 1437-1441.

Boros, D. L., and Whitfield, J. R. (1999). Enhanced Th1 and dampened Th2 responses synergize to inhibit acute granulomatous and fibrotic responses in murine schistosomiasis mansoni. Inf. Imm. *67*, 1187-1193.

Bosshardt, S. C., Freeman, G. L., Secor, W. E., and Colley, D. G. (1997). IL-10 deficit correlates with chronic, hypersplenomegaly syndrome in male CBA/J mice infected with *Schistosoma mansoni*. Par. Immunol. *19*, 347-353.

Boyer, M. H., Kalfayan, L.J., and Ketchum, D.G. (1977). The host antigen phenomenon in experimental murine schistosomiasis. III. Destruction of parasites transferred from mice to hamsters. Am. J. Trop. Med. Hyg. *26*, 254-257.

Brattig, N. W., Rathjens, U., Ernst, M., Geisinger, F., Renz, A., and Tischendorf, F.W. (2000). Lipopolysaccharide-like molecules derived from *Wolbachia endobacteria* of the filaria *Onchocerca volvulus* are candidate mediators in the sequence of inflammatory and antiinflammatory responses of human monocytes. Micr. Inf. *10*, 1147-1157.

Brindley, P. J., and Dobson, C. (1981). Genetic control of liability to infection with *Nematospiroides dubius* in mice: selection of refractory and liable populations of mice. Parasitology *83*, 51-65.

Brown, M., Kizza, M., Watera, C., Quigley, M. A., Rowland, S., Hughes, P., Whitworth, J. A. G., and Elliott, A. M. (2004). Helminth infection is not associated with faster progression of HIV disease in coinfected adults in Uganda. J. Inf. Dis. *190*, 1869-1879.

Brunet, L. R., Beall, M., Dunne, D. W., and Pearce, E. J. (1999). Nitric oxide and the Th2 response combine to prevent severe hepatic damage during *Schistosoma mansoni* infection. J. Immunol. *163*, 4976-4984.

Brunet, L. R., Finkelman, F. D., Cheever, A. W., Kopf, M. A., and Pearce, E. J. (1997). IL-4 protects against TNF-alpha-mediated cachexia and death during acute schistosomiasis. J. Immunol. *159*, 777-785.

Bruschi, F., and Lucchi, N. W. (2001). Enzymatic antioxidant systems in helminth parasites: no doubt on their evasive role. Acta Parasitol. *46*, 233-241.

Bundy, D. A. (1988). Population ecology of intestinal helminth infections in human communities. Phil. Trans. Roy. Soc. Lond. Ser. B Biol. Sci. *31*, 405-420.

Bundy, D. A. P. (1986). Epidemiologic aspects of Trichuris and Trichuriasis in Caribbean communities. Trans. Roy. Soc. Trop. Med.Hyg. *80*, 706-718.

Bundy, D. A. P., and Medley, G. F. (1992). Immuno-epidemiology of human geohelminthiasis - ecological and immunological determinants of worm burden. Parasitology *104*, S105-S119.

Bundy, D. A. P., Grenfell, B. T., and Rajagopalan, P. K. (1991). Immunoepidemiology of lymphatic filariasis - the relationship between infection and disease. Immunoparasitol. Today- A Combined Issue of Immunology Today and Parasitol. Today, A71-A75.

Bundy, D. A. P., Kan, S. P., and Rose, R. (1988). Age-related prevalence, intensity and frequency-distribution of gastrointestinal helminth infection in urban slum children from Kuala-Lumpur, Malaysia. Trans. Roy. Soc. Trop. Med.Hyg. *82*, 289-294.

Butterworth, A., Dunne, D., Fulford, A., Capron, M., Khalife, J., Capron, A., Koech, D., Ouma, J., and Sturrock, R. (1988). Immunity in human schistosomiasis mansoni: cross-reactive IgM and IgG2 anti-carbohydrate antibodies block the expression of immunity. Biochimie *70*, 1053-1063.

Butterworth, A. E., and Hagan, P. (1987). Immunity in human schistosomiasis. Parasitol. Today *3*, 11-16.

Butterworth, A. E., Benstedsmith, R., Capron, A., Capron, M., Dalton, P. R., Dunne, D. W., Grzych, J. M., Kariuki, H. C., Khalife, J., Koech, D., Mugambe, M., Ouma, J.H., Siongok, T.K.A., and Sturrock, R.F. (1987). Immunity in human schistosomiasis mansoni: prevention by blocking antibodies of the expression of immunity in young children. Parasitology *94*, 281-300.

Byram, J. E., and Vn Lichtenberg, F. (1977). Altered schistosome granuloma formation in nude mice. Am. J. Trop. Med. Hyg. *26*, 944-956.

Byram, J. E., Doenhoff, M.J., Musallam, R., Brink, L.H., and Von Lichtenberg, F. (1979). *Schistosoma mansoni* infections in T-cell deprived mice, and the ameliorating effect of administering homologous chronic infection serum. II. Pathology. Am. J. Trop. Med. Hyg. *28*, 274-285.

Calandra, T., and Roger, T. (2003). Macrophage migration inhibitory factor: A regulator of innate immunity. Nature Rev. Immunol. *3*, 791-800.

Callahan, H. L., Crouch, R. K., and James, E. R. (1988). Helminth anti-oxidant enzymes: a protective mechanism against host oxidants. Parasitol. Today *4*, 218-225.

Capron, A., Capron, M., Dombrowicz, D., and Riveau, G. (2001). Vaccine strategies against schistosomiasis: From concepts to clinical trials. Int. Arch. All. App. Immunol. *124*, 9-15.

Capron, M., Carlier, Y., Nzeyimana, H., Minoprio, P., Santoro, F., Sellin, B., and Capron, A. (1980). *In vitro* study of immunological events in human and experimental schistosomiasis: relationships between cytotoxic antibodies and circulating Schistosoma antigens. Par. Immunol. 2, 223-235.

Carlow, C. K. S., Spielman, A., and Philipp, M. (1987). A developmentally regulated surface epitope expressed by the infective larvae of *Brugia malayi* which is rapidly lost after infection. J. Cell. Biochem. 165-165.

Carmona, C., Dowd, A.J., Smith, A.M., and Dalton, J.P. (1993). Cathepsin L proteinase secreted by *Fasciola hepatica in vitro* prevents antibody-mediated eosinophil attachment to newly excysted juveniles. Mol. Biochem. Parasitol. *62*, 9-17.

Cervi, L., MacDonald, A.S., Kane, C., Dzierszinski, F., and Pearce, E.J. (2004). Cutting edge: dendritic cells copulsed with microbial and helminth antigens undergo modified maturation, segregate the antigens to distinct intracellular compartments, and concurrently induce microbe-specific Th1 and helminth-specific Th2 responses. J. Immunol. *172*, 2016-2220.

Chabasse, D., Bertrand, B., Leroux, J.P., Gauthey, N., and Hocquet, P. (1985). A case of schistosomiasis mansoni detected 37 years after infection. Bull. Soc. Path. Exot. Fil. *78*, 643-647.

Chan, L., Bundy, D. A. P., and Kan, S. P. (1994). Aggregation and predisposition to *Ascaris lumbricoides* and *Trichuris trichiura* at the familial level. Trans. Roy. Soc. Trop. Med. Hyg. *88*, 46-48.

Chan, M. S., and Isham, V. S. (1998). A stochastic model of schistosomiasis immuno-epidemiology. Math. Biosci. *151*, 179-198.

Chandrashekar, R., Curtis, K.C., Lu, W., and Weil, G.J. (1998). Molecular cloning of an enzymatically active thioredoxin peroxidase from *Onchocerca volvulus*. Mol. Biochem. Parasitol. *93*, 309-312.

Chandrashekar, R., Ogunrinade, A. F., Henry, R. W., Lustigman, S., and Weil, G. J. (1993). *Onchocerca volvulus*: monoclonal antibodies to immune complex-associated parasite antigens. Exp. Parasitol. *77*, 224-234.

Cheever, A. W. (1965). A comparative study of *Schistosoma mansoni* infections in mice, gerbils and multimammate rats: I. The relation of portal hypertension to size of hepatic granulomas. Am. J. Trop. Med. Hyg. *14*, 211-226.

Cheever, A. W., Byram, J. E., Hieny, S., Vonlichtenberg, F., Lunde, M. N., and Sher, A. (1985). Immunopathology of *Schistosoma japonicum* and *S. mansoni* infection in B-cell depleted mice. Par. Immunol. *7*, 399-413.

Cheever, A. W., Hoffmann, K. F., and Wynn, T. A. (2000). Immunopathology of schistosomiasis mansoni in mice and men. Immunol. Today *21*, 465-466.

Cheever, A. W., Lewis, F. A., and Wynn, T. A. (1997). *Schistosoma mansoni*: unisexual infections sensitize mice for granuloma formation around intravenously injected eggs. Parasitol. Res. *83*, 57-59.

Cheever, A. W., Poindexter, R. W., and Wynn, T. A. (1999). Egg laying is delayed but worm fecundity is normal in SCID mice infected with *Schistosoma japonicum* and *S. mansoni* with or without recombinant tumor necrosis factor alpha treatment. Inf. Imm. *67*, 2201-2208.

Chen, L., Rao, K. V. N., He, Y. X., and Ramaswamy, K. (2002). Skin-stage schistosomula of *Schistosoma mansoni* produce an apoptosis-inducing factor that can cause apoptosis of T cells. J. Biol. Chem. *277*, 34329-34335.

Chevillard, C., Moukoko, C. E., Elwali, N., Bream, J. H., Kouriba, B., Argiro, L., Rahoud, S., Mergani, A., Henri, S., Gaudart, J., Mohamed-Ali, Q., Hall, H.A., and Dessein, A.J. (2003). IFN-gamma polymorphisms (IFN-gamma+2109 and IFN-gamma+3810) are associated with severe hepatic fibrosis in human hepatic schistosomiasis (*Schistosoma mansoni*). J. Immunol. *171*, 5596-5601.

Chiaramonte, M. G., Cheever, A. W., Malley, J. D., Donaldson, D. D., and Wynn, T. A. (2001). Studies of murine schistosomiasis reveal interleukin-13 blockade as a treatment for established and progressive liver fibrosis. Hepatology *34*, 273-282.

Chiaramonte, M. G., Mentink-Kane, M., Jacobson, B. A., Cheever, A. W., Whitters, M. J., Goad, M. E. P., Wong, A., Collins, M., Donaldson, D. D., Grusby, M. J., and Wynn, T. A. (2003). Regulation and function of the interleukin 13 receptor alpha 2 during a T helper cell type 2-dominant immune response. J. Exp. Med. *197*, 687-701.

Chiaramonte, M. G., Schopf, L. R., Neben, T. Y., Cheever, A. W., Donaldson, D. D., and Wynn, T. A. (1999). IL-13 is a key regulatory cytokine for Th2 cell-mediated pulmonary granuloma formation and IgE responses induced by *Schistosoma mansoni* eggs. J. Immunol. *162*, 920-930.

Chow, S. C., Brown, A., and Pritchard, D. (2000). The human hookworm pathogen *Necator americanus* induces apoptosis in T lymphocytes. Par. Immunol. *22*, 21-29.

Cioli, D. (1976). Transfer of *Schistosoma mansoni* into the mesenteric veins of hamsters. Int. J. Parasitol. *6*, 349-354.

Clegg, J. A. (1974). Host antigens and the immune response in schistosomiasis. In Parasites in the Immunized Host: Mechanisms of Survival, R. Porter, and J. Knight, ed. (Elsevier, North Holland.), pp. 161-176.

Coker, C. M. (1957). Effect of cortisone on natural immunity to *Schistosoma mansoni* in mice. Proc. Soc. Exp. Biol. Med. *96*, 1-3.

Colley, D. G. (1976). Adoptive suppression of granuloma formation. J. Exp. Med. *143*, 696-700.

Colley, D. G., Lewis, F.A., and Todd, C.W. (1979). Adoptive suppression of granuloma formation by T lymphocytes and by lymphoid cells sensitive to cyclophosphamide. Cell. Immunol. *46*, 192-200.

Colwell, D. A., and Wescott, R.B. (1973). Prolongation of egg production of *Nippostrongylus brasiliensis* in mice concurrently infected with *Nematospiroides dubius*. J. Parasitol. *59*, 216.

Constant, S. L., and Bottomly, K. (1997). Induction of TH1 and TH2 CD4+ T cell responses: the alternative approaches. Ann. Rev. Immunol. *15*, 297-322.

Cooke, A., Zaccone, P., Raine, T., Phillips, J. M., and Dunne, D. W. (2004). Infection and autoimmunity: are we winning the war, only to lose the peace? Tr. Parasitol. *20*, 316-321.

Cookson, E., Blaxter, M.L., and Selkirk, M.E. (1992). Identification of the major soluble cuticular glycoprotein of lymphatic filarial nematode parasites (gp29) as a secretory homolog of glutathione peroxidase. Proc. Nat. Acad. Sci. U.S.A. *89*, 5837-5841.

Cooper, P. J. (2004). The potential impact of early exposures to geohelminth infections on the development of atopy. Clin. Rev. All. Immunol. *26*, 5-14.

Cooper, P. J., Chico, M. E., Sandoval, C., Espinel, I., Guevara, A., Kennedy, M. W., Urban, J. F., Griffin, G. E., and Nutman, T. B. (2000). Human infection with *Ascaris lumbricoides* is associated with a polarized cytokine response. J. Inf. Dis. *182*, 1207-1213.

Cooper, P. J., Espinel, I., Paredes, W., Guderian, R. H., and Nutman, T. B. (1998). Impaired tetanus-specific cellular and humoral responses following tetanus vaccination in human onchocerciasis: a possible role for interleukin-10. J. Inf. Dis. *178*, 1133-1138.

Cooper, P. J., Mancero, T., Espinel, M., Sandoval, C., Lovato, R., Guderian, R. H., and Nutman, T. B. (2001). Early human infection with *Onchocerca volvulus* is associated with an enhanced parasite-specific cellular immune response. J. Inf. Dis. *183*, 1662-1668.

Crawford, C., Behnke, J. M., and Pritchard, D. I. (1989). Suppression of heterologous immunity by *Nematospiroides dubius* antigens *in vitro*. Int. J. Parasitol. *19*, 29-34.

Croll, N. A., and Ghadirian, E. (1981). Wormy persons: cxontributions to the nature and patterns of overdispersion with *Ascaris lumbricoides*, *Ancylosotma duodenale*, *Necator americanus* and *Trichuris trichiura*. Trop.Geogr. Med. *33*, 241-248.

Crompton, D. W. T. (1984). "Parasites and people." Macmillan Publishers, London.

Culley, F. J., Brown, A., Conroy, D. M., Sabroe, I., Pritchard, D. I., and Williams, T. J. (2000). Eotaxin is specifically cleaved by hookworm metalloproteases preventing its action *in vitro* and *in vivo*. J. Immunol. *165*, 6447-6453.

Currie, R. M., Needham, C. S., Drake, L. J., Cooper, E. S., and Bundy, D. A. P. (1998). Antigenic variability in *Trichuris trichiura* populations. Parasitology *117*, 347-353.

Dai, W. J., Waldvogel, A., Jungi, T., Stettler, M., and Gottstein, B. (2003). Inducible nitric oxide synthase deficiency in mice increases resistance to chronic infection with *Echinococcus multilocularis*. Immunology *108*, 238-244.

Dainichi, T., Maekawa, Y., Ishii, K., Zhang, T. Q., Nashed, B. F., Sakai, T., Takashima, M., and Himeno, K. (2001). Nippocystatin, a cysteine protease inhibitor from *Nippostrongylus brasiliensis*, inhibits antigen processing and modulates antigen-specific immune response. Inf. Imm. *69*, 7380-7386.

Damian, R. T. (1964). Molecular mimicry: antigen sharing by parasite and host and its consequences. Am. Nat. *98*, 129-149.

Damian, R. T., Greene, N.D., and Hubbard, W.J. (1973). Occurrence of mouse alpha 2 - macroglobulin antigenic determinants on *Schistosoma mansoni* adults, with evidence on their nature. J. Parasitol. *59*, 64-73.

Damian, R. T. (1987). Molecular mimicry revisited. Parasitol. Today *3*, 263-266.

Damian, R. T. (1987). The exploitation of immune responses by parasites: presidential address. J. Parasitol. *73*, 3-13.

Darani, H. Y., Curtis, R. H. C., McNeice, C., Price, H. P., Sayers, J. R., and Doenhoff, M. J. (1997). *Schistosoma mansoni*: anomalous immunogenic properties of a 27 kDa larval serine protease associated with protective immunity. Parasitology *115*, 237-247.

Das, P. K., Manoharan, A., Srividya, A., Grenfell, B. T., Bundy, D. A. P., and Vanamail, P. (1990). Frequency-distribution of *Wuchereria-bancrofti* microfilariae in human populations and its relationships with age and sex. Parasitology *101*, 429-434.

Dasgupta, A., Bala, S., and Dutta, S. N. (1987). Cellular unresponsiveness in patients with soluble circulating antigens in Bancroftian filariasis. Ind. J. Med. Res. *85*, 136-139.

Davies, A. J., Hall, J.G., Targett, G.A., and Murray, M. (1980). The biological significance of the immune response with special reference to parasites and cancer. J. Parasitol. *66*, 705-721.

Davies, S. J., Grogan, J. L., Blank, R. B., Lim, K. C., Locksley, R. M., and McKerrow, J. H. (2001). Modulation of blood fluke development in the liver by hepatic CD4(+) lymphocytes. Science *294*, 1358-1361.

Davies, S. J., Lim, K. C., Blank, R. B., Kim, J. H., Lucas, K. D., Hernandez, D. C., Sedgwick, J. D., and McKerrow, J. H. (2004). Involvement of TNF in limiting liver pathology and promoting parasite survival during schistosome infection. Int. J. Parasitol. *34*, 27-36.

Day, K. P., Grenfell, B., Spark, R., Kazura, J. W., and Alpers, M. P. (1991a). Age specific patterns of change in the dynamics of *Wuchereria bancrofti* infection in Papua New Guinea. Am. J. Trop. Med. Hyg. *44*, 518-527.

Day, K. P., Gregory, W. F., and Maizels, R. M. (1991b). Age-specific acquisition of immunity to infective larvae in a Bancroftian filariasis endemic area of Papua New Guinea. Par. Immunol. *13*, 277-290.

Day, K. P., Gregory, W. F., and Maizels, R. M. (1991). Age-specific acquisition of immunity to infective larvae in a Bancroftian filariasis endemic area of Papua New Guinea. Par. Immunol. *13*, 277-290.

De Jesus, A. R., Silva, A., Santana, L.B., Magalhaes, A., De Jesus, A.A., De Almeida, R.P., Rego, M.A., Burattini, M.N., Pearce, E.J., and Carvalho, E.M. (2002). Clinical and immunologic evaluation of 31 patients with acute schistosomiasis mansoni. J. Inf. Dis. *185*, 98-105.

De Jong, E. C., Vieira, P.L., Kalinski, P., Schuitemaker, J.H., Tanaka, Y., Wierenga, E.A., Yazdanbakhsh, M., and Kapsenberg, M.L. (2002). Microbial compounds selectively induce Th1 cell-promoting or Th2 cell-promoting dendritic cells *in vitro* with diverse th cell-polarizing signals. J. Immunol. *168*, 1704-1709.

De Vijver, K. K. V., Hokke, C. H., van Remoortere, A., Jacobs, W., Deelder, A. M., and Van Marck, E. A. (2004). Glycans of *Schistosoma mansoni* and keyhole limpet haemocyanin induce hepatic granulomas *in vivo*. Int. J. Parasitol. *34*, 951-961.

Dean, D. A. (1977). Decreased binding of cytotoxic antibody by developing *Schistosoma mansoni*: evidence for a surface change independent of host antigen adsorption and membrane turnover. J. Parasitol. *63*, 418-426.

Dehlawi, M. S., and Wakelin, D. (1988). Suppression of mucosal mastocytosis by *Nematospiroides dubius* results from an adult worm-mediated effect upon host lymphocytes. Par. Immunol. *10*, 85-95.

Dell, A., Haslam, S. M., Morris, H. R., and Khoo, K. H. (1999). Immunogenic glycoconjugates implicated in parasitic nematode diseases. Biochim. Biophys. Acta *1455*, 353-362.

DeMatteis, S., Pirotto, F., Marques, J., Nieto, A., Orn, A., and Baz, A. (2001). Modulation of the cellular immune response by a carbohydrate rich fraction from *Echinococcus granulosus* protoscoleces in infected or immunized Balb/c mice. Par. Immunol. *23*, 1-9.

Demeure, C. E., Rihet, P., Abel, L., Ouattara, M., Bourgois, A., and Dessein, A. J. (1993). Resistance to *Schistosoma mansoni* in humans - influence of the IgE/IgG4 balance and IgG2 in immunity to reinfection after chemotherapy. J. Inf. Dis. *168*, 1000-1008.

Demeure, C. E., Rihet, P., Abel, L., Ouattara, M., Bourgois, A., and Dessein, A. J. (1993). Resistance to *Schistosoma mansoni* in humans: influence of the IgE/IgG4 balance and IgG2 in immunity to reinfection after chemotherapy. J. Inf. Dis. *168*, 1000-1008.

Deng, J. S., Gold, D., LoVerde, P. T., and Fishelson, Z. (2003). Inhibition of the complement membrane attack complex by *Schistosoma mansoni* paramyosin. Inf. Imm. *71*, 6402-6410.

Despommier, D. D. (1993). *Trichinella spiralis* and the concept of niche. J. Parasitol. *79*, 472-482.

Despommier, D. D. (1998). How does *Trichinella spiralis* make itself at home? Parasitol. Today *14*, 318-323.

Dessaint, J. P., Camus, D., Fischer, E., and Capron, A. (1977). Inhibition of lymphocyte proliferation by factor(s) produced by *Schistosoma mansoni*. Eur. J. Immunol. *7*, 624-629.

Dessein, A., Kouriba, B., Eboumbou, C., Dessein, H., Argiro, L., Marquet, S., Elwali, N., Rodrigues, V., Li, Y. S., Doumbo, O., and Chevillard, C. (2004). Interleukin-13 in the skin and interferon-gamma in the liver are key players in immune protection in human schistosomiasis. Imm. Rev. *201*, 180-190.

Devaney, E., Martin, S. A. M., and Thompson, F. J. (1996). Stage-specific gene expression in lymphatic filarial nematodes. Parasitol. Today *12*, 418-424.

Devaney, E., and Osborne, J. (2000). The third-stage larva (L3) of Brugia: its role in immune modulation and protective immunity. Micr. Inf. *2*, 1363-1371.

Dimock, K. A., Eberhard, M. L., and Lammie, P. J. (1996). Th1-like antifilarial immune responses predominate in antigen- negative persons. Inf. Imm. *64*, 2962-2967.

Dineen, J. K. (1963). Antigenic relationships between host and parasite. Nature *197*, 471-472.

Dissanayake, S., Amith, R. S., and Shahin, A. (2004). *Taenia crassiceps* carbohydrates stimulate IL-6 expression in naive murine macrophages via Toll-like receptors (TLRs). Mol. Immunol. *41*, 391-398.

Doenhoff, M., Musallam, R., Bain, J., and McGregor, A. (1978). Studies on the host-parasite relationship in *Schistosoma mansoni*-infected mice: the immunological dependence of parasite egg excretion. Immunology *35*, 771-778.

Doenhoff, M., Musallam, R., Bain, J., and McGregor, A. (1979). *Schistosoma mansoni* infections in T-cell deprived mice, and the ameliorating effect of administering homologous chronic infection serum. I. Pathogenesis. Am. J. Trop. Med. Hyg. *28*, 260-263.

Doenhoff, M. J. (1997). A role for granulomatous inflammation in the transmission of infectious disease: schistosomiasis and tuberculosis. Parasitology *115*, S113-S125.

Doenhoff, M. J. (1998). Granulomatous inflammation and the transmission of infection: schistosomiasis - and TB too? Immunol. Today *19*, 462-467.

Doenhoff, M. J., and Bain, J. (1978). The immune-dependence of schistosomicidal chemotherapy: relative lack of efficacy of an antimonial in *Schistosoma mansoni*-infected mice deprived of their T-cells and the demonstration of drug-antiserum synergy. Clin. Exp. Immunol. *33*, 232-238.

Doenhoff, M. J., Chiodini, P. L., and Hamilton, J. V. (2004). Specific and sensitive diagnosis of schistosome infection: can it be done with antibodies? Tr. Parasitol. *20*, 35-39.

Doenhoff, M. J., Hassounah, O., Murare, H., Bain, J., and Lucas, S. (1986). The schistosome egg granuloma: immunopathology in the cause of host protection or parasite survival. Trans. Roy. Soc. Trop. Med.Hyg. *80*, 503-514.

Doenhoff, M. J., Hassounah, O. A., and Lucas, S. B. (1985). Does the immunopathology induced by schistosome eggs potentiate parasite survival. Immunol. Today *6*, 203-206.

Doenhoff, M. J., Modha, J., and Lambertucci, J. R. (1988). Anti-schistosome chemotherapy enhanced by antibodies specific for a parasite esterase. Immunology *65*, 507-510.

Doenhoff, M. J., Pearson, S., Dunne, D. W., Bickle, Q., Lucas, S., Bain, J., Musallam, R., and Hassounah, O. (1981). Immunological control of hepatotoxicity and parasite egg excretion in *Schistosoma mansoni* infections - stage specificity of the reactivity of immune serum in T-cell deprived mice. Trans. Roy. Soc. Trop. Med.Hyg. *75*, 41-53.

Doenhoff, M. J., Stanley, R. G., Griffiths, K., and Jackson, C. L. (2002). An anti-atherogenic effect of *Schistosoma mansoni* infections in mice associated with a parasite-induced lowering of blood total cholesterol. Parasitology *125*, 415-421.

Doetze, A., Satoguina, J., Burchard, G., Rau, T., Loliger, C., Fleischer, B., and Hoerauf, A. (2000). Antigen-specific cellular hyporesponsiveness in a chronic human helminth infection is mediated by T(h)3/T(r)1-type cytokines IL-10 and transforming growth factor-beta but not by a T(h)1 to T(h)2 shift. Int. Immunol. *12*, 623-630.

Domingo, E. O., and Warren, K.S. (1968). Endogenous desensitization: changing host granulomatou response to schistosome eggs at different stages of infection with *Schistosoma mansoni*. Am. J. Pathol. 52, 369-379.

Donaldson, J., van Houtert, M. F. J., and Sykes, A. R. (1998). The effect of nutrition on the periparturient parasite status of mature ewes. Animal Science *67*, 523-533.

Douch, P. G. C., Harrison, G. B. L., Buchanan, L. L., and Greer, K. S. (1988). Relationship of nematode cholinesterase activity and nematode burdens to the development of resistance to Trichostrongyle infections in sheep. Vet. Parasitol. 27, 291-308.

Druillhe, P., Tall, A., and Sokhna, C. (2005). Worms can worsen malaria: towards a new means to roll back malaria. Tr. Parasitol. *21*, 359-362.

Dunne, D., Mountford, A. (2001). Resistance to infection in humans and animal models. In Schistosomiasis, A. A. F. Mahmoud, ed. (London, Imperial College Press).

Dunne, D. W., Bickle, Q. D., Butterworth, A. E., and Richardson, B. A. (1987). The blocking of human antibody-dependent, eosinophil-mediated killing of *Schistosoma mansoni* schistosomula by monoclonal antibodies which cross-react with a polysaccharide-containing egg antigen. Parasitology *94*, 269-280.

Dunne, D. W., Butterworth, A. E., Fulford, A. J. C., Kariuki, H. C., Langley, J. G., Ouma, J. H., Capron, A., Pierce, R. J., and Sturrock, R. F. (1992). Immunity after treatment of human schistosomiasis: association between IgE antibodies to adult worm antigens and resistance to reinfection. Eur. J. Immunol. *22*, 1483-1494.

Dunne, D. W., Hassounah, O., Musallam, R., Lucas, S., Pepys, M. B., Baltz, M., and Doenhoff, M. (1983). Mechanisms of *Schistosoma mansoni* egg excretion: parasitological observations in immunosuppressed mice reconstituted with immune serum. Par. Immunol. *5*, 47-60.

Dunne, D. W., Jones, F. M., and Doenhoff, M. J. (1991). The purification, characterization, serological activity and hepatotoxic properties of two cationic glycoproteins (alpha-1 and omega-1) from *Schistosoma mansoni* eggs. Parasitology *103*, 225-236.

Dunne, D. W., Lucas, S., Bickle, Q., Pearson, S., Madgwick, L., Bain, J., and Doenhoff, M. J. (1981). Identification and partial purification of an antigen (omega-1) from *Schistosoma mansoni* eggs which is putatively hepatotoxic in T-cell deprived mice. Trans. Roy. Soc. Trop. Med.Hyg. *75*, 54-71.

Eberl, M., Langermans, J. A. M., Vervenne, R. A., Nyame, A. K., Cummings, R. D., Thomas, A. W., Coulson, P. S., and Wilson, R. A. (2001). Antibodies to glycans dominate the host response to schistosome larvae and eggs: Is their role protective or subversive? J. Inf. Dis. *183*, 1238-1247.

Edwards, A. J., Burt, J.S., and Ogilvie, B.M. (1971). The effect of immunity upon some enzymes of the parasitic nematode, *Nippostrongylus brasiliensis*. Parasitology *62*, 339-347.

Edwards, M. K., Busto, P., James, E. R., Carlow, C. K. S., and Philipp, M. (1990). Antigenic and dynamic properties of the surface of Onchocerca microfilariae. Trop. Med. Parasitol. *41*, 174-180.

Elkins, D. B., Haswellelkins, M., and Anderson, R. M. (1986). The epidemiology and control of intestinal helminths in the Pulicat Lake region of southern India .1. Study design and pretreatment and post-treatment observations on *Ascaris lumbricoides* infection. Trans. Roy. Soc. Trop. Med.Hyg. *80*, 774-792.

Elliott, D. E., Setiawan, T., Metwali, A., Blum, A., Urban, J. F., and Weinstock, J. V. (2004). *Heligmosomoides polygyrus* inhibits established colitis in IL- 10-deficient mice. Eur. J. Immunol. *34*, 2690-2698.

Else, K. J., Entwistle, G. M., and Grencis, R. K. (1993). Correlations between worm burden and markers of Th1 and Th2 cell subset induction in an inbred strain of mouse infected with *Trichuris muris*. Par. Immunol. *15*, 595-600.

Else, K. J., and Finkelman, F. D. (1998). Intestinal nematode parasites, cytokines and effector mechanisms. Int. J. Parasitol. *28*, 1145-1158.

Else, K. J., Finkelman, F. D., Maliszewski, C. R., and Grencis, R. K. (1994). Cytokine-mediated regulation of chronic intestinal helminth infection. J. Exp. Med. *179*, 347-351.

Else, K. J., and Grencis, R. K. (1991). Cellular immune responses to the murine nematode parasite *Trichuris muris*.1. Differential cytokine production during acute or chronic infection. Immunology *72*, 508-513.

Elson, L. H., Calvopina, M., Paredes, W., Araujo, E., Bradley, J. E., Guderian, R. H., and Nutman, T. B. (1995). Immunity to onchocerciasis: putative immune persons produce a Th1-like response to *Onchocerca volvulus*. J. Inf. Dis. *171*, 652-658.

Ey, P. L. (1988). *Heligmosomoides polygyrus*: retarded development and stunting of larvae by antibodies specific for excretory secretory antigens. Exp. Parasitol. *65*, 232-243.

Fallon, P. G. (2000). Immunopathology of schistosomiasis: a cautionary tale of mice and men. Immunol. Today *21*, 29-35.

Fallon, P. G., Cooper, R. O., Probert, A. J., and Doenhoff, M. J. (1992). Immune-dependent chemotherapy of schistosomiasis. Parasitology *105*, S41-S48.

Fallon, P. G., and Dunne, D. W. (1999). Tolerization of mice to *Schistosoma mansoni* egg antigens causes elevated type 1 and diminished type 2 cytokine responses and increased mortality in acute infection. J. Immunol. *162*, 4122-4132.

Fallon, P. G., Teixeira, M. M., Neice, C. M., Williams, T. J., Hellewell, P. G., and Doenhoff, M. J. (1996). Enhancement of *Schistosoma mansoni* infectivity by intradermal injections of larval extracts: a putative role for larval proteases. J. Inf. Dis. *173*, 1460-1466.

Faquim-Mauro, E. L., and Macedo, M. S. (1998). The immunosuppressive activity of *Ascaris suum* is due to high molecular weight components. Clin. Exp. Immunol. *114*, 245-251.

Fatima, M., Horta, M., and Ramalhopinto, F. J. (1991). Role of human decay-accelerating factor in the evasion of *Schistosoma mansoni* from the complement-mediated killing *in vitro*. J. Exp. Med. *174*, 1399-1406.

Faulkner, H., Renauld, J. C., Van Snick, J., and Grencis, R. K. (1998). Interleukin-9 enhances resistance to the intestinal nematode *Trichuris muris*. Inf. Imm. *66*, 3832-3840.

Faulkner, H., Turner, J., Kamgno, J., Pion, S. D., Boussinesq, M., and Bradley, J. E. (2002). Age- and infection intensity-dependent cytokine and antibody production in human trichuriasis: the importance of IgE. J. Inf. Dis. *185*, 665-672.

Faveeuw, C., Angeli, V., Fontaine, J., Maliszewski, C., Capron, A., Van Kaer, L., Moser, M., Capron, M., and Trottein, F. (2002). Antigen presentation by CD1d contributes to the amplification of Th2 responses to *Schistosoma mansoni* glycoconjugates in mice. J. Immunol. *169*, 906-912.

Fenner, F. (1983). The Florey lecture, 1983. Biological control, as exemplified by smallpox eradication and myxomatosis. Proc. Roy. Soc. Lond. Ser. B *218*, 259-285.

Ferreira, A. M., Irigoin, F., Breijo, M., Sim, R. B., and Diaz, A. (2000). How *Echinococcus granulosus* deals with complement. Parasitol. Today *16*, 168-172.

File, S. (1995). Interaction of schistosome eggs with vascular endothelium. J. Parasitol. *81*, 234-238.

Finkelman, F. D., Shea-Donohue, T., Morris, S. C., Gildea, L., Strait, R., Madden, K. B., Schopf, L., and Urban, J. F. (2004). Interleukin-4-and interleukin-13-mediated host protection against intestinal nematode parasites. Imm. Rev. *201*, 139-155.

Fletcher, M., Loverde, P. T., and Woodruff, D. S. (1981). Genetic-variation in *Schistosoma mansoni*: enzyme polymorphisms in populations from Africa, Southwest Asia, South America, and the West Indies. Am. J. Trop. Med. Hyg. *30*, 406-421.

FloresVillanueva, P. O., Zheng, X. X., Strom, T. B., and Stadecker, M. J. (1996). Recombinant IL-10 and IL-10/Fc treatment down-regulate egg antigen-specific delayed hypersensitivity reactions and egg granuloma formation in schistosomiasis. J. Immunol. *156*, 3315-3320.

Forsyth, K. P., Copeman, D. B., and Mitchell, G. F. (1984). Differences in the surface radioiodinated proteins of skin and uterine microfilariae of *Onchocerca gibsoni*. Mol. Biochem. Parasitol. *10*, 217-229.

Fraser, E. M., and Kennedy, M. W. (1991). Heterogeneity in the expression of surface-exposed epitopes among larvae of *Ascaris lumbricoides*. Par. Immunol. *13*, 219-225.

Freedman, D. O., and Ottesen, E. A. (1988). Eggs of *Schistosoma mansoni* stimulate endothelial cell proliferation *in vitro*. J. Inf. Dis. *158*, 556-562.

Fulford, A. J. C., Butterworth, A. E., Ouma, J. H., and Sturrock, R. F. (1995). A statistical approach to schistosome population-dynamics and estimation of the life-span of *Schistosoma mansoni* in man. Parasitology *110*, 307-316.

Fulford, A. J. C., Butterworth, A. E., Sturrock, R. F., and Ouma, J. H. (1992). On the use of age-intensity data to detect immunity to parasitic infections, with special reference to *Schistosoma mansoni* in Kenya. Parasitology *105*, 219-227.

Gallin, M., Edmonds, K., Ellner, J. J., Erttmann, K. D., White, A. T., Newland, H. S., Taylor, H. R., and Greene, B. M. (1988). Cell-mediated immune responses in human infection with *Onchocerca volvulus*. J. Immunol. *140*, 1999-2007.

Galvani, A. P. (2003). Immunity, antigenic heterogeneity, and aggregation of helminth parasites. J. Parasitol. *89*, 232-241.

Gatrill, A. J., Kee, J., Behnke, J. M., and Wakelin, D. (1991). *Acanthocheilonema viteae* (*Dipetalonema viteae*) in mice: differences in the relative binding of microfilarial surface-specific antibody may explain the contrasting response phenotypes of Balb/C and C57BL/10. J. Helm. *65*, 211-218.

Gems, D. (2000). Longevity and ageing in parasitic and free-living nematodes. Biogerontol. *1*, 289-307.

Gems, D., and Maizels, R. M. (1996). An abundantly expressed mucin-like protein from *Toxocara canis* infective larvae: The precursor of the larval surface coat glycoproteins. Proc. Nat. Acad. Sci. U.S.A. *93*, 1665-1670.

Gerencer, M., Marinculic, A., Rapic, D., Frankovic, M., and Valpotic, I. (1992). Immunosuppression of *in vivo* and *in vitro* lymphocyte-responses in swine induced by *Trichinella spiralis* or excretory secretory antigens of the parasite. Vet. Parasitol. *44*, 263-273.

Gershon, R. K., Carter. RL., and Kondo, K. (1967). On concomitant immunity in tumour-bearing hamsters. Nature *11213*, 674-676.

Ghendler, Y., Arnon, R., and Fishelson, Z. (1994). *Schistosoma mansoni*: isolation and characterization of Smpi56, a novel protease inhibitor. Exp. Parasitol. *78*, 121-131.

Ghosh, I., Eisinger, S.W., Raghavan, N., and Scott, A.L. (1998). Thioredoxin peroxidases from *Brugia malayi*. Mol. Biochem. Parasitol. *91*, 207-220.

Gibbs, H. C. (1986). Hypobiosis in parasitic nematodes: an update. Adv. Parasitol. *25*, 129-174.

Godot, V., Harraga, S., Beurton, I., Deschaseaux, M., Sarciron, E., Gottstein, B., and Vuitton, D. A. (2000). Resistance/susceptibility to *Echinococcus multilocularis* infection and cytokine profile in humans. I. Comparison of patients with progressive and abortive lesions. Clin. Exp. Immunol. *121*, 484-490.

Godot, V., Harraga, S., Podoprigora, G., Liance, M., Bardonnet, K., and Vuitton, D. A. (2003). IFN alpha-2a protects mice against a helminth infection of the liver and modulates immune responses. Gastroenterol. *124*, 1441-1450.

Goldring, O. L., Clegg, J.A., Smithers, S.R., and Terry, R.J. (1976). Acquisition of human blood group antigens by *Schistosoma mansoni*. Clin. Exp. Immunol. *26*, 181-187.

Gomez-Munoz, M. T., Canals-Caballero, A., Almeria, S., Pasquali, P., Zarlenga, D. S., and Gasbarre, L. C. (2004). Inhibition of bovine T lymphocyte responses by extracts of the stomach worm *Ostertagia ostertagi*. Vet. Parasitol. *120*, 199-214.

Goodridge, H. S., Marshall, F.A., Wilson, E.H., Houston, K.M., Liew, F.Y., Harnett, M.M., and Harnett, W. (2004). *In vivo* exposure of murine dendritic cell and macrophage bone marrow progenitors to the phosphorylcholine-containing filarial nematode glycoprotein ES-62 polarizes their differentiation to an anti-inflammatory phenotype. Immunology *113*, 491-498.

Goodridge, H. S., Marshall, F.A., Else, K.J., Houston, K.M., Egan, C., Al-Riyami, L., Liew, F.Y., Harnett, W., and Harnett, M.M. (2005a). Immunomodulation via novel use of TLR4 by the filarial nematode phosphorylcholine-containing secreted product, ES-62. J. Immunol. *174*, 284-293.

Goodridge, H. S., Deehan, M.R., Harnett, W., and Harnett, M.M. (2005b). Subversion of immunological signalling by a filarial nematode phosphorylcholine-containing secreted product. Cell. Sign. *17*, 11-16.

Gordon, S. (2003). Alternative activation of macrophages. Nature Rev. Immunol. *3*, 23-35.

Goyal, P. K., and Wakelin, D. (1993). Influence of variation in host strain and parasite isolate on inflammatory and antibody responses to *Trichinella spiralis* in mice. Parasitology *106*, 371-378.

Gray, C. A., and Lawrence, R. A. (2002). A role for antibody and Fc receptor in the clearance of *Brugia malayi* microfilariae. Eur. J. Immunol. *32*, 1114-1120.

Greene, B. M., Fanning, M. M., and Ellner, J. J. (1983). Non-specific suppression of antigen-induced ymphocyte blastogenesis in *Onchocerca volvulus* infection in man. Clin. Exp. Immunol. *52*, 259-265.

Grencis, R. K., and Entwistle, G. M. (1997). Production of an interferon-gamma homologue by an intestinal nematode: functionally significant or interesting artefact. Parasitology *115*, S101-S106.

Grencis, R. K., Hultner, L., and Else, K. J. (1991). Host protective immunity to *Trichinella spiralis* in mice: activation of Th-cell subsets and lymphokine secretion in mice expressing different response phenotypes. Immunology *74*, 329-332.

Grieve, R. B. (1990). Immunological relevance of the cuticle and epicuticle of larval *Dirofilaria immitis* and *Toxocara canis*. Acta Tropica *47*, 399-402.

Gruden-Movsesijan, A., Petrovic, M., and Sofronic-Milosavljevic, L. (2003). Interaction of mannan-binding lectin with *Trichinella spiralis* glycoproteins, a possible innate immune mechanism. Par. Immunol. *25*, 545-552.

Gryseels, B. (1994). Human resistance to schistosoma infections: age or experience. Parasitol. Today *10*, 380-384.

Haig, D. M., Lima, G.C., and Mota, I. (1980). Antibody suppression in mice infected with *Nippostrongylus brasiliensis*. Par. Immunol. *2*, 175-187.

Haisch, K., Schramm, G., Falcone, F. H., Alexander, C., Schlaak, M., and Haas, H. (2001). A glycoprotein from *Schistosoma mansoni* eggs binds non-antigen- specific immunoglobulin E and releases interleukin-4 from human basophils. Par. Immunol. *23*, 427-434.

Hamajima, F., Yamamoto, M., Tsuru, S., Yamakami, K., Fujino, T., Hamajima, H., and Katsura, Y. (1994). Immunosuppression by a neutral thiol protease from parasitic helminth larvae in mice. Par. Immunol. *16*, 261-273.

Harn, D. A., Cianci, C. M. L., and Caulfield, J. P. (1989). *Schistosoma mansoni*: immunization with cercarial glycocalyx preparation increases the adult worm burden. Exp. Parasitol. *68*, 108-110.

Harnett, W., Worms, M.J., Kapil, A., Grainger, M., and Parkhouse, R.M. (1989). Origin, kinetics of circulation and fate *in vivo* of the major excretory-secretory product of *Acanthocheilonema viteae*. Parasitology *99*, 229-239.

Harnett, W., Houston, K.M., Tate, R., Garate, T., Apfel, H., Adam, R., Haslam, S.M., Panico, M., Paxton, T., Dell, A., Morris, H., and Brzeski, H. (1999). Molecular cloning and demonstration of an aminopeptidase activity in a filarial nematode glycoprotein. Mol. Biochem. Parasitol. *104*, 11-23.

Harnett, W., McInnes, I. B., and Harnett, M. M. (2004). ES-62, a filarial nematode-derived immunomodulator with anti- inflammatory potential. Immunol. Lett. *94*, 27-33.

Harris, A. R. C., Russell, R. J., and Charters, A. D. (1984). A review of schistosomiasis in immigrants in western Australia, demonstrating the unusual longevity of *Schistosoma mansoni*. Trans. Roy. Soc. Trop. Med.Hyg. *78*, 385-388.

Harrison, R. A., Bickle, Q., and Doenhoff, M. J. (1982). Factors affecting the acquisition of resistance against *Schistosoma mansoni* in the mouse: evidence that the mechanisms which mediate resistance during early patent infections may lack immunological specificity. Parasitology *84*, 93-110.

Harrison, R. A., and Doenhoff, M. J. (1983). Retarded development of *Schistosoma mansoni* in immunosuppressed mice. Parasitology *86*, 429-438.

Hartmann, S., and Lucius, R. (2003). Modulation of host immune responses by nematode cystatins. Int. J. Parasitol. *33*, 1291-1302.

Hassounah, O. A., and Doenhoff, M. J. (1993). Comparison of the hepatoprotective effects of immune cells and serum in *Schistosoma mansoni*-infected immunosuppressed mice. Par. Immunol. *15*, 657-661.

Haswell-Elkins, M., Elkins, D., and Anderson, R. M. (1989). The influence of individual, social group and household factors on the distribution of *Ascaris lumbricoides* within a community and implications for control strategies. Parasitology *98*, 125-134.

Haswell-Elkins, M. R., Elkins, D. B., and Anderson, R. M. (1987). Evidence for predisposition in humans to infection with Ascaris, hookworm, Enterobius and Trichuris in a South Indian fishing community. Parasitology *95*, 323-337.

Hayes, K. S., Bancroft, A. J., and Grencis, R. K. (2004). Immune-mediated regulation of chronic intestinal nematode infection. Imm. Rev. *201*, 75-88.

Helmby, H., and Grencis, R. K. (2003). Contrasting roles for IL-10 in protective immunity to different life cycle stages of intestinal nematode parasites. Eur. J. Immunol. *33*, 2382-2390.

Henderson, G. S., Nix, N. A., Montesano, M. A., Gold, D., Freeman, G. L., McCurley, T. L., and Colley, D. G. (1993). Two distinct pathological syndromes in Male CBA/J inbred mice with chronic *Schistosoma mansoni* infections. Am. J. Pathol. *142*, 703-714.

Henri, S., Chevillard, C., Mergani, A., Paris, P., Gaudart, J., Camilla, C., Dessein, H., Montero, F., Elwali, N., Saeed, O. K., Magzoub, M., and Dessein, A.J. (2002). Cytokine regulation of periportal fibrosis in humans infected with *Schistosoma mansoni*: IFN-gamma is associated with protection against fibrosis and TNF-alpha with aggravation of disease. J. Immunol. *169*, 929-936.

Herbert, D. R., Holscher, C., Mohrs, M., Arendse, B., Schwegmann, A., Radwanska, M., Leeto, M., Kirsch, R., Hall, P., Mossmann, H., Claussen, B., Forster, I., and Brombacher, F. (2004). Alternative macrophage activation is essential for survival during schistosomiasis and down modulates T helper 1 responses and immunopathology. Immunity *20*, 623-635.

Herve, M., Angeli, V., Pinzar, E., Wintjens, R., Faveeuw, C., Narumiya, S., Capron, A., Urade, Y., Capron, M., Riveau, G., and Trottein, F. (2003). Pivotal roles of the parasite PGD(2) synthase and of the host D prostanoid receptor 1 in schistosome immune evasion. Eur.J. Immunol. *33*, 2764-2772.

Hesse, M., Piccirillo, C.A., Belkaid, Y., Prufer, J., Mentink-Kane, M., Leusink, M., Cheever, A.W., Shevach, E.M., and Wynn, T.A. (2004). The pathogenesis of schistosomiasis is controlled by cooperating IL-10-producing innate effector and regulatory T cells. J. Immunol. *172*, 3157-3166.

Himmi, E. H., Auriault, C., Pierce, R. J., Joseph, M., Tartar, A., and Capron, A. (1985). Regulatory role of a peptide from the 2nd constant domain of imunoglobulin-G. 2. *In vitro* effect on granuloma-formation around *S. mansoni* eggs. Int. J. Immunopharm. *7*, 231-237.

Hockley, D. J., and McLaren, D.J. (1973). *Schistosoma mansoni*: changes in the outer membrane of the tegument during development from cercariae to adult worm. Int. J. Parasitol. *3*, 13-25.

Hoffmann, K. F., Wynn, T.A., and Dunne, D.W. (2002). Cytokine-mediated host responses during schistosome infections; walking the fine line between immunological control and immunopathology. Adv. Parasitol. *52*, 265-307.

Hoffmann, K. F., Caspar, P., Cheever, A. W., and Wynn, T. A. (1998). IFN-gamma, IL-12, and TNF-alpha are required to maintain reduced liver pathology in mice vaccinated with *Schistosoma mansoni* eggs and IL-12. J. Immunol. *161*, 4201-4210.

Hoffmann, K. F., Cheever, A. W., and Wynn, T. A. (2000). IL-10 and the dangers of immune polarization: excessive type 1 and type 2 cytokine responses induce distinct forms of lethal immunopathology in murine schistosomiasis. J. Immunol. *164*, 6406-6416.

Hoffmann, K. F., James, S. L., Cheever, A. W., and Wynn, T. A. (1999). Studies with double cytokine-deficient mice reveal that highly polarized Th1-and Th2-type cytokine and antibody responses contribute equally to vaccine-induced immunity to *Schistosoma mansoni*. J. Immunol. *163*, 927-938.

Hoffmann, W. H., Pfaff, A. W., Schulz-Key, H., and Soboslav, P. T. (2001). Determinants for resistance and susceptibility to microfilaraemia in *Litomosoides sigmodontis* filariasis. Parasitology *122*, 641-649.

Hogarth, P. J., Folkard, S. G., Taylor, M. J., and Bianco, A. E. (1995). Accelerated clearance of Onchocerca microfilariae and resistance to reinfection in Interleukin-4 gene knockout mice. Par. Immunol. *17*, 653-657.

Hogg, K. G., Kumkate, S., and Mountford, A. P. (2003). IL-10 regulates early IL-12-mediated immune responses induced by the radiation-attenuated schistosome vaccine. Int. Immunol. *15*, 1451-1459.

Holcman, B., and Heath, D. D. (1997). The early stages of *Echinococcus granulosus* development. Acta Tropica *64*, 5-17.

Holland, M. J., Harcus, Y. M., Riches, P. L., and Maizels, R. M. (2000). Proteins secreted by the parasitic nematode *Nippostrongylus brasiliensis* act as adjuvants for Th2 responses. Eur. J. Immunol. *30*, 1977-1987.

Houdijk, J. G. M., Jessop, N. S., Knox, D. P., and Kyriazakis, I. (2003). Breakdown of immunity to *Nippostrongylus brasiliensis* in lactating rats. Br. J. Nutr. *90*, 809-814.

Hussain, R., Poindexter, R. W., and Ottesen, E. A. (1992). Control of allergic reactivity in human filariasis: predominant localization of blocking antibody to the IgG4 subclass. J. Immunol. *148*, 2731-2737.

Imai, S., and Fujita, K. (2004). Molecules of parasites as immunomodulatory drugs. Curr. Top. Med. Chem. *4*, 539-552.

Jackson, J. A., Turner, J. D., Rentoul, L., Faulkner, H., Behnke, J. M., Hoyle, M., Grencis, R. K., Else, K. J., Kamgno, J., Boussinesq, M., and Bradley, J. E. (2004). T helper cell type 2 responsiveness predicts future susceptibility to gastrointestinal nematodes in humans. J. Inf. Dis. *190*, 1804-1811.

Jacobs, W., Deelder, A., Bogers, J., Van de Vijver, K., and Van Marck, E. (1999a). Schistosomal granuloma modulation III. *Schistosoma haematobium* worms accelerate *S. mansoni* soluble egg antigen-induced hepatic granuloma formation *in vivo*. Parasitol. Res. *85*, 905-909.

Jacobs, W., van Dam, G., Bogers, J., Deelder, A., and Van Marck, E. (1999b). Schistosomal granuloma modulation. I. *Schistosoma mansoni* worm antigens CAA and CCA prime egg-antigen-induced hepatic granuloma formation. Parasitol. Res. *85*, 7-13.

Janeway, C. A., and Medzhitov, R. (2002). Innate immune recognition. Ann. Rev. Immunol. *20*, 197-216.

Jankovic, D., Cheever, A. W., Kullberg, M. C., Wynn, T. A., Yap, G., Caspar, P., Lewis, F. A., Clynes, R., Ravetch, J. V., and Sher, A. (1998). CD4+ T cell-mediated granulomatous pathology in schistosomiasis is downregulated by a B cell-dependent mechanism requiring Fc receptor signaling. J. Exp. Med. *187*, 619-629.

Jenkins, D. C., and Phillipson, R.F. (1971). The kinetics of repeated low-level infections of *Nippostronylus brasiliensis* in the laboratory rat. Parasitology *62*, 457-465.

Jenkins, D. C. (1972). *Nippostrongylus brasiliensis*: observations on the comparative immunogenicity of adult worms from primary and immune-adapted infections. Parasitology *65*, 547-550.

Jenkins, D. C., and Phillipson, R.F. (1972). Evidence that the nematode *Nippostrongylus brasiliensis* can adapt to and overcome the effects of host immunity. Int. J. Parasitol. *2*, 353-359.

Jenkins, S. N., and Behnke, J.M. (1977). Impairment of primary expulsion of *Trichuris muris* in mice concurrently infected with *Nematospiroides dubius*. Parasitology *75*. 71-78.

Jenson, J. S., O'Connor, R., Osborne, J., and Devaney, E. (2002). Infection with Brugia microfilariae induces apoptosis of CD4(+) T lymphocytes: a mechanism of immune unresponsiveness in filariasis. Eur. J. Immunol. *32*, 858-867.

Joachim, A., Ruttkowski, B., and Daugschies, A. (1999). Changing surface antigen and carbohydrate patterns during the development of *Oesophagostomum dentatum*. Parasitology *119*, 491-501.

Jones, J. F., Crandall, C.A., and Crandall, R.B. (1984). *In vivo* and *in vitro* responses to sheep erythrocytes by lymph node cells from mice with trichinellosis. Clin. Exp. Immunol. *57*, 301-306.

Jwo, J., and Loverde, P. T. (1989). The ability of fractionated sera from animals vaccinated with irradiated cercariae of *Schistosoma mansoni* to transfer immunity to mice. J. Parasitol. *75*, 252-260.

Kabatereine, N. B., Vennervald, J. B., Ouma, J. H., Kemijumbi, J., Butterworth, A. E., Dunne, D. W., and Fulford, A. J. C. (1999). Adult resistance to schistosomiasis mansoni: age-dependence of reinfection remains constant in communities with diverse exposure patterns. Parasitology *118*, 101-105.

Kalkofen, U. P. (1970). Attachment and feeding behavior of *Ancylostoma caninum*. Zeit. Parasitenk. *33*, 339-354.

Kar, S. K., Mania, J., Baldwin, C. I., and Denham, D. A. (1993). The sheath of the microfilaria of *Wuchereria bancrofti* has albumin and immunoglobulin on its surface. Par. Immunol. *15*, 297-300.

Karanja, D. M. S., Colley, D. G., Nahlen, B. L., Ouma, J. H., and Secor, W. E. (1997). Studies on schistosomiasis in western Kenya .1. Evidence for immune-facilitated excretion of schistosome eggs from patients with *Schistosoma mansoni* and human immunodeficiency virus coinfections. Am. J. Trop. Med. Hyg. *56*, 515-521.

Keiser, P. B., and Nutman, T.B. (2004). *Strongyloides stercoralis* in the immunocompromised population. Clin. Microbiol. Rev. *17*, 208-217.

Kemp, W.M., Damian, R.T., Greene, N.D., and Lushbaugh, W.B. (1976). Immunocytochemical localization of mouse alpha 2-macroglobulinlike antigenic determinants on *Schistosoma mansoni* adults. J. Parasitol. *62*, 413-419.

Kemp, W. M., Merritt, S.C., Bogucki, M.S., Rosier, J.G., and Seed, J.R. (1977). Evidence for adsorption of heterospecific host immunoglobulin on the tegument of *Schistosoma mansoni*. J. Immunol. *119*, 1849-1845.

Kennedy, M. W., and Qureshi, F. (1986). Stage-specific secreted antigens of the parasitic larval stages of the nematode ascaris. Immunology *58*, 515-522.

Khalife, J., Capron, M., Capron, A., Grzych, J. M., Butterworth, A. E., Dunne, D. W., and Ouma, J. H. (1986). Immunity in human schistosomiasis mansoni: regulation of protective immune mechanisms by IgM blocking antibodies. J. Exp. Med. *164*, 1626-1640.

King, C. L., Kumaraswami, V., Poindexter, R. W., Kumari, S., Jayaraman, K., Alling, D. W., Ottesen, E. A., and Nutman, T. B. (1992). Immunological tolerance in lymphatic filariasis: diminished parasite-specific lymphocyte-T and lymphocyte-B precursor frequency in the microfilaremic state. J. Clin. Inv. *89*, 1403-1410.

King, C. L., Mahanty, S., Kumaraswami, V., Abrams, J. S., Regunathan, J., Jayaraman, K., Ottesen, E. A., and Nutman, T. B. (1993). Cytokine control of parasite-specific anergy in human lymphatic filariasis: preferential induction of a regulatory T-helper type-2 lymphocyte subset. J. Clin. Inv. *92*, 1667-1673.

King, C. L., Medhat, A., Malhotra, I., Nafeh, M., Helmy, A., Khaudary, J., Ibrahim, S., ElSherbiny, M., Zaky, S., Stupi, R. J., Brustoski, K., Shehata, M., and Shata, M.T. (1996). Cytokine control of parasite-specific anergy in human urinary schistosomiasis: IL-10 modulates lymphocyte reactivity. J. Immunol. *156*, 4715-4721.

Knox, D. P., and Jones, D. G. (1992). A comparison of superoxide dismutase (SOD, EC-1.15.1.1) distribution in gastrointestinal nematodes. Int. J. Parasitol. *22*, 209-214.

Kong, Y., Chung, Y. B., Cho, S. Y., and Kang, S. Y. (1994). Cleavage of immunoglobulin-G by excretory-secretory cathepsin S-like protease of *Spirometra mansoni* plerocercoid. Parasitology *109*, 611-621.

Koski, K. G., and Scott, M. E. (2001). Gastrointestinal nematodes, nutrition and immunity: breaking the negative spiral. Annual Review of Nutrition *21*, 297-321.

Koyama, K., and Ito, Y. (1996). Comparative studies on immune responses to infection in susceptible B10.BR mice infected with different strains of the murine nematode parasite *Trichuris muris*. Par. Immunol. *18*, 257-263.

Kurniawan, A., Yazdanbakhsh, M., Vanree, R., Aalberse, R., Selkirk, M. E., Partono, F., and Maizels, R. M. (1993). Differential expression of IgE and IgG4 specific antibody responses in asymptomatic and chronic human filariasis. J. Immunol. *150*, 3941-3950.

Lammie, P. J., Eberhard, M. L., Leiva, L. E., Lowrie, R. C., and Katz, S. P. (1988). The effect of diethylcarbamazine treatment of Bancroftian filariasis on the immunological reactivity of microfilaremic individuals. Trans. Roy. Soc. Trop. Med.Hyg. *82*, 726-729.

Landa, A., Laclette, J. P., Nicholsonweller, A., and Shoemaker, C. B. (1993). cDNA cloning and recombinant expression of collagen-binding and complement inhibitor activity of *Taenia solium* paramyosin (AgB). Mol. Biochem. Parasitol. *60*, 343-348.

Langley, J. G., Kariuki, H. C., Hammersley, A. P., Ouma, J. H., Butterworth, A. E., and Dunne, D. W. (1994). Human-IgG subclass responses and subclass restriction to *Schistosoma mansoni* egg antigens. Immunology *83*, 651-658.

Langrish, C. L., McKenzie, B. S., Wilson, N. J., Malefyt, R. D., Kastelein, R. A., and Cua, D. J. (2004). IL-12 and IL-23: master regulators of innate and adaptive immunity. Imm. Rev. *202*, 96-105.

Lawrence, C. E. (2003). Is there a common mechanism of gastrointestinal nematode expulsion? Par. Immunol. *25*, 271-281.

Lawrence, C. E., Paterson, Y. Y. W., Wright, S. H., Knight, P. A., and Miller, H. R. P. (2004). Mouse mast cell protease-1 is required for the enteropathy induced by gastrointestinal helminth infection in the mouse. Gastroenterol. *127*, 155-165.

Lawrence, R. A., Allen, J. E., Osborne, J., and Maizels, R. M. (1994). Adult and microfilarial stages of the filarial parasite *Brugia malayi* stimulate contrasting cytokine and Ig isotype responses in Balb/C mice. J. Immunol. *153*, 1216-1224.

Lawrence, R. A., and Devaney, E. (2001). Lymphatic filariasis: parallels between the immunology of infection in humans and mice. Par. Immunol. *23*, 353-361.

Le Goff, L., Lamb, T. J., Graham, A. L., Harcus, Y., and Allen, J. E. (2002). IL-4 is required to prevent filarial nematode development in resistant but not susceptible strains of mice. Int. J. Parasitol. *32*, 1277-1284.

Le Hesran, J. Y., Akiana, J., El Ndiaye, H. M., Dia, M., Senghor, P., and Konate, L. (2004). Severe malaria attack is associated with high prevalence of *Ascaris lumbricoides* infection among children in rural Senegal. Trans. Roy. Soc. Trop. Med.Hyg. *98*, 397-399.

Lee, D. L. (1996). Why do some nematode parasites of alimentary tract secrete acetylcholinesterase? Int. J. Parasitol. *26*, 499-508.

Lee, T. D. G., and Wright, K.A. (1978). The morphology and attachment and probable feeding site of the nematode *Trichuris muris* (Schrank, 1788) Hall, 1916. Can. J. Zool. *56*, 1889-1905.

Lehmnan, J. S. J., Mott, K.E., Morrow, R.H.Jr, Muniz, T.M. and Boyer, M.H. (1976). The intensity and effects of infection with *Schistosoma mansoni* in a rural community in northeast Brazil. Am. J. Trop. Med. Hyg. *25*, 285-294.

Leiva, L. E., Lammie, P.J. (1989). Modulation of lymphocyte activation by soluble *Brugia pahangi* extracts. Trop. Med. Parasitol. *40*, 327-331.

Lewis, F. A., Hieny, S., and Sher, A. (1985). Evidence against the existence of specific *Schistosoma mansoni* subpopulations which are resistant to irradiated vaccine-induced immunity. Am. J. Trop. Med. Hyg. *34*, 86-91.

Lewis, J. W., and Bryant, V. (1976). The distribution of *Nematospiroides dubius* within the small intestine of laboratory mice. J. Helm. *50*, 163-171.

Lightowlers, M. W., and Rickard, M. D. (1988). Excretory secretory products of helminth parasites: effects on host immune responses. Parasitology *96*, S123-S166.

Liu, Z. G., Liu, Q., Pesce, J., Whitmire, J., Ekkens, M. J., Foster, A., VanNoy, J., Sharpe, A. H., Urban, J. F., and Gause, W. C. (2002). *Nippostrongylus brasiliensis* can induce B7-independent antigen- specific development of IL-4-producing T cells from naive CD4 T cells *in vivo*. J. Immunol. *169*, 6959-6968.

Liwski, R., Zhou, J., McAlister, V., and Lee, T. D. G. (2000). Prolongation of allograft survival by *Nippostrongylus brasiliensis* is associated with decreased allospecific cytotoxic T lymphocyte activity and development of T cytotoxic cell type 2 cells. Transplantation *69*, 1912-1922.

Loeffler, D. A., Lundy, S. K., Singh, K. P., Gerard, H. C., Hudson, A. P., and Boros, D. L. (2002). Soluble egg antigens from *Schistosoma mansoni* induce angiogenesis-related processes by up-regulating vascular endothelial growth factor in human endothelial cells. J. Inf. Dis. *185*, 1650-1656.

Loke, P., Nair, M.G., Parkinson, J., Guiliano, D., Blaxter, M., and Allen, J.E. (2002). IL-4 dependent alternatively-activated macrophages have a distinctive *in vivo* gene expression phenotype. BMC Immunology *3*, 7.

Loke, P., MacDonald, A. S., Robb, A., Maizels, R. M., and Allen, J. E. (2000). Alternatively activated macrophages induced by nematode infection inhibit proliferation via cell-to-cell contact. Eur. J. Immunol. *30*, 2669-2678.

Lolis, E., and Bucala, R. (2003). Macrophage migration inhibitory factor. Exp. Opin. Ther. Targ. *7*, 153-164.

Loukas, A., Hintz, M., Linder, D., Mullin, N. P., Parkinson, J., Tetteh, K. K. A., and Maizels, R. M. (2000). A family of secreted mucins from the parasitic nematode *Toxocara canis* bears diverse mucin domains but shares similar flanking six-cysteine repeat motifs. J. Biol. Chem. *275*, 39600-39607.

Loukas, A., Jones, M. K., King, L. T., Brindley, P. J., and McManus, D. P. (2001). Receptor for Fc on the surfaces of schistosomes. Inf. Imm. *69*, 3646-3651.

Loukas, A., and Maizels, R. M. (2000). Helminth C-type lectins and host-parasite interactions. Parasitol. Today *16*, 333-339.

Loukas, A., Mullin, N. P., Tetteh, K. K. A., Moens, L., and Maizels, R. M. (1999). A novel C-type lectin secreted by a tissue-dwelling parasitic nematode. Curr. Biol. *9*, 825-828.

LoVerde, P. T. (1998). Do antioxidants play a role in schistosome host-parasite interactions? Parasitol. Today *14*, 284-289.

Loverde, P. T., Dewald, J., and Minchella, D. J. (1985a). Further studies of genetic variation in *Schistosoma mansoni*. J. Parasitol. *71*, 732-734.

Loverde, P. T., Dewald, J., Minchella, D. J., Bosshardt, S. C., and Damian, R. T. (1985b). Evidence for host-induced selection in *Schistosoma mansoni*. J. Parasitol. *71*, 297-301.

Lu, W., Egerton, G.L., Bianco, A.E., and Williams, S.A. (1998). Thioredoxin peroxidase from *Onchocerca volvulus*: a major hydrogen peroxide detoxifying enzyme in filarial parasites. Mol. Biochem. Parasitol. *91*, 221-235.

Lubiniecki, A. S., and Cypess, R.H. (1975). Immunological sequelae of *Trichinella spiralis* infection in mice: effect on the antibody responses to sheep erythrocytes and Japanse B encephalitis virus. Inf. Imm. *11*, 1306-1311.

Lucas, S., Musallam, R., Bain, J., Hassounah, O., Bickle, Q., and Doenhoff, M. (1980). The pathological effects of immunosuppression of *Schistosoma mansoni*-infected mice, with particular reference to survival and hepatotoxicity after thymectomy and treatment with antithymocyte serum, and treatment with hydrocortisone acetate. Trans. Roy. Soc. Trop. Med.Hyg. *74*, 633-643.

Lunde, M. N., Paranjape, R., Lawley, T. J., and Ottesen, E. A. (1988). Filarial antigen in circulating immune-complexes from patients with *Wuchereria bancrofti* filariasis. Am. J. Trop. Med. Hyg. *38*, 366-371.

MacDonald, A. J., Turaga, P. S. D., Harmon-Brown, C., Tierney, T. J., Bennett, K. E., McCarthy, M. C., Simonek, S. C., Enyong, P. A., Moukatte, D. W., and Lustigman, S. (2002). Differential cytokine and antibody responses to adult and larval stages of *Onchocerca volvulus* consistent with the development of concomitant immunity. Inf. Imm. *70*, 2796-2804.

MacDonald, A. S., Straw, A.D., Bauman, B., Pearce, E.J. (2001). CD8-dendritic cell activation status plays an integral role in influencing Th2 response development. J. Immunol. *167*, 1982-1988.

MacDonald, A. S., Maizels, R. M., Lawrence, R. A., Dransfield, I., and Allen, J. E. (1998). Requirement for *in vivo* production of IL-4, but not IL-10, in the induction of proliferative suppression by filarial parasites. J. Immunol. *160*, 4124-4132.

Machado, D. C., Horton, D., Harrop, R., Peachell, P. T., and Helm, B. A. (1996). Potential allergens stimulate the release of mediators of the allergic response from cells of mast cell lineage in the absence of sensitization with antigen-specific IgE. Eur. J. Immunol. *26*, 2972-2980.

Mahanty, S., King, C. L., Kumaraswami, V., Regunathan, J., Maya, A., Jayaraman, K., Abrams, J. S., Ottesen, E. A., and Nutman, T. B. (1993). Il-4-secreting and Il-5-secreting lymphocyte populations are preferentially stimulated by parasite-derived antigens in human tissue invasive nematode infections. J. Immunol. *151*, 3704-3711.

Mahanty, S., Luke, H. E., Kumaraswami, V., Narayanan, P. R., Vijayshekaran, V., and Nutman, T. B. (1996a). Stage-specific induction of cytokines regulates the immune response in lymphatic filariasis. Exp. Parasitol. *84*, 282-290.

Mahanty, S., Mollis, S. N., Ravichandran, M., Abrams, J. S., Kumaraswami, V., Jayaraman, K., Ottesen, E. A., and Nutman, T. B. (1996b). High levels of spontaneous and parasite antigen-driven interleukin-10 production are associated with antigen-specific hyporesponsiveness in human lymphatic filariasis. J. Inf. Dis. *173*, 769-773.

Mahanty, S., and Nutman, T. B. (1995). Immunoregulation in human lymphatic filariasis: the role of interleukin-10. Par. Immunol. *17*, 385-392.

Mahanty, S., Ravichandran, M., Raman, U., Jayaraman, K., Kumaraswami, V., and Nutman, T. B. (1997). Regulation of parasite antigen-driven immune responses by interleukin-10 (IL-10) and IL-12 in lymphatic filariasis. Inf. Imm. *65*, 1742-1747.

Maizels, R., Burke, J., Sutanto, I., Purnomo, and Partono, F. (1986). Secreted and surface antigens from larval stages of *Wuchereria bancrofti*, the major human lymphatic filarial parasite. Mol. Biochem. Parasitol. *19*, 27-34.

Maizels, R. M., Bundy, D. A. P., Selkirk, M. E., Smith, D. F., Anderson, R. M. (1993). Immunological modulation and evasion by helminth parasites in human populations. Nature *365*, 797-805.

Maizels, R. M., Balic, A., Gomez-Escobar, N., Nair, M., Taylor, M. D., and Allen, J. E. (2004). Helminth parasites - masters of regulation. Imm. Rev. *201*, 89-116.

Maizels, R. M., Blaxter, M. L., and Scott, A. L. (2001a). Immunological genomics of *Brugia malayi*: filarial genes implicated in immune evasion and protective immunity. Par. Immunol. *23*, 327-344.

Maizels, R. M., Denham, D. A., and Sutanto, I. (1985). Secreted and circulating antigens of the filarial parasite *Brugia pahangi*: analysis of *in vitro* released components and detection of parasite products *in vivo*. Mol. Biochem. Parasitol. *17*, 277-288.

Maizels, R. M., Gomez-Escobar, N., Gregory, W. F., Murray, J., and Zang, X. X. (2001b). Immune evasion genes from filarial nematodes. Int. J. Parasitol. *31*, 889-898.

Maizels, R. M., and Holland, M. J. (1998). Parasite immunity: pathways for expelling intestinal helminths. Curr. Biol. *8*, R711-R714.

Maizels, R. M., and Kurniawan-Atmadja, A. (2002). Variation and polymorphism in helminth parasites. Parasitology *125*, S25-S37.

Maizels, R. M., and Lawrence, R. A. (1991). Immunological tolerance: the key feature in human filariasis. Parasitol. Today *7*, 271-276.

Maizels, R. M., Philipp, M., Dasgupta, A., and Partoni, F. (1984). Human serum albumin is a major component on the surface of microfilariae of *Wuchereria bancrofti*. Par. Immunol. *6*, 185-190.

Malhotra, I., Mungai, P., Wamachi, A., Kioko, J., Ouma, J. H., Kazura, J. W., and King, C. L. (1999). Helminth- and bacillus Calmette Guerin-induced immunity in children sensitized in utero to filariasis and schistosomiasis. J. Immunol. *162*, 6843-6848.

Mania, J., and Kar, S. K. (1994). Detection of albumin and immunoglobulin on surface of *Brugia malayi* mf by immunofluorescence. Ind. J. Med. Res. *99*, 65-67.

Manoury, B., Gregory, W.F., Maizels, R.M., and Watts, C. (2001). Bm-CPI-2, a cystatin homolog secreted by the filarial parasite *Brugia malayi*, inhibits class II MHC-restricted antigen processing. Curr. Biol. *11*, 447-451.

Marikovsky, M., Arnon, R., and Fishelson, Z. (1988). Proteases secreted by transforming schistosomula of *Schistosoma mansoni* promote resistance to killing by complement. J. Immunol. *141*, 273-278.

Marikovsky, M., Parizade, M., Arnon, R., and Fishelson, Z. (1990) Complement regulation on the surface of cultured schistosomula and adult worms of *Schistosoma mansoni*. Eur. J. Immunol. *20*, 221-227.

Maron, R., Palanivel, V., Weiner, H. L., and Harn, D. A. (1998). Oral administration of schistosome egg antigens and insulin B-chain generates and enhances Th2-type responses in NOD mice. Clin. Immunol. Immunopathol. *87*, 85-92.

Marquet, S., Abel, L., Hillaire, D., Dessein, H., Kalil, J., Feingold, J., Weissenbach, J., and Dessein, A. J. (1996). Genetic localization of a locus controlling the intensity of infection by *Schistosoma mansoni* on chromosome 5q31-q33. Nature Genet. *14*, 181-184.

Marshall, E., and Howells, R. E. (1986). Turnover of the surface proteins of adult and 3rd-stage and 4th-stage larval *Brugia pahangi*. Mol. Biochem. Parasitol. *18*, 17-24.

Marshall, J. S. (2004). Mast-cell responses to pathogens. Clin. Immunol. Immunopathol.re Reviews Immunology *4*, 787-799.

Mazingue, C., Stadler, B. M., Quatannens, B., Capron, A., and Deweck, A. (1986). Schistosome-derived inhibitory factor: an immunosuppressive agent preferentially active on lymphocytes-T. Int. Arch. All. App. Immunol. *80*, 347-354.

McElroy, P. J., and Befus, A. D. (1987). *Nippostrongylus brasiliensis*-induced depression of heterologous immune responses: effects of the histamine H2 receptor antagonist, cimetidine. J. Parasitol. *73*, 1158-1163.

McGreevy, P. B., Ratiwayanto, S., Tuti, S., McGreevy, M.M., and Dennis, D.T. (1980). *Brugia malayi*: relationship between anti-sheath antibodies and amicrofilaremia in natives living in an endemic area of South Kalimantan, Borneo. Am. J. Trop. Med. Hyg. *29*, 553-562.

McKee, A. S., and Pearce, E.J. (2004). CD25+CD4+ cells contribute to Th2 polarization during helminth infection by suppressing Th1 response development. J. Immunol. *173*, 1224-1331.

McKenzie, G. J., Bancroft, A., Grencis, R. K., and McKenzie, A. N. J. (1998). A distinct role for interleukin-13 in Th2-cell-mediated immune responses. Curr. Biol. *8*, 339-342.

McKerrow, J. H., and Salter, J. (2002). Invasion of skin by schistosoma cercariae. Tr. Parasitol. *18*, 193-195.

McLaren, D. J., and Hockley, D.J. (1977). Blood flukes have a double outer membrane. Nature *269*, 147-149.

Medhat, A., Shehata, M., Bucci, K., Mohamed, S., Dief, A. D. E., Badary, S., Galal, H., Nafeh, M., and King, C. L. (1998). Increased interleukin-4 and interleukin-5 production in response to *Schistosoma haematobium* adult worm antigens correlates with lack of reinfection after treatment. J. Inf. Dis. *178*, 512-519.

Mentink-Kane, M. M., Cheever, A. W., Thompson, R. W., Hari, D. M., Kabatereine, N. B., Vennervald, B. J., Ouma, J. H., Mwatha, J. K., Jones, F. M., Donaldson, D. D., Grusby, M.J., Dunne, D.W., and Wynn, T.A. (2004). IL-13 receptor alpha 2 down-modulates granulomatous inflammation and prolongs host survival in schistosomiasis. Proc. Nat. Acad. Sci. U.S.A. *101*, 586-590.

Michel, J. F., Lancaster, M.B., and Hong, C. (1978). The length of *Ostertagia ostertagi* in populations of uniform age. Int. J. Parasitol. *8*, 437-441.

Miller, H. R. P. (1984). The protective mucosal response against gastrointestinal nematodes in ruminants and laboratory animals. Vet. Immunol. Immunopathol. *6*, 167-259.

Miller, K. M., Lustigman, S., Brotman, B., Andrus, L., Weiss, N., Hotze, C., and Prince, A. M. (1991). *Onchocerca volvulus*: characterization of monoclonal antibodies reactive with surface components of 3rd-stage larvae. Trop. Med. Parasitol. *42*, 85-90.

Min, B., Prout, M., Hu-Li, J., Zhu, J. F., Jankovic, D., Morgan, E. S., Urban, J. F., Dvorak, A. M., Finkelman, F. D., LeGros, G., and Paul, W. E. (2004). Basophils produce IL-4 and accumulate in tissues after infection with a Th2-inducing parasite. J. Exp. Med. *200*, 507-517.

Mitchell, G. F. (1979). Effector cells, molecules and mechanisms in host-protective immunity to parasites. Immunology *38*, 209-223.

Mitre, E., Taylor, R. T., Kubofcik, J., and Nutman, T. B. (2004). Parasite antigen-driven basophils are a major source of IL-4 in human filarial infections. J. Immunol. *172*, 2439-2445.

Modha, J., and Doenhoff, M. J. (1994). *Schistosoma mansoni* host-parasite relationship: interaction of contrapsin with adult worms. Parasitology *109*, 487-495.

Modha, J., Parikh, V., Gauldie, J., and Doenhoff, M. J. (1988). An association between schistosomes and contrapsin, a mouse serine protease inhibitor (serpin). Parasitology *96*, 99-109.

Montenegro, S. M., Miranda, P., Mahanty, S., Abath, F.G., Teixeira, K.M., Coutinho, E.M., Brinkman, J., Goncalves, I., Domingues, L.A., Domingues, A.L., Sher, A., and Wynn, T.A. (1999). Cytokine production in acute versus chronic human *Schistosomiasis mansoni*: the cross-regulatory role of interferon-gamma and interleukin-10 in the responses of peripheral blood mononuclear cells and splenocytes to parasite antigens. J. Inf. Dis. *179*, 1502-1514.

Montesano, M. A., Colley, D. G., Willard, M. T., Freeman, G. L., and Secor, W. E. (2002). Idiotypes expressed early in experimental *Schistosoma mansoni* infections predict clinical outcomes of chronic disease. J. Exp. Med. *195*, 1223-1228.

Montesano, M. A., Freeman, G. L., Secor, W. E., and Colley, D. G. (1997). Immunoregulatory idiotypes stimulate T helper 1 cytokine responses in experimental *Schistosoma mansoni* infections. J. Immunol. *158*, 3800-3804.

More, S. J., and Copeman, D. B. (1990). A highly specific and sensitive monoclonal antibody-based ELISA for the detection of circulating antigen in Bancroftian filariasis. Trop. Med. Parasitol. *41*, 403-406.

Mosmann, T. R., and Coffman, R. L. (1989). Th1-cell and Th2-cell: different patterns of lymphokine secretion lead to different functional properties. Ann. Rev. Immunol. *7*, 145-173.

Mosmann, T. R., and Moore, K. W. (1991). The role of IL-10 in crossregulation of Th1 and Th2 responses. Immunoparasitol. Today-a Combined Issue of Immunology Today and Parasitol. Today, A49-A53.

Mota-Santos, T. A., Tavares, C.A., Gazzinelli, G., and Pellegrino, J. (1977). Immunosuppression mediated by adult worms in chronic schistosomiasis mansoni. Am. J. Trop. Med. Hyg. *26*, 727-731.

Mountford, A. P., Coulson, P. S., Saunders, N., and Wilson, R. A. (1989). Characteristics of protective immunity in mice induced by drug-attenuated larvae of *Schistosoma mansoni*: antigen localization and antibody responses. J. Immunol. *143*, 989-995.

Mountford, A. P., and Trottein, F. (2004). Schistosomes in the skin: a balance between immune priming and regulation. Tr. Parasitol. *20*, 221-226.

Murare, H. M., Agnew, A., Baltz, M., Lucas, S. B., and Doenhoff, M. J. (1987). The response to *Schistosoma bovis* in normal and T-cell deprived mice. Parasitology *95*, 517-530.

Murare, H. M., and Doenhoff, M. J. (1987). Parasitological observations of *Schistosoma bovis* in normal and T-cell deprived mice. Parasitology *95*, 507-516.

Murray, J., Manoury, B., Balic, A., Watts, C., Maizels, R.M. (2005) Bm-CPI-2, a cystatin from *Brugia malayi* nematode parasites, differs from *Caenorhabditis elegans* cystatins in a specific site mediating inhibition of the antigen-processing enzyme AEP. Mol. Biochem. Parasitol. *139*, 197-203.

Mwatha, J. K., Kimani, G., Kamau, T., Mbugua, G. G., Ouma, J. H., Mumo, J., Fulford, A. J. C., Jones, F. M., Butterworth, A. E., Roberts, M. B., and Dunne, D. W. (1998). High levels of TNF, soluble TNF receptors, soluble ICAM-1, and IFN-gamma, but low levels of IL-5, are associated with hepatosplenic disease in human schistosomiasis mansoni. J. Immunol. *160*, 1992-1999.

Nacher, M., Singhasivanon, P., Traore, B., Vannaphan, S., Gay, F., Chindanond, D., Franetich, J. F., Mazier, D., and Looareesuwan, S. (2002a). Helminth infections are associated with protection from cerebral malaria and increased nitrogen derivatives concentrations in Thailand. Am. J. Trop. Med. Hyg. *66*, 304-309.

Nacher, M., Singhasivanon, P., Yimsamran, S., Manibunyong, W., Thanyavanich, N., Wuthisen, P., and Looareesuwan, S. (2002b). Intestinal helminth infections are associated with increased incidence of *Plasmodium falciparum* malaria in Thailand. J. Parasitol. *88*, 55-58.

Nair, M. G., Gallagher, L. J., Taylor, M. D., Loke, P., Coulson, P. S., Wilson, R. A., Maizels, R. M., and Allen, J. E. (2005). Chitinase and Fizz family members are a generalized feature of nematode infection with selective upregulation of Ym1 and F10.1 by antigen-presenting cells. Inf. Imm. *73*, 385-394.

Ngaiza, J. R., and Doenhoff, M. J. (1990). Blood platelets and schistosome egg excretion. Proc. Soc. Exp. Biol. Med. *193*, 73-79.

Ngaiza, J. R., Doenhoff, M. J., and Jaffe, E. A. (1993). *Schistosoma mansoni* egg attachment to cultured human umbilical vein endothelial cells: an *in vitro* model of an early step of parasite egg excretion. J. Inf. Dis. *168*, 1576-1580.

Noel, W., Raes, G., Ghassabeh, G. H., De Baetselier, P., and Beschin, A. (2004). Alternatively activated macrophages during parasite infections. Tr. Parasitol. *20*, 126-133.

Nookala, S., Srinivasan, S., Kaliraj, P., Narayanan, R. B., and Nutman, T. B. (2004). Impairment of tetanus-specific cellular and humoral responses following tetanus vaccination in human lymphatic filariasis. Inf. Imm. *72*, 2598-2604.

Nutman, T. B., and Kumaraswami, V. (2001). Regulation of the immune response in lymphatic filariasis: perspectives on acute and chronic infection with *Wuchereria bancrofti* in South India. Par. Immunol. *23*, 389-399.

Nutman, T. B., Kumaraswami, V., and Ottesen, E. A. (1987b). Parasite-specific anergy in human filariasis: insights after analysis of parasite antigen-driven lymphokine production. J. Clin. Invest. *79*, 1516-1523.

Nutman, T. B., Kumaraswami, V., Pao, L., Narayanan, P. R., and Ottesen, E. A. (1987a). An analysis of *in vitro* B-cell immune responsiveness in human lymphatic filariasis. J. Immunol. *138*, 3954-3959.

O'Connor, R. A., Jenson, J. S., and Devaney, E. (2000). NO contributes to proliferative suppression in a murine model of filariasis. Inf. Imm. *68*, 6101-6107.

O'Connor, R. A., Jenson, J. S., Osborne, J., and Devaney, E. (2003). An enduring association? Microfilariae and immunosupression in lymphatic filariasis. Tr. Parasitol. *19*, 565-570.

Ogilvie, B. M. (1974). Antigenic variation in the nematode *Nippostrongylus brasiliensis.*, Associated Scientific Publishers, Amsterdam and London).

Okano, M., Satoskar, A. R., Nishizaki, K., Abe, M., and Harn, D. A. (1999). Induction of Th2 responses and IgE is largely due to carbohydrates functioning as adjuvants on *Schistosoma mansoni* egg antigens. J. Immunol. *163*, 6712-6717.

Okano, M., Satoskar, A. R., Nishizaki, K., and Harn, D. A. (2001). Lacto-N-fucopentaose III found on *Schistosoma mansoni* egg antigens functions as adjuvant for proteins by inducing Th2- type response. J. Immunol. *167*, 442-450.

Oliver-Gonzalez, J., Torregrossa, M.V. (1944). A substance in animal parasites related to human isoagglutinins. J. Inf. Dis. *74*, 173-177.

Oliver-Gonzalez, J. (1949). Functional antigens in helminths. J. Inf. Dis. *78*, 232-237.

Omer Ali, P., Smithers, S. R., Bickle, Q., Phillips, S. M., Harn, D., and Simpson, A. J. G. (1988). Analysis of the anti-*Schistosoma mansoni* surface antibody response during murine infection and its potential contribution to protective immunity. J. Immunol. *140*, 258-264.

Omer Ali, P. O., Mansour, M., Woody, J. N., Smithers, S. R., and Simpson, A. J. G. (1989). Antibody to carbohydrate and polypeptide epitopes on the surface of schistosomula of *Schistosoma mansoni* in Egyptian patients with acute and chronic schistosomiasis. Parasitology *98*, 417-424.

Opal, S. M., and DePalo, V. A. (2000). Anti-inflammatory cytokines. Chest *117*, 1162-1172.

Osborne, J., and Devaney, E. (1998). The L3 of Brugia induces a T(h)2-polarized response following activation of an IL-4-producing CD4-CD8- alpha beta T cell population. Int. Immunol. *10*, 1583-1590.

Osborne, J., and Devaney, E. (1999). Interleukin-10 and antigen-presenting cells actively suppress Th1 gels in BALB/c mice infected with the filarial parasite *Brugia pahangi*. Inf. Imm. *67*, 1599-1605.

Oshiro, T. M., Rafael, A., Enobe, C. S., Fernandes, I., and Macedo-Soares, M. F. (2004). Comparison of different monoclonal antibodies against immunosuppressive proteins of *Ascaris suum*. Braz. J. Med. Biol. Res. *37*, 223-226.

Ottesen, E. A., Weller, P.F., and Heck, L. (1977). Specific cellular immune unresponsiveness in human filariasis. Immunology *33*, 413-421.

Ottesen, E. A. (1992). Infection and disease in lymphatic filariasis: an immunological perspective. Parasitology *104*, S71-S79.

Ottesen, E. A., Skvaril, F., Tripathy, S. P., Poindexter, R. W., and Hussain, R. (1985). Prominence of IgG4 in the IgG antibody response to human filariasis. J. Immunol. *134*, 2707-2712.

Ouma, J. H., Fulford, A. J. C., Kariuki, H. C., Kimani, G., Sturrock, R. F., Muchemi, G., Butterworth, A. E., and Dunne, D. W. (1998). The development of Schistosomiasis mansoni in an immunologically naive immigrant population in Masongaleni, Kenya. Parasitology *117*, 123-132.

Page, A. P., Rudin, W., Fluri, E., Blaxter, M. L., and Maizels, R. M. (1992). *Toxocara canis*: a labile antigenic surface-coat overlying the epicuticle of infective larvae. Exp. Parasitol. *75*, 72-86.

Parizade, M., Arnon, R., Lachmann, P. J., and Fishelson, Z. (1994). Functional and antigenic similarities between a 94-Kd protein of *Schistosoma mansoni* (SCIP-1) and human CD59. J. Exp. Med. *179*, 1625-1636.

Pastrana, D. V., Raghavan, N., FitzGerald, P., Eisinger, S.W., Metz, C., Bucala, R., Schleimer, R.P., Bickel, C., and Scott, A.L. (1998). Filarial nematode parasites secrete a homologue of the human cytokine macrophage migration inhibitory factor. Inf. Imm. *66*, 5955-5963.

Paterson, J. C. M., Garside, P., Kennedy, M. W., and Lawrence, C. E. (2002). Modulation of a heterologous immune response by the products of *Ascaris suum*. Inf. Imm. *70*, 6058-6067.

Patton, E. A., Brunet, L. R., La Flamme, A. C., Pedras-Vasconcelos, J., Kopf, M., and Pearce, E. J. (2001). Severe schistosomiasis in the absence of interleukin-4 (IL-4) is IL-12 independent. Inf. Imm. *69*, 589-592.

Pearce, E. J., Caspar, P., Grzych, J.M., Lewis, F.A., and Sher, A. (1991). Downregulation of Th1 cytokine production accompanies induction of Th2 responses by a parasitic helminth, *Schistosoma mansoni*. J. Exp. Med. *173*, 159-166.

Pearce, E. J., and MacDonald, A.S. (2002). The immunobiology of schistosomiasis. Nature Rev.Immunol. *2*, 499-511.

Pearce, E. J., Basch, P. F., and Sher, A. (1986). Evidence that the reduced surface antigenicity of developing *Schistosoma mansoni* schistosomula is due to antigen shedding rather than host molecule acquisition. Par. Immunol. *8*, 79-94.

Pearce, E. J., Hall, B. F., and Sher, A. (1990). Host-specific evasion of the alternative complement pathway by schistosomes correlates with the presence of a phospholipase C-sensitive surface molecule resembling human decay accelerating factor. J. Immunol. *144*, 2751-2756.

Pemberton, A. D., Knight, P. A., Gamble, J., Colledge, W. H., Lee, J. K., Pierce, M., and Miller, H. R. P. (2004). Innate BALB/c enteric epithelial responses to *Trichinella spiralis*: Inducible expression of a novel goblet cell lectin, intelectin-2, and its natural deletion in C57BL/10 mice. J. Immunol. *173*, 1894-1901.

Pestka, S., Krause, C. D., Sarkar, D., Walter, M. R., Shi, Y. F., and Fisher, P. B. (2004). Interleukin-10 and related cytokines and receptors. Ann. Rev. Immunol. *22*, 929-979.

Pfaff, A. W., Schulz-Key, H., Soboslay, P.T., Taylor, D.W., MacLennan, K., and Hoffmann, W.H. (2002). *Litomosoides sigmodontis* cystatin acts as an immunomodulator during experimental filariasis. Int. J. Parasitol. *32*, 171-178.

Philipp, M., Parkhouse, R.M.E., and Ogilvie, B.M. (1980). Changing proteins on the surface of a parasitic nematode. Nature *287*, 538-540.

Philipp, M., and Rumjanek, F.D. (1984). Antigenic and dynamic properties of helminth surface structures. Mol. Biochem. Parasitol. *10*, 245-268.

Philipp, M. (1984). Acetylcholinesterase secreted by intestinal nematodes: a reinterpretation of its putative role of biochemical holdfast. Trans. Roy. Soc. Trop. Med.Hyg. *78*, 138-139.

Philipp, M., Worms, M. J., McLaren, D. J., Ogilvie, B. M., Parkhouse, R. M., and Taylor, P. M. (1984). Surface proteins of a filarial nematode: a major soluble antigen and a host component on the cuticle of *Litomosoides carinii*. Par. Immunol. *6*, 63-82.

Piessens, W. F., McGreevy, P.B., Piessens, P.W., McGreevy, M., Koiman, I., Saroso, J.S., and Dennis, D.T. (1980a). Immune responses in human infections with *Brugia malayi*: specific cellular unresponsiveness to filarial antigens. J. Clin. Invest. *65*, 172-179.

Piessens, W. F., Ratiwayanto, S., Tuti, S., Palmieri, J.H., Piessens, P.W., Koiman, I., and Dennis, D.T. (1980b). Antigen-specific suppressor cells and suppressor factors in human filariasis with *Brugia malayi*. N. Eng. J. Med. *302*, 833-837.

Piessens, W. F., Partono, F., Hoffman, S. L., Ratiwayanto, S., Piessens, P. W., Palmieri, J. R., Koiman, I., Dennis, D. T., and Carney, W. P. (1982). Antigen-specific suppressor lymphocyte-T in human lymphatic filariasis. N. Eng. J. Med. *307*, 144-148.

Pleass, R. J., Kusel, J. R., and Woof, J. M. (2000). Cleavage of human IgE mediated by *Schistosoma mansoni*. Int. Arch. All. App. Immunol. *121*, 194-204.

Polderman, A. M. (1979). Transmission dynamics of endemic schistosomiasis. Trop.Geogr. Med. *31*, 465-475.

Porto, A. F., Santos, S. B., Alcantara, L., Guerreiro, J. B., Passos, J., Gonzalez, T., Neva, F., Gonzalez, D., Ho, J. L., and Carvalho, E. M. (2004). HTLV-1 modifies the clinical and immunological response to schistosomiasis. Clin. Exp. Immunol. *137*, 424-429.

Potter, A., Stephens, D., and De Keulenaer, B. (2003). Strongyloides hyper-infection: a case for awareness. Ann. Trop. Med. Parasitol. *97*, 855-860.

Prier, R. C., and Lammie, P. J. (1988). Differential regulation of *in vitro* humoral and cellular immune responsiveness in *Brugia pahangi*-infected jirds. Inf. Imm. *56*, 3052-3057.

Pritchard, D. I., Ali, N. M. H., and Behnke, J. M. (1984). Analysis of the mechanism of immunodepression following heterologous antigenic stimulation during concurrent infection with *Nematospiroides dubius*. Immunology *51*, 633-642.

Pritchard, D. I., and Behnke, J. M. (1985). The suppression of homologous immunity by soluble adult antigens of *Nematospiroides dubius*. J. Helm. *59*, 251-256.

Pritchard, D. I., Lawrence, C. E., Appleby, P., Gibb, I. A., and Glover, K. (1994). Immunosuppressive proteins secreted by the gastrointestinal nematode parasite *Heligmosomoides polygyrus*. Int. J. Parasitol. *24*, 495-500.

Prost, A., Nebout, M., Rougemont, A. (1979). Lepromatous leprosy and onchocerciasis. Br. Med. J. *1*, 589-590.

Quinnell, R. J. (1992). The Population-dynamics of *Heligmosomoides polygyrus* in an enclosure population of wood mice. J. Anim. Ecol. *61*, 669-679.

Quinnell, R. J. (2003). Genetics of susceptibility to human helminth infection. Int. J. Parasitol. *33*, 1219-1231.

Raleigh, J. M., Brandon, M. R., and Meeusen, E. (1996). Stage-specific expression of surface molecules by the larval stages of *Haemonchus contortus*. Par. Immunol. *18*, 125-132.

Ramaswamy, K., Kumar, P., and He, Y. X. (2000). A role for parasite-induced PGE(2) in IL-10-mediated host immunoregulation by skin stage schistosomula of *Schistosoma mansoni*. J. Immunol. *165*, 4567-4574.

Ramaswamy, K., Salafsky, B., Potluri, S., He, Y. X., Li, J. W., and Shibuya, T. (1996). Secretion of an anti-inflammatory, immunomodulatory factor by schistosomulae of *Schistosoma mansoni*. J. Inflamm. *46*, 13-22.

Rathaur, S., Muller, S., Maizels, R. M., and Walter, R. D. (1992). Identification of circulating parasite acetylcholinesterase in human and rodent filariasis. Parasitol. Res. *78*, 671-676.

Ravindran, B., Satapathy, A. K., and Sahoo, P. K. (1994). Bancroftian filariasis: differential reactivity of anti-sheath antibodies in microfilariae carriers. Par. Immunol. *16*, 321-323.

Ravindran, B., Satapathy, A. K., Sahoo, P. K., and Geddam, J. J. B. (2000). Protective immunity in human Bancroftian filariasis: inverse relationship between antibodies to microfilarial sheath and circulating filarial antigens. Par. Immunol. *22*, 633-637.

Rezende, S. A., SilvaTeixeira, D. N., Drummond, S. C., and Goes, A. M. (1997). IL-10 plays a role in the modulation of human granulomatous hypersensitivity against *Schistosoma mansoni* eggs induced by immune complexes. Scand. J. Immunol. *46*, 96-102.

Riffkin, M. C., Seow, H. F., Wood, P. R., Brown, L. E., Jackson, D. C., and Scheerlinck, J. P. Y. (2000). *Trichostrongylus colubriformis* extract upregulates TNF-alpha receptor expression and enhances TNF-alpha sensitivity of L929 cells. Immunol. Cell Biol. *78*, 575-579.

Rigano, R., Profumo, E., Bruschi, F., Carulli, G., Azzara, A., Ioppolo, S., Buttari, B., Ortona, E., Margutti, P., Teggi, A., and Siracusano, A. (2001). Modulation of human immune response by *Echinococcus granulosus* antigen B and its possible role in evading host defenses. Inf. Imm. *69*, 288-296.

Rihet, P., Demeure, C. E., Bourgois, A., Prata, A., and Dessein, A. J. (1991). Evidence for an association between human resistance to *Schistosoma mansoni* and high anti-larval IgE levels. Eur. J. Immunol. *21*, 2679-2686.

Rodrigues, V., Piper, K., Couissinier-Paris, P., Bacelar, O., Dessein, H., and Dessein, A. J. (1999). Genetic control of schistosome infections by the SM1 locus of the 5q31-q33 region is linked to differentiation of type 2 helper T lymphocytes. Inf. Imm. *67*, 4689-4692.

Rodriguez-Sosa, M., Satoskar, A. R., David, J. R., and Terrazas, L. I. (2003). Altered T helper responses in CD40 and interleukin-12 deficient mice reveal a critical role for Th1 responses in eliminating the helminth parasite *Taenia crassiceps*. Int. J. Parasitol. *33*, 703-711.

Rogers, S. H., and Bueding, E. (1975). Anatomical localization of glucose uptake by *Schistosoma mansoni* adults. Int. J. Parasitol. *5*, 369-371.

Rook, G. A. W., and Stanford, J. L. (1998). Give us this day our daily germs. Immunol. Today *19*, 113-116.

Roye, O., Delacre, M., Williams, I. R., Auriault, C., and Wolowczuk, I. (2001). Cutaneous interleukin-7 transgenic mice display a propitious environment to *Schistosoma mansoni* infection. Par. Immunol. *23*, 133-140.

Rutitzky, L. I., Hernandez, H. J., and Stadecker, M. J. (2001). Th1-polarizing immunization with egg antigens correlates with severe exacerbation of immunopathology and death in schistosome infection. Proc. Nat. Acad. Sci. U.S.A. *98*, 13243-13248.

Rutitzky, L. I., Mirkin, G. A., and Stadecker, M. J. (2003). Apoptosis by neglect of CD4(+) Th cells in granulomas: a novel effector mechanism involved in the control of egg-induced immunopathology in murine schistosomiasis. J. Immunol. *171*, 1859-1867.

Sabin, E. A., Araujo, M. I., Carvalho, E. M., and Pearce, E. J. (1996). Impairment of tetanus toxoid-specific Th1-like immune responses in humans infected with *Schistosoma mansoni*. J. Inf. Dis. *173*, 269-272.

Sabin, E. A., and Pearce, E. J. (1995). Early IL-4 production by non-CD4(+) cells at the site of antigen deposition predicts the development of a T-helper-2 cell response to *Schistosoma mansoni* eggs. J. Immunol. *155*, 4844-4853.

Sadun, E. H. (1963). Fluorescent antibody technique for helminth infections. Exp. Parasitol. *13*, 72-82.

Saeftel, M., Arndt, M., Specht, S., Volkmann, L., and Hoerauf, A. (2003). Synergism of gamma interferon and interleukin-5 in the control of murine filariasis. Inf. Imm. *71*, 6978-6985.

Saeftel, M., Volkmann, L., Korten, S., Brattig, N., Al-Qaoud, K., Fleischer, B., and Hoerauf, A. (2001). Lack of interferon-gamma confers impaired neutrophil granulocyte function and imparts prolonged survival of adult filarial worms in murine filariasis. Micr. Inf. *3*, 203-213.

Sandor, M., Weinstock, J. V., and Wynn, T. A. (2003). Granulomas in schistosome and mycobacterial infections: a model of local immune responses. Tr. Immunol. *24*, 44-52.

Sartono, E., Kruize, Y. C. M., Kurniawan, A., Vandermeide, P. H., Partono, F., Maizels, R. M., and Yazdanbakhsh, M. (1995b). Elevated cellular immune responses and interferon-gamma release after long-term diethylcarbamazine treatment of patients with human lymphatic filariasis. J. Inf. Dis. *171*, 1683-1687.

Sartono, E., Kruize, Y. C. M., Partono, F., Kurniawan, A., Maizels, R. M., and Yazdanbakhsh, M. (1995a). Specific T cell unresponsiveness in human filariasis: diversity in underlying mechanisms. Par. Immunol. *17*, 587-594.

Sartono, E., Lopriore, C., Kruize, Y. C. M., Kurniawan-Atmadja, A., Maizels, R. M., and Yazdanbakhsh, M. (1999). Reversal in microfilarial density and T cell responses in human lymphatic filariasis. Par. Immunol. *21*, 565-571.

Satapathy, A. K., Sahoo, P. K., Geddam, J. J. B., Mohanty, M. C., and Ravindran, B. (2001). Human Bancroftian filariasis: loss of patent microfilaraemia is not associated with production of antibodies to microfilarial sheath. Par. Immunol. *23*, 163-167.

Satoguina, J., Mempel, M., Larbi, J., Badusche, M., Loliger, C., Adjei, O., Gachelin, G., Fleischer, B., and Hoerauf, A. (2002). Antigen-specific T regulatory-1 cells are associated with immunosuppression in a chronic helminth infection (onchocerciasis). Micr. Inf. *4*, 1291-1300.

Saunders, N., Wilson, R. A., and Coulson, P. S. (1987). The outer bilayer of the adult schistosome tegument surface has a low turnover rate *in vitro* and *in vivo*. Mol. Biochem. Parasitol. *25*, 123-131.

Schad, G. A., and Anderson, R. M. (1985). Predisposition to hookworm infection in humans. Science *228*, 1537-1540.

Schierack, P., Lucius, R., Sonnenburg, B., Schilling, K., and Hartmann, S. (2003). Parasite-specific immunomodulatory functions of filarial cystatin. Inf. Imm. *71*, 2422-2429.

Schmidt, G.D., and Roberts, L.S. (2000) Foundations of Parasitology, 6th Ed. McGraw Hill, Boston.

Schonemeyer, A., Lucius, R., Sonnenburg, B., Brattig, N., Sabat, R., Schilling, K., Bradley, J., and Hartmann, S. (2001). Modulation of human T cell responses and macrophage functions by onchocystatin, a secreted protein of the filarial nematode *Onchocerca volvulus*. J. Immunol. *167*, 3207-3215.

Schopf, L. R., Hoffmann, K. F., Cheever, A. W., Urban, J. F., and Wynn, T. A. (2002). IL-10 is critical for host resistance and survival during gastrointestinal helminth infection. J. Immunol. *168*, 2383-2392.

Schramm, G., Falcone, F. H., Gronow, A., Haisch, K., Mamat, U., Doenhoff, M. J., Oliveira, G., Galle, J., Dahinden, C. A., and Haas, H. (2003). Molecular characterization of an interleukin-4-inducing factor from *Schistosoma mansoni* eggs. J. Biol. Chem. *278*, 18384-18392.

Secor, W. E., Karanja, D. M. S., and Colley, D. G. (2004). Interactions between schistosomiasis and human immunodeficiency virus in western Kenya. Mem. Inst. Osw. Cruz *99*, 93-95.

Selby, G. R., and Wakelin, D. (1975). Suppression of the immune response to *Trichuris muris* in lactating mice. Parasitology *71*, 77-85.

Semnani, R. T., Law, M., Kubofcik, J., and Nutman, T. B. (2004). Filaria-induced immune evasion: Suppression by the infective stage of *Brugia malayi* at the earliest host-parasite interface. J. Immunol. *172*, 6229-6238.

Semnani, R. T., Liu, A. Y., Sabzevari, H., Kubofcik, J., Zhou, J., Gilden, J. K., and Nutman, T. B. (2003). *Brugia malayi* microfilariae induce cell death in human dendritic cells, inhibit their ability to make IL-12 and IL-10, and reduce their capacity to activate CD4(+) T cells. J. Immunol. *171*, 1950-1960.

Shaw, D. J., Grenfell, B. T., and Dobson, A. P. (1998). Patterns of macroparasite aggregation in wildlife host populations. Parasitology *117*, 597-610.

Sher, A., Hall, B.F., and Vadas, M.A. (1978). Acquisition of murine major histocompatibility complex gene products by schistosomula of *Schistosoma mansoni*. J. Exp. Med. *148*, 46-57.

Sher, A., Fiorentino, D., Caspar, P., Pearce, E., and Mosmann, T. (1991). Production of IL-10 by CD4+ T lymphocytes correlates with down-regulation of Th1 cytokine synthesis in helminth infection. J. Immunol. *147*, 2713-2716.

Sher, A., and Coffman, R. L. (1992). Regulation of immunity to parasites by T-cells and T-cell derived cytokines. Ann. Rev. Immunol. *10*, 385-409.

Sher, A., Jankovic, D., Cheever, A., and Wynn, T. (1996). An IL-12-based vaccine approach for preventing immunopathology in schistosomiasis. In Inerleukin 12: Cellular and Mol. Immunol. of an Important Regulatory Cytokine, pp. 202-207.

Shinkai, K., Mohrs, M., and Locksley, R. M. (2002). Helper T cells regulate type-2 innate immunity *in vivo*. Nature *420*, 825-829.

Simpson, A. J. G. (1990). Schistosome surface antigens: developmental expression and immunological function. Parasitol. Today *6*, 40-45.

Simpson, A. J. G., Correaoliveira, R., Smithers, S. R., and Sher, A. (1983). The exposed carbohydrates of schistosomula of *Schistosoma mansoni* and their modification during maturation *in vivo*. Mol. Biochem. Parasitol. *8*, 191-205.

Simpson, A. J. G., Singer, D., McCutchan, T. F., Sacks, D. L., and Sher, A. (1983). Evidence that schistosome MHC antigens are not synthesized by the parasite but are acquired from the host as intact glycoproteins. J. Immunol. *131*, 962-965.

Skelly, P. J. (2004). Immunoparasitology series: intravascular schistosomes and complement. Tr. Parasitol. *20*, 370-374.

Smith, H. V., Kusel, J. R., and Girdwood, R. W. A. (1983). The production of human A-blood and B-blood group like substances by *in vitro* maintained second stage *Toxocara canis* larvae: their presence on the outer larval surfaces and in their excretions secretions. Clin. Exp. Immunol. *54*, 625-633.

Smith, N. C., and Bryant, C. (1986). The role of host-generated free radicals in helminth infections: *Nippostrongylus brasiliensis* and *Nematospiroides dubius* compared. Int. J. Parasitol. *16*, 617-622.

Smith, P., Walsh, C. M., Mangan, N. E., Fallon, R. E., Sayers, J. R., McKenzie, A. N. J., and Fallon, P. G. (2004). *Schistosoma mansoni* worms induce anergy of T cells via selective up-regulation of programmed death ligand 1 on macrophages. J. Immunol. *173*, 1240-1248.

Smithers, S. R., and Terry, R.J. (1967). Resistance to experimental infection with *Schistosoma mansoni* in rhesus monkeys induced by the transfer of adult worms. Trans. Roy. Soc. Trop. Med.Hyg. *61*, 517-533.

Smithers, S. R., and Terry, R.J. (1969). Immunity in schstosomiasis. Ann. N. Y. Acad. Sci. *160*, 826-840.

Smithers, S. R., Terry, R.J., and Hockley, D.J. (1969). Host antigens in schistosomiasis. Proc. Roy. Soc. Lond. Ser. B *171*, 483-494.

Snary, D., Smith, M.A., and Clegg, J.A. (1980). Surface proteins of *Schistosoma mansoni* and their expression during morphogenesis. Eur. J. Immunol. *10*, 573-575.

Soboslay, P. T., Luder, C. G. K., Hoffmann, W. H., Michaelis, I., Helling, G., Heuschkel, C., Dreweck, C. M., Blanke, C. H., Pritze, S., Banla, M., and Schulzkey, H. (1994). Ivermectin-facilitated immunity in onchocerciasis: activation of parasite-specific Th1-type responses with subclinical *Onchocerca volvulus* infection. Clin. Exp. Immunol. *96*, 238-244.

Specht, S., Volkmann, L., Wynn, T., and Hoerauf, A. (2004). Interleukin-10 (IL-10) counterregulates IL-4-dependent effector mechanisms in murine filariasis. Inf. Imm. *72*, 6287-6293.

Spolski, R. J., Corson, J., Thomas, P. G., and Kuhn, R. E. (2000). Parasite-secreted products regulate the host response to larval *Taenia crassiceps*. Par. Immunol. *22*, 297-305.

Sprent, J. F. A. (1959). The evolution of living organisms. Melbourne University Press, Victoria.).

Srivatsan, J., Smith, D. F., and Cummings, R. D. (1992). The human blood fluke *Schistosoma mansoni* synthesizes glycoproteins containing the Lewis-X antigen. J. Biol. Chem. *267*, 20196-20203.

Stadecker, M. J., Asahi, H., Finger, E., Hernandez, H. J., Rutitzky, L. I., and Sun, J. (2004). The immunobiology of Th1 polarization in high-pathology schistosomiasis. Imm. Rev. *201*, 168-179.

Stanley, P., and Stein, P.E. (2003). BmSPN2, a serpin secreted by the filarial nematode *Brugia malayi*, does not inhibit human neutrophil proteinases but plays a noninhibitory role. Biochemistry *42*, 6241-6248.

Stavitsky, A. B. (2004). Regulation of granulomatous inflammation in experimental models of schistosomiasis. Inf. Imm. *72*, 1-12.

Stavitsky, A. B., Metz, C., Liu, S. F., Jia, X. L., and Bucala, R. (2003). Blockade of macrophage migration inhibitory factor (MIF) in *Schistosoma japonicum*-infected mice results in an increased adult worm burden and reduced fecundity. Par. Immunol. *25*, 369-374.

Steel, C., Guinea, A., McCarthy, J. S., and Ottesen, E. A. (1994). Long-term effect of prenatal exposure to maternal microfilaremia on immune responsiveness to filarial parasite antigens. Lancet *343*, 890-893.

Steel, C., and Nutman, T. B. (2003). CTLA-4 in filarial infections: Implications for a role in diminished T cell reactivity. J. Immunol. *170*, 1930-1938.

Stewart, G. R., Boussinesq, M., Coulson, T., Elson, L., Nutman, T., and Bradley, J. E. (1999). Onchocerciasis modulates the immune response to mycobacterial antigens. Clin. Exp. Immunol. *117*, 517-523.

Strachan, D. P. (1989). Hay-fever, hygiene, and household size. British Medical Journal *299*, 1259-1260.

Subramanian, S., Stolk, W. A., Ramaiah, K. D., Plaisier, A. P., Krishnamoorthy, K., Van Oortmarssen, G. J., Amalraj, D. D., Habbema, J. D. F., and Das, P. K. (2004). The dynamics of *Wuchereria bancrofti* infection: a model-based analysis of longitudinal data from Pondicherry, India. Parasitology *128*, 467-482.

Summers, R. W., Elliott, D. E., Urban, J. F., Thompson, R., and Weinstock, J. V. (2005). *Trichuris suis* therapy in Crohn's disease. Gut *54*, 87-90.

Tang, J. M., Dobson, C., and McManus, D. P. (1995). Antigens in phenotypes of *Heligmosomoide polygyrus* raised selectively from different strains of mice. Int. J. Parasitol. *25*, 847-852.

Tang, L., Ou, X., Henkle-Duhrsen, K., and Selkirk, M.E. (1994). Extracellular and cytoplasmic CuZn superoxide dismutases from Brugia lymphatic filarial nematode parasites. Inf. Imm. *62*, 961-967.

Tanguay, G. V., and Scott, M. E. (1992). Factors generating aggregation of *Heligmosomoides polygyrus* (Nematoda) in laboratory mice. Parasitology *104*, 519-529.

Tarleton, R. L., and Kemp, W.M. (1981). Demonstration of IgG-Fc and C3 receptors on adult *Schistosoma mansoni*. J. Immunol. *126*, 379-384.

Tawill, S., Le Goff, L., Ali, F., Blaxter, M., and Allen, J. E. (2004). Both free-living and parasitic nematodes induce a characteristic Th2 response that is dependent on the presence of intact glycans. Inf. Imm. *72*, 398-407.

Taylor, D. W., Goddard, J. M., and McMahon, J. E. (1986). Surface components of *Onchocerca volvulus*. Mol. Biochem. Parasitol. *18*, 283-300.

Taylor, M. D., LeGoff, L., Harris, A., Malone, E., Allen, J.E., and Maizels, R.M. (2005) Removal of regulatory T cell activity reverses hyporesponsiveness and leads to filarial parasite clearance *in vivo*. J. Immunol. ***In Press***.

Taylor, M. G., Bickle, Q.D. (1986). Irradiated schistosome vaccines. Parasitol. Today 2, 132-134.

Taylor, M. J. (2002). Wolbachia endosymbiotic bacteria of filarial nematodes: a new insight into disease pathogenesis and control. Arch. Med. Res. *33*, 422-424.

Telford, G., Wheeler, D. J., Appleby, P., Bowen, J. G., and Pritchard, D. I. (1998). *Heligmosomoides polygyrus* immunomodulatory factor (IMF), targets T-lymphocytes. Par. Immunol. *20*, 601-611.

Terrazas, L. I., Bojalil, R., Govezensky, T., and Larralde, C. (1998). Shift from an early protective Th1-type immune response to a late permissive Th2-type response in murine cysticercosis (*Taenia crassiceps*). J. Parasitol. *84*, 74-81.

Tezuka, H., Imai, S., Hidano, S., Tsukidate, S., and Fujita, K. (2003). Various types of *Dirofilaria immitis* polyproteins selectively induce a Th2-type immune response. Inf. Imm. *71*, 3802-3811.

Thambiah, G., Whitworth, J., Hommel, M., and Devaney, E. (1992). Identification and characterization of a parasite antigen in the circulating immune-complexes of *Onchocerca volvulus* infected patients. Trop. Med. Parasitol. *43*, 271-276.

Thomas, P. G., and Harn, D. A. (2004). Immune biasing by helminth glycans. Cell. Microbiol. *6*, 13-22.

Timmann, C., Fuchs, S., Thoma, C., Lepping, B., Brattig, N. W., Sievertsen, J., Thye, T., Muller-Myhsok, B., and Horstmann, R. (2004). Promoter haplotypes of the interleukin-10 gene influence proliferation of peripheral blood cells in response to helminth antigen. Genes Imm. *5*, 256-260.

Tingley, G. A., Butterworth, A. E., Anderson, R. M., Kariuki, H. C., Koech, D., Mugambi, M., Ouma, J. H., Siongok, T. K. A., and Sturrock, R. F. (1988). Predisposition of humans to infection with *Schistosoma mansoni*: evidence from the reinfection of individuals following chemotherapy. Trans. Roy. Soc. Trop. Med.Hyg. *82*, 448-452.

Torpier, G., Capron, A., and Ouaissi, M.A. (1979). Receptor for IgG(Fc) and human beta2-microglobulin on *S. mansoni* schistosomula. Nature *278*, 447-449.

Trizio, D., Della Bruna, C., and Isetta, A.M. (1980). Modification of immune responsiveness in murine *Schistosomiasis mansoni*. I. Time course after cercarial exposure. Immunology *40*, 353-358.

Trottein, F., Descamps, L., Nutten, S., Dehouck, M. P., Angeli, V., Capron, A., Cecchelli, R., and Capron, M. (1999a). *Schistosoma mansoni* activates host microvascular endothelial cells to acquire an anti-inflammatory phenotype. Inf. Imm. *67*, 3403-3409.

Trottein, F., Nutten, S., Angeli, V., Delerive, P., Teissier, E., Capron, A., Staels, B., and Capron, M. (1999b). *Schistosoma mansoni* schistosomula reduce E-selectin and VCAM-1 expression in TNF-alpha-stimulated lung microvascular endothelial cells by interfering with the NF-kappa B pathway. Eur. J. Immunol. *29*, 3691-3701.

Trottein, F., Pavelka, N., Vizzardelli, C., Angeli, V., Zouain, C. S., Pelizzola, M., Capozzoli, M., Urbano, M., Capron, M., Belardelli, F.,Granucci, F., and Riciardi-Castagnoli, P. (2004). A type IIFN-dependent pathway induced by *Schistosoma mansoni* eggs in mouse myeloid dendritic cells generates an inflammatory signature. J. Immunol. *172*, 3011-3017.

Turner, J. D., Faulkner, H., Kamgno, J., Cormont, F., Van Snick, J., Else, K. J., Grencis, R. K., Behnke, J. M., Boussinesq, M., and Bradley, J. E. (2003). Th2 cytokines are associated with reduced worm burdens in a human intestinal helminth infection. J. Inf. Dis. *188*, 1768-1775.

Tweardy, D. J., Osman, G. S., Elkholy, A., and Ellner, J. J. (1987). Failure of immunosuppressive mechanisms in human *Schistosoma mansoni* infection with hepatosplenomegaly. J. Clin. Microbiol. *25*, 768-773.

Van Dam, G. J., Bergwerff, A. A., Thomasoates, J. E., Rotmans, J. P., Kamerling, J. P., Vliegenthart, J. F. G., and Deelder, A. M. (1994). The immunologically reactive O-linked polysaccharide chains derived from circulating cathodic antigen isolated from the human blood fluke *Schistosoma mansoni* have Lewis-X as repeating unit. Eur. J. Biochem. *225*, 467-482.

Van Dam, G. J., Bogitsh, B. J., vanZeyl, R. J. M., Rotmans, J. P., and Deelder, A. M. (1996a). *Schistosoma mansoni*: *in vitro* and *in vivo* excretion of CAA and CCA by developing schistosomula and adult worms. J. Parasitol. *82*, 557-564.

Van Dam, G. J., Claas, F. H. J., Yazdanbakhsh, M., Kruize, Y. C. M., vanKeulen, A. C. I., Ferreira, S., Rotmans, J. P., and Deelder, A. M. (1996b). *Schistosoma mansoni* excretory circulating cathodic antigen shares Lewis-x epitopes with a human granulocyte surface antigen and evokes host antibodies mediating complement-dependent lysis of granulocytes. Blood *88*, 4246-4251.

Van Dam, G. J., Seino, J., Rotmans, J. P., Daha, M. R., and Deelder, A. M. (1993). *Schistosoma mansoni* circulating anodic antigen but not circulating cathodic antigen interacts with complement component C1q. Eur. J. Immunol. *23*, 2807-2812.

Van den Biggelaar, A. H. J., van Ree, R., Rodrigues, L. C., Lell, B., Deelder, A. M., Kremsner, P. G., and Yazdanbakhsh, M. (2000). Decreased atopy in children infected with *Schistosoma haematobium*: a role for parasite-induced interleukin-10. Lancet *356*, 1723-1727.

Van der Kleij, D., Latz, E., Brouwers, J., Kruize, Y. C. M., Schmitz, M., Kurt-Jones, E. A., Espevik, T., de Jong, E. C., Kapsenberg, M. L., Golenbock, D. T., Tielens, A.G., and Yazdanbakhsh, M. (2002). A novel host-parasite lipid cross-talk: schistosomal lyso-phosphatidylserine activates Toll-like receptor 2 and affects immune polarization. J. Biol. Chem. *277*, 48122-48129.

Van der Kleij, D., van den Biggelaar, A. H. J., Kruize, Y. C. M., Retra, K., Fillie, Y., Schmitz, M., Kremsner, P. G., Tielens, A. G. M., and Yazdanbakhsh, M. (2004). Responses to toll-like receptor ligands in children living in areas where schistosome infections are endemic. J. Inf. Dis. *189*, 1044-1051.

Van der Kleij, D., and Yazdanbakhsh, M. (2003). Control of inflammatory diseases by pathogens: lipids and the immune system. Eur. J. Immunol. *33*, 2953-2963.

Van Die, I., Engering, A., and Van Kooyk, Y. (2004). C-type lectins in innate immunity to pathogens. Trends in Glycoscience and Glycotechnology *16*, 265-279.

Vanamail, P., Ramaiah, K. D., Pani, S. P., Das, P. K., Grenfell, B. T., and Bundy, D. A. P. (1996). Estimation of the fecund life span of *Wuchereria bancrofti* in an endemic area. Trans. Roy. Soc. Trop. Med.Hyg. *90*, 119-121.

Vanamail, P., Subramanian, S., Das, P. K., Pani, S. P., Rajagopalan, P. K., Bundy, D. A. P., and Grenfell, B. T. (1989). Estimation of age-specific rates of acquisition and loss of *Wuchereria bancrofti* infection. Trans. Roy. Soc. Trop. Med.Hyg. *83*, 689-693.

Vella, A. T., and Pearce, E. J. (1992). CD4+ Th2 response induced by *Schistosoma mansoni* eggs develops rapidly, through an early, transient, Th0-like stage. J. Immunol. *148*, 2283-2290.

Velupillai, P., and Harn, D.A. (1994). Oligosaccharide-specific induction of interleukin 10 production by B220+ cells from schistosome-infected mice: a mechanism for regulation of CD4+ T-cell subsets. Proc. Nat. Acad. Sci. U.S.A. *91*, 18-22.

Velupillai, P., dos Reis, E. A., dos Reis, M. G., and Harn, D. A. (2000). Lewis(x)-containing oligosaccharide attenuates schistosome egg antigen-induced immune depression in human schistosomiasis. Human Immunol. *61*, 225-232.

Verwaerde, C., Damonneville, M., Capron, A., and Auriault, C. (1986). Role of serine proteases of *Schistosoma mansoni* in the regulation of IgE synthesis. Immunobiology *173*, 254-255.

Viney, M. E., Brown, M., Omoding, N. E., Bailey, J. W., Gardner, M. P., Roberts, E., Morgan, D., Elliott, A. M., and Whitworth, J. A. G. (2004). Why does HIV infection not lead to disseminated strongyloidiasis? J. Inf. Dis. *190*, 2175-2180.

Volkmann, L., Bain, O., Saeftel, M., Specht, S., Fischer, K., Brombacher, F., Matthaei, K. I., and Hoerauf, A. (2003). Murine filariasis: interleukin 4 and interleukin 5 lead to containment of different worm developmental stages. Medical Microbiol. Immunol. *192*, 23-31.

Von Lichtenberg, F. (1964). Studies on granuloma formation: III. Antigen sequestration and destruction in the schistosome pseudotubercle. Am. J. Pathol. *45*, 75-94.

Vuitton, D. A. (2003). The ambiguous role of immunity in echinococcosis: protection of the host or of the parasite? Acta Trop. *85*, 119-132.

Wahid, F. N., and Behnke, J. M. (1996). Genetic control of acquired resistance to *Heligmosomoides polygyrus*: overcoming genetically determined weak responder status by strategic immunization with ivermectin-abbreviated infections. J. Helm. *70*, 159-168.

Wakelin, D. (1975). Genetic control of immune responses to parasites: selection for responsiveness and non-responsiveness to *Trichuris muris* in random-bred mice. Parasitology *71*, 377-384.

Wakelin, D., and Lloyd, M. (1976). Immunity to primary and challenge infections of *Trichinella spiralis* in mice: a re-examination of conventional parameters. Parasitology *72*, 173-182.

Wakelin, D., Farias, S. E., and Bradley, J. E. (2002). Variation and immunity to intestinal worms. Parasitology *125*, S39-S50.

Wallerstein, R. S. (1949). Longevity of *Schistosoma mansoni*, observations based on a case. Am. J. Trop. Med. *29*, 717-722.

Ward, D. J., Nutman, T. B., Zeaflores, G., Portocarrero, C., Lujan, A., and Ottesen, E. A. (1988). Onchocerciasis and immunity in humans: enhanced T-cell responsiveness to parasite antigen in putatively immune individuals. J. Inf. Dis. *157*, 536-543.

Warren, K. (1975). Hepatosplenic schistosomiasis mansoni: an immunologic disease. Bull. N. Y. Acad. Med. *51*, 545-550.

Warren, K. S. (1964). Correlation between experimental and human infection with *Schistosoma mansoni*. Nature *201*, 899-901.

Warren, K. S. (1966). The pathogenesis of "clay-pipe stem cirrhosis" in mice with chronic schistosomiasis mansoni, with a note on the longevity of the schistosomes. Am. J. Pathol. *49*, 477-489.

Warren, K. S., and Domingo, E.O. (1970a). *Schistosoma mansoni*: stage specificity of granuloma formation around eggs after exposure to irradiated cercariae, unisexual infections, or dead worms. Exp. Parasitol. *27*, 60-66.

Warren, K. S., and Domingo, E.O. (1970b). Granuloma formation around *Schistosoma mansoni*, *S. haematobium*, and *S. japonicum* eggs. Size and rate of development, cellular composition, cross-sensitivity, and rate of egg destruction. Am. J. Trop. Med. Hyg. *19*, 292-304.

Warren, K. S., and Berry, E.G. (1972). Induction of hepatosplenic disease by single pairs of the Philippine, Formosan, Japanese, and Chinese strains of *Schistosoma japonicum*. J. Inf. Dis. *126*, 482-491.

Warren, K. S. (1973). Regulation of the prevalence and intensity of schistosomiasis in man: immunology or ecology. J. Inf. Dis. *127*, 595-609.

Wassom, D. L., Dougherty, D. A., and Dick, T. A. (1988). *Trichinella spiralis* infections of inbred mice: immunologically specific responses induced by different Trichinella isolates. J. Parasitol. *74*, 283-287.

Webster, M., Fulford, A. J. C., Braun, G., Ouma, J. H., Kariuki, H. C., Havercroft, J. C., Gachuhi, K., Sturrock, R. F., Butterworth, A. E., and Dunne, D. W. (1996). Human immunoglobulin E responses to a recombinant 22.6- kilodalton antigen from *Schistosoma mansoni* adult worms are associated with low intensities of reinfection after treatment. Inf. Imm. *64*, 4042-4046.

Weil, G. J., Hussain, R., Kumaraswami, V., Tripathy, S. P., Phillips, K. S., and Ottesen, E. A. (1983). Prenatal allergic sensitization to helminth antigens in offspring of parasite-infected mothers. J. Clin. Invest. *71*, 1124-1129.

Weinmann, C. J., and Hunter, G.W.3rd (1960). Studies on schistosomiasis. 14. Effects of cortisone upon the *Schistosoma mansoni* burden in mice. Exp. Parasitol. *9*, 239-242.

Wilkins, H. A., Blumenthal, U. J., Hagan, P., Hayes, R.J., and Tulloch, S. (1987). Resistance to reinfection after treatment of urinary schistosomiasis. Trans. Roy. Soc. Trop. Med. Hyg. *81*, 29-35.

Wills-Karp, M., Santeliz, J., and Karp, C. L. (2001). The germless theory of allergic disease: revisiting the hygiene hypothesis. Nature Rev. Immunol. *1*, 69-75.

Wilson, E. H., Katz, E., Goodridge, H.S., Harnett, M.M., and Harnett, W. (2003). *In vivo* activation of murine peritoneal B1 cells by the filarial nematode phosphorylcholine-containing glycoprotein ES-62. *25*, 463-466.

Wilson, M. S., and Maizels, R. M. (2004). Regulation of allergy and autoimmunity in helminth infection. Clinical Reviews in Allergy and Immunology *26*, 35-50.

Wilson, R. A., and Barnes, P.E. (1977). The formation and turnover of the membranocalyx on the tegument of *Schistosoma mansoni*. Parasitology *74*, 61-71.

Wilson, R. A., Coulson, P. S., and McHugh, S. M. (1983). A significant part of the concomitant immunity of mice to *Schistosoma mansoni* Is the consequence of a leaky hepatic portal system, not immune killing. Par. Immunol. *5*, 595-601.

Wolday, D., Mayaan, S., Mariam, Z. G., Berhe, N., Seboxa, T., Britton, S., Galai, N., Landay, A., and Bentwich, Z. (2002). Treatment of intestinal worms is associated with decreased HIV plasma viral load. J. AIDS *31*, 56-62.

Wolowczuk, I., Nutten, S., Roye, O., Delacre, M., Capron, M., Murray, R. M., Trottein, F., and Auriault, C. (1999). Infection of mice lacking interleukin-7 (IL-7) reveals an unexpected role for IL-7 in the development of the parasite *Schistosoma mansoni*. Inf. Imm. *67*, 4183-4190.

Woolhouse, M. E. J., Taylor, P., Matanhire, D., and Chandiwana, S. K. (1991). Acquired-immunity and epidemiology of *Schistosoma haematobium*. Nature *351*, 757-759.

Woolhouse, M. E. J. (1998). Patterns in parasite epidemiology: the peak shift. Parasitol. Today *14*, 428-434.

Wu, Z., Boonmars, T., Nagano, I., Nakada, T., and Takahashi, Y. (2003). Molecular expression and characterization of a homologue of host cytokine macrophage migration inhibitory factor from Trichinella spp. J. Parasitol. *89*, 507-515.

Wuhrer, M., Dennis, R. D., Doenhoff, M. J., and Geyer, R. (2000). A fucose-containing epitope is shared by keyhole limpet haemocyanin and *Schistosoma mansoni* glycosphingolipids. Mol. Biochem. Parasitol. *110*, 237-246.

Wynn, T. A., Cheever, A. W., Williams, M. E., Hieny, S., Caspar, P., Kuhn, R., Muller, W., and Sher, A. (1998). IL-10 regulates liver pathology in acute murine *Schistosomiasis mansoni* but is not required for immune down-modulation of chronic disease. J. Immunol. *160*, 4473-4480.

Wynn, T. A., Thompson, R. W., Cheever, A. W., and Mentink-Kane, M. M. (2004). Immunopathogenesis of schistosomiasis. Imm. Rev. *201*, 156-167.

Yazdanbakhsh, M., Kremsner, P. G., and van Ree, R. (2002). Immunology: Allergy, parasites, and the hygiene hypothesis. Science *296*, 490-494.

Yazdanbakhsh, M., and Matricardi, P. M. (2004). Parasites and the hygiene hypothesis: regulating the immune system? Clin. Rev. All. Immunol. *26*, 15-23.

Yazdanbakhsh, M., Paxton, W. A., Kruize, Y. C. M., Sartono, E., Kurniawan, A., Wout, A. V., Selkirk, M. E., Partono, F., and Maizels, R. M. (1993). T-cell responsiveness correlates differentially with antibody isotype levels in clinical and asymptomatic filariasis. J. Inf. Dis. *167*, 925-931.

Yazdanbakhsh, M., Sartono, E., Kruize, Y. C. M., Kurniawan, A., Vanderpouwkraan, T., Vandermeide, P. H., Selkirk, M. E., Partono, F., Hintzen, R. Q., Vanlier, R. A. W., and Maizels, R. M. (1993). Elevated levels of T-cell activation antigen CD27 and increased interleukin-4 production in human lymphatic filariasis. Eur. J. Immunol. *23*, 3312-3317.

Yi, X. Y., Omerall, P., Kelly, C., Simpson, A. J. G., and Smithers, S. R. (1986). IgM antibodies recognizing carbohydrate epitopes shared between schistosomula and miracidia of *Schistosoma mansoni* that block *in vitro* killing. J. Immunol. *137*, 3946-3954.

Young, A. R., and Meeusen, E. N. (2002). Galectins in parasite infection and allergic inflammation. Glycoconj. J. *19*, 601-606.

Zaccone, P., Fehervari, Z., Jones, F. M., Sidobre, S., Kronenberg, M., Dunne, D. W., and Cooke, A. (2003). *Schistosoma mansoni* antigens modulate the activity of the innate immune response and prevent onset of type 1 diabetes. Eur. J. Immunol. *33*, 1439-1449.

Zang, X., Yazdanbakhsh, M., Jiang, H., Kanost, M.R., and Maizels, R.M. (1999). A novel serpin expressed by blood-borne microfilariae of the parasitic nematode *Brugia malayi* inhibits human neutrophil serine proteinases. Blood *94*, 1418-1428.

Zang, X., Taylor, P., Wang, J.M., Meyer, D.J., Scott. A.L., Walkinshaw, M.D., and Maizels, R.M. (2002). Homologues of human macrophage migration inhibitory factor from a parasitic nematode. Gene cloning, protein activity, and crystal structure. J. Biol. Chem. *277*, 44261-44267.

Zang, X. X., and Maizels, R. M. (2001). Serine proteinase inhibitors from nematodes and the arms race between host and pathogen. Tr. Biochem. Sci. *26*, 191-197.

7

Viral Subversion of Humoral Immune Responses

Mansun Law, Pietro Paolo Sanna and Dennis R. Burton*

Abstract

Viruses are a group of hugely diverse organisms. Many viruses have developed strategies to subvert host immunity and enhance their ability to propagate in a specific host. This review focuses on the strategies viruses use to evade humoral immune responses and virus neutralizing antibodies (NAbs) in particular. Viral strategies to subvert NAb responses to the virus surface (SU) proteins include (i) mutation by antigenic drift and shift, and recombination; (ii) masking of conserved epitopes; (iii) glycan shielding; (iv) SU protein decoys; (iv) original antigenic sin; (v) irregular SU protein spacing; and (vi) multiple SU proteins and virus forms. Viruses also evade humoral responses by spreading directly between cells, up-regulating and hijacking host complement regulatory proteins, encoding antagonistic Fc and complement receptors, and inducing immunosuppression. Understanding these mechanisms will provide valuable information for the design of anti-viral therapies and vaccines.

1. Introduction

Humoral immunity is defined by the anti-microbial activities present in the body fluids and blood. These activities are mediated by components from both the innate and adaptive immune responses including complement, antibodies (Abs), cytokines and microbicidal inhibitors such as lipoproteins and peptides. Amongst these, Abs play a central anti-viral role. They can act directly on viruses to prevent their attachment, fusion and entry to cells, or to aggregate virus particles to reduce infectivity; alternatively Abs can act in concert with complement or immune cells such as macrophages and natural killer cells to destroy virus or infected cells by lysis or phagocytosis (Burton, 2002; Zinkernagel et al., 2001).

*For correspondence email mlaw@scripps.edu

Neutralizing Abs (NAbs) have been demonstrated to be crucial factors in protection against virus challenge by Ab passive transfer experiments and by the effects of maternal Ab transfer through lactation and they are the most reliable efficacy index for all available vaccines (Hilleman, 2004; Zinkernagel, 2003; Zinkernagel et al., 2001). The primary role of Ab is often associated with the neutralization of virus to control virus replication and prevent spreading early in infection. In contrast, cell-mediated immunity, including CD8+ cytotoxic T lymphocytes (CTLs), is thought to be more relevant in virus clearance, once infection is established, and recovery. However there is increasing evidence that B lymphocytes, the immune cells that make Abs, may also be important in some cases in virus clearance and recovery although the mechanisms are still unclear (Gerhard, 2001; Mozdzanowska et al., 2000; Topham and Doherty, 1998). In addition the increasing interest in natural Abs in hampering infections (Baumgarth, 2000; Ochsenbein and Zinkernagel, 2000) and the suggested requirement for Abs to sustain memory T cells (Bachmann et al., 2004) have added a new dimension to the possible roles of Abs and B cells in viral clearance and immune regulation.

NAbs are conventionally defined as Abs that can block virus infectivity in *in vitro* assays by targeting epitopes exposed on the surface (SU) proteins of viruses. Although NAbs are generally associated with protective activity *in vivo*, non-neutralizing Abs can fulfill this role in some cases. Examples include Abs that target the neuraminidase (NA) and matrix protein 2 (M2) of influenza virus (Gerhard, 2001) and the A33R protein of vaccinia virus (Galmiche et al., 1999).

Viruses are enormously diverse organisms with either RNA or DNA genomes. Control of different viruses may require very different strategies. RNA viruses generally have a high mutation rate due to the poor fidelity of the RNA-dependent RNA polymerase or reverse transcriptase whereas DNA viruses are more stable genetically (Anderson et al., 2004). Genetic mutations are the principal mechanism for virus evasion for RNA viruses (see below). Different viruses have different modes of infection (Hilleman, 2004). Some infect, replicate and spread swiftly to another new host before the host mounts a vigorous immune responses to clear the infection; these viruses can be considered "hit-and-run" viruses, *e.g.* variola virus (causes smallpox), measles virus, foot-and-mouth disease virus and Dengue virus. These viruses normally cause an acute infection and are cytolytic in nature. Some viruses infect the host and establish persistence, or chronic infection, and can be considered "hit-and-stay" viruses that include viruses from the families of herpesvirus, lymphocytic choriomeningitis virus (LCMV), adenoviruses, human immunodeficiency virus (HIV), hepatitis C virus (HCV). Persistence of viruses in the face of potential immune responses can be achieved by different methods: (1) Some viruses, *e.g.* LCMV, poliovirus, alphaviruses, Japanese encephalitis virus and tick-borne encephalitis virus, infect sites that are immune privileged such as the central nervous system (CNS), which is separated and protected by the blood-brain barrier from normal immune surveillance by the lymphatic system. In addition, neurons do not express class I major histocompatibility complex (MHC) and therefore viruses infecting the CNS will not be subjected to CTL responses (Whitton and Oldstone, 2001). (2) Some viruses, *e.g.* herpesviruses, establish persistence by latent infection where the viruses shut down the expression of most viral genes and refrain from replication in order to avoid the immune system. Often, they also encode viral immunomodulators to suppress inflammatory responses and to interfere with class I MHC presentation to evade CTLs and NK cells (Tortorella et al., 2000). (3) Some viruses, such as HIV and HCV, replicate actively and persist in the presence of potent immune responses. These viruses have an extraordinarily high replication rate, in addition to a high mutation rate, and maintain a population of many closely related viral variants known as quasispecies within the host. The fast kinetic of mutations and virus replication have ensured that escape mutants arise a step ahead of immune surveillance (Hahn, 2003; Johnson and Desrosiers, 2002; Richman et al., 2003; Wei et al., 2003). Persistent viruses are probably the biggest challenge in infectious diseases in terms of public health and vaccine development.

This chapter deals with the strategies used by viruses for subverting the host humoral responses. The focus of the first part of the chapter is how the intrinsic properties of viral surface (SU) proteins allow viruses to avoid Ab responses. The second part of the chapter discusses other mechanisms that viruses use to escape humoral immunity.

2. Evasion via properties of SU proteins

The virus SU proteins, *e.g.* hemagglutinin (HA) of influenza virus, gp120 and gp41 of HIV and G protein of vesicular stomatitis virus (VSV), are responsible for the attachment and/or penetration of cells. These proteins are the most important targets for humoral immunity because they are highly exposed on the virus surface and, for many enveloped viruses, also on the infected cell surface. Several viruses are known to actively down regulate the expression of SU proteins on the infected cell surface, a mechanism probably important for the conservation of SU proteins for virus assembly but also of importance in reducing the chance of immune recognition by Abs. Latent herpesviruses actively silence the expression of the majority of the genome including the SU glycoproteins (Tortorella et al., 2000). The transmembrane SU protein gp41 of HIV contains multiple motifs in the cytoplasmic tail for the endocytosis of the spikes of HIV (Ohno et al., 1997; Wyss et al., 2001) whereas the neutralizing determinant of the extracellular enveloped form of vaccinia virus, B5R glycoprotein, contains retrieval sequences for recycling from the plasma membrane by clathrin-coated vesicles (Ward and Moss, 2000). Nevertheless for proper function in virus attachment and entry, the SU proteins should be exposed on virions and potentially available for NAb recognition. However, many viral SU proteins have evolved special features for the evasion of NAbs, either during a persistent infection or for re-infecting a host with existing immunity against the virus. The general principles are summarized below.

2.1. Mutation of SU proteins

Genetic mutation is probably the most powerful mechanism of viral subversion of immune responses. It facilitates the survival of virus within the host during persistent infection (*e.g.* HIV and HCV) and allows the virus to re-infect a population even in the presence of herd immunity (*e.g.* influenza virus). Recent understanding of cell-mediated immunity suggests that immune escape mutations can also occur in non-SU proteins and much effort has concentrated on defining the major CTL epitopes in different viruses.

Genetic mutations are particularly prominent for RNA- and retroviruses due to the poor fidelity of the viral RNA-dependent RNA polymerases or reverse transcriptases (RTs) in virus replication (reviewed by Anderson et al., 2004). These enzymes introduce errors including base-mismatches, deletions and insertions into the virus genome but are unable to remove the errors because of the lack of 3'- to 5'-exonucleolytic proofreading activity in the enzyme. The errors introduced are retained in the replication intermediates (RNA-RNA duplex for RNA viruses and RNA-DNA duplex for retroviruses) and will not be repaired as in other DNA organisms. RNA polymerases and RTs can also jump between templates and result in "recombination". The cumulative change of the antigenicity of viral proteins because of the replication error and immune selection is known as "antigenic drift". In addition, some RNA viruses have a segmented genome and co-infection of a host with multiple viruses of some species can potentially result in the reassortment of segments from different viruses and the creation of new virus hybrids. This mechanism is known as "antigenic shift".

2.1.1. Genetic recombination

In the replication of RNA- and retroviruses, errors are randomly introduced into the genome. The high error rate in replication will produce a huge population of virus variants, known as quasispecies, with a closely related genome. Within the quasispecies population, many will be non-viable and defective due to lethal mutations. Genetic recombination can ameliorate lethal mutations by swapping viable fragments between defective genomes (Mikkelsen and Pedersen, 2000). The nature of genetic recombination in these viruses is rather different from that of other organisms which have a special mechanism for the cross-over of genetic materials (Heyer et al., 2003). Genetic recombination of RNA- and retroviruses occur when RNA polymerase or RTs jump from the replicating genome to an adjacent template and continue duplicating the genetic materials (Arnold and Cameron, 1999; Kirkegaard and Baltimore, 1986; Mikkelsen and Pedersen, 2000; Robertson et al., 1995). In the case of multiple viruses infecting the same cell, exchange of genetic materials by recombination will allow the formation of new virus hybrids with mosaic genomes from parental genomes. In this case genetic recombination can significantly change the virulence of a virus and its sensitivity to NAbs.

2.1.2. Antigenic shift

The terms antigenic drift and shift are derived from the study of the escape of influenza virus from host immunity. Influenza virus is an RNA virus with a segmented genome and influenza A and B viruses contain 8 RNA segments whereas influenza C virus has 7 (Wright and Webster, 2001). Most knowledge of influenza virus, particularly of its epidemiology and mechanisms of immune escape, come from the study of influenza A virus which is responsible for the re-occurrence of pandemic flu. Three virus SU proteins, the virus HA, NA and M2 proteins, are implicated as targets for protective Abs although Abs to HA are the major determinant in virus neutralization and protection (Gerhard, 2001). Abs to NA and M2 predispose the host to less severe infection and illness but do not block infection (Colman, 1994; Neirynck et al., 1999).

Antigenic shift is a mechanism for immune evasion that is only possible for virus with a segmented genome, such as influenza virus. When a cell is infected by multiple influenza viruses, gene segments from the different viruscs can reassort into a virion and result in a new virus containing a new combination of genes from the contributing viruses. The capacity for genetic reassortment requires the independent segregation of the influenza virus gene segments into the virions during replication (Wright and Webster, 2001). Reassortment can also occur in viruses having diploid genomes such as HIV. Instead of immune evasion of whole virus, this mechanism is probably more important in the conservation of individual gene segments. Genetic reassortment can greatly affect the property of an influenza virus such as host range and pathogenicity. For instance, the replacement of the nucleocapsid protein (NP) from an avian influenza virus by a human strain produced a new virus that failed to replicate in chicken cells but replicated in standard mammalian cells (MDCK) for influenza culture (Scholtissek et al., 1978).

Antigenic shift is an extremely dangerous event because this can potentially convert an avirulent influenza virus into a highly pathogenic killer (Steinhauer, 1999). The devastating effects have been demonstrated repeatedly in several flu pandemics in history that have claimed untold millions of lives (Wright and Webster, 2001). Fifteen HA and nine NA subtypes have been identified and >30% difference by polypeptide sequence are found between the subtypes (Rohm et al., 1996). Co-infection of animals in the wild can potentially introduce exotic HA or NA, *e.g.* from avian species, into an influenza virus backbone that can infect humans and results in a virus that possesses an HA or NA not previously encountered by humans. With no NAbs or immunity to the new HA or NA, the new virus can spread relatively unchecked through the naïve human population. The outbreak of the H5N1 bird flu in Hong Kong in late 1997 is highly alarming because the new virus is highly pathogenic and killed >30% of the infected patients (Sims et al., 2003). Antigenic shift had brought the HA subtype 5, an avian HA that was found in the virus causing outbreaks in poultry farms earlier in the year, into the human population (Shortridge et al., 1998; Xu et al., 1999). Fortunately this virus was unable to transmit efficiently between infected humans and the epidemic was brought under control by a massive culling of the local poultry. However, recent antigenic shift between 2001 to 2002 in aquatic birds in south China may have given rise to an even more pathogenic H5N1 influenza A virus (genotype Z) (Govorkova et al., 2005; Li et al., 2004; Tran et al., 2004), and by the time of this writing, 69 cases have been reported in Vietnam and Thailand since January 2004, of which 46 were fatal (http://www.who.int/csr/don/2005_03_11/en/). This virus is also not highly transmissible between humans (Liem and Lim, 2005), however, the perpetual gene reassortment of influenza virus in wild hosts may one day create a highly contagious killer.

2.1.3. Antigenic Drift

Antigenic drift is the change of antigenicity of viral proteins due to point mutations. RNA- and retroviruses have a high error rate during replication and therefore drift much more rapidly than DNA viruses. The error rate, or frequency of mutations per nucleotide per cycle of genome replication, of these viruses ranges from 10^{-3} to 10^{-5} and is 10- to 10^5-fold higher than other DNA-based organisms (Anderson et al.,

2004). The infidelity in the duplication of genomes results in the virus existing as quasispecies during an infection and the popularity of certain quasispecies would be determined by their fitness under the selective pressure of the immune system.

HIV and HCV have an extraordinarily high replication rate. It is estimated that HCV produces 10^{12} virions per day with a half-life about 2-3 hours (Neumann et al., 1998; Ramratnam et al., 1999). HIV also produces about 10^{10} to 10^{11} virions per day with a half-life of about 5.8 hours (Ho et al., 1995; Perelson et al., 1996; Wei et al., 1995; Xu et al., 1999). This results in huge population of quasispecies that can harbor mutations at any location of the population genome. The mutation rate of the HIV genome is estimated to be of the order of 10^{-5} per base per replication (Mansky and Temin, 1995; O'Neil et al., 2002). Therefore, for a genome of 10^4 base pairs long, $>10^8$ virus variants should be produced every day and the immune system is challenged constantly with a huge array of viral variants (Coffin, 1995; McMichael and Phillips, 1997).

Although errors are introduced into the genome randomly, hot spots of mutation, the hypervariable regions, are found in all virus genomes due to selective pressure of the host immune responses. In many cases, the hypervariable regions in the SU proteins are associated with NAb pressure.

The hypervariable region 1 (HVR1) of HCV envelope glycoprotein E2 appears to be the major determinant for NAb recognition (Farci et al., 1996). Changes in this region are associated with viral persistence and escape from NAbs during the course of HCV infection (van Doorn et al., 1995). In HIV-1 gp120, five variable regions are identified because of their heterogeneity among HIV isolates (Starcich et al., 1986). The first 4 variable regions are called variable (V) loops because each of them is flanked by a pair of conserved cysteine residues which form a disulfide bond at the base of the linear sequences (Leonard et al., 1990). The V loops of gp120 are likely to be exposed on the surface of gp120 (Kwong et al., 1998) although the level of exposure of different loops is debated (Hartley et al., 2005). Mutations in the V2 and V3 have been shown to allow escape from NAbs (Masuda et al., 1990; Schonning et al., 1996; Yoshiyama et al., 1994).

In influenza virus, hypermutated residues are also found on the surface of HA and NA (reviewed by Bizebard et al., 2001). The influenza virus HA is a homo-trimer made of subunits consisting of a membrane-proximal helix rich stem structure and a membrane-distal receptor-binding globular domain (Wiley and Skehel, 1987; Wilson and Cox, 1990; Wilson et al., 1981). The globular domain is composed of mainly beta-structure with an eight-stranded Swiss-roll type structure. The receptor (sialic acid) binding site contains residues highly conserved in both influenza A and B viruses and is located in a shallow, concave pocket at the distal end of the globular head. The conserved receptor binding site is surrounded by highly variable protruding loops and escape from neutralizing monoclonal Abs (mAbs) in tissue culture is generally associated with amino acid changes clustered around the conserved receptor binding site (Bizebard et al., 1995; Caton et al., 1982; Wiley et al., 1981; Wilson and Cox, 1990). Study of the substitutions of amino acids in HA between epidemic strains reveals that substitutions of surface but not buried residues are retained in subsequent epidemics, suggesting that these surface substitutions are antigenically significant (Bizebard et al., 2001).

In summary, antigenic drift caused by the selective pressure of NAbs is likely associated with exposed residues found in the hypervariable regions of the SU proteins.

2.2. Masking of Conserved Epitopes of SU Proteins

Conserved regions of the SU proteins are likely to be functionally important, *e.g.* structure and folding of the protein and receptor binding. Mutations in these regions will likely reduce the fitness of the virus. Receptor sites in particular are, by definition, conserved and generally exposed and therefore potential targets for NAbs. Many viruses have evolved mechanisms to reduce NAb access to receptor sites.

The canyon hypothesis was originally proposed using human rhinovirus 14 (HRV14) as a model to explain how viruses evade NAbs by hiding the receptor binding sites in deep depressions of the virus surface known as "canyons" (Rossmann, 1989). In HRV14, a canyon of 20 Å deep and 20 Å wide encircles each of the icosahedral fivefold axes of the capsid is identified. While the conserved residues important for function are located at the canyon floor, mutations for NAb escape are mapped to the rim of canyon (Rossmann, 1989; Rossmann et al., 1985). However the general validity of the "canyon" hypothesis to explain evasion from NAbs is disputed because many viruses, such as foot-and-mouth disease virus (FMDV) and alphaviruses, have their conserved receptor binding residues highly exposed on their surface (reviewed by Smith, 2001). Indeed even in the HRV14 model, NAbs that can penetrate the canyon and interact with the conserved residues have also been identified (Smith et al., 1996).

The "Ab footprint" hypothesis was proposed to explain NAb escape of viruses having a small receptor binding site, such as the HA and NA of influenza virus that bind to sialic acids on glycan moieties on the cell surface (Bizebard et al., 1995; Colman et al., 1983; Weis et al., 1988). The hypothesis suggests that, when the epitopes of NAbs are larger than the viral functional sites, the interface between Ab and antigen (Ag), the "footprint" will unavoidably include residues that are not involved in function even when the Abs target the functional sites. Mutations at these residues would allow the virus to escape Ab neutralization without compromising viability. This proposal is supported by the fact that influenza virus can escape NAbs by changing only a few residues (3 or 4) in the Ab/Ag interface (13 to 20 contacting residues) (Bizebard et al., 1995; Nuss et al., 1993). The interface between Ab and Ag is typically 400 to 900Å^2 and involves multiple contact residues (Colman et al., 1983; Janin and Chothia, 1990; Stanfield et al., 1990). It is suggested that the disruption of only a few interactions can reduce the affinity of an Ab by 3 orders of magnitude (Bizebard et al., 2001; Colman et al., 1983). Fleury et al. showed that virus escape in tissue culture is associated with a significant reduction of the affinity of the selecting Ab (Fleury et al., 1998). In addition, the substitution of a bulkier or electro-incompatible residue at or close to the sites of contact can severely inhibit Ab/Ag interactions due to steric hindrance or electro-repulsion (Lawrence and Colman, 1993; McCoy et al., 1997).

In other viruses that use protein receptors, the "Ab footprint" theory may not be appropriate to explain escape because the Ab/Ag interface can be as large as that of viral attachment proteins and receptors, *e.g.* the interface between HIV gp120 and CD4 is about 800Å^2 (Kwong et al., 1998). Kwong et al. suggested that the complementarity of gp120 and CD4 is not perfect and many gp120 residues contact CD4 via main-chain atoms. Therefore gp120 probably can tolerate mutations that allow escape from NAbs targeting the CD4 binding site (CD4bs) but without affecting CD4 binding. Indeed, the HIV envelope spike is a trimer formed by an equal number of gp41 and gp120 molecules and possesses many special features for the evasion of NAbs (Johnson and Desrosiers, 2002; Kwong et al., 1998; Parren et al., 1999; Wyatt et al., 1998; Zwick et al., 2004). The structure of gp120 is defined as an inner domain, outer domain and bridging sheet (Kwong et al., 1998; Wyatt et al., 1998). The CD4bs is located at the interface of these 3 domains whereas the chemokine receptor binding site is located at an approximately 90° angle to the CD4bs (Rizzuto and Sodroski, 2000; Rizzuto et al., 1998). In terms of immunogenicity, gp120 can be divided into 3 faces: neutralizing, non-neutralizing, and silent (Wyatt et al., 1998). The neutralizing face is believed to contain the CD4bs and coreceptor binding site; the non-neutralizing face overlaps with the inner domain that interacts non-covalently with gp41 for the formation of the trimeric spike; and the silent face overlaps with the outer domain and is heavily glycosylated and appears to be "immunosilent" (see Glycosylation of SU protein below). Despite being found on the neutralizing face, the CD4bs is suggested to be partially masked by the variable loops (V1-3) thereby subverting NAb binding (Wyatt and Sodroski, 1998). The CD4bs also appears to be relatively recessed and more easily recognized by CD4 (one Ig domain in width) then the Fab arms of an Ab (2 Ig domains). Furthermore, recent structural data suggest that CD4bs is not properly formed until CD4 binds (Chen et al., 2005). The other functional receptor site on gp120 is the CCR5 binding site. This site is largely masked until after CD4 binding. After CD4 binding, scFv and Fab fragments of NAbs can, in fact, gain access to the site but whole IgG molecules cannot, presumably because of the proximity of the host cell membrane (Labrijn et al., 2003). This dramatically illustrates efficient evasion of the humoral response by HIV.

2.3. Glycosylation of SU proteins

Glycosylation is important for the proper folding and function of many viral SU proteins (Lee et al., 1992; Li et al., 1993; Lu et al., 1995; Meunier et al., 1999; Pal et al., 1989; Wright et al., 1989). In some viruses, such as HIV, SIV, HCV and Ebola virus, glycosylation is extensive, accounting for 40 to 50% of the molecular weight (reviewed by Pikora, 2004; Zhang et al., 2004). Apparently many of the glycans are redundant and non-essential for function. Indeed some of these glycans are dispensable for SU protein functions but have an important role in immune evasion (Cheng-Mayer et al., 1999; Ohgimoto et al., 1998; Schonning et al., 1996; Skehel et al., 1984; Wei et al., 2003; Wyatt et al., 1998).

Errors in replication can lead to substitutions of amino acid residues which in turn can affect the post-translational modification, such as glycosylation, of the proteins. The addition or removal of glycans on SU proteins can have a great effect on NAb activity. *N*-linked glycans are about 2 kDa (10- to 25-fold larger than an amino acid residue) and can extend 15-20 Å away from the protein surface for a complex carbohydrate (Wei et al., 2003; Zhang et al., 2004). We have discussed above that the normal Ab/Ag interface is about 400-900 Å^2 and the addition of a glycan moiety within or close to the Ab/Ag binding sites can disrupt the interactions. Reports of escape mutants involving glycosylation changes are found in influenza virus (Skehel et al., 1984; Skehel and Wiley, 2000), respiratory syncytial virus (Zlateva et al., 2004), HIV (Back et al., 1994; Cheng-Mayer et al., 1999; Losman et al., 2001; Ly and Stamatatos, 2000; Schonning et al., 1996; Wei et al., 2003), SIV (Chackerian et al., 1997; Ohgimoto et al., 1998; Ye et al., 2000). In HCV the removal of glycans can enhance Ab responses (Fournillier et al., 2001), suggesting that glycans on HCV may play a role in NAb escape (Zhang et al., 2004).

A highly exposed face of HIV-1 gp120, the outer domain, is considered "immunosilent" because of the densely packed glycans that are attached by the host cell glycosylation machinery and are not expected to elicit Abs. The "evolving glycan shield" was proposed to describe the mechanism of NAb subversion by the constant evolution of the glycan packing on HIV gp120 as a result of natural mutations and immune selection to prevent NAb but not receptor binding (Wei et al., 2003). Despite the lack of general immunogenicity of the silent face of gp120, an interesting NAb, 2G12, can neutralize many primary isolates of HIV-1 by recognizing a pattern of oligomannoses on the silent face of gp120 (Calarese et al., 2003; Sanders et al., 2002; Scanlan et al., 2002; Trkola et al., 1996). The NAb 2G12 is unusual because it targets carbohydrate moieties on a virus which is uncommon, although protective Abs to bacterial capsules are well characterized (Paradiso and Lindberg, 1996). The structure of 2G12 is unusual because the 2 VH domains are interlocked to form VH domain-swapped dimers that create a new binding site for the high-affinity recognition of a carbohydrate cluster (Calarese et al., 2003). Abs with 2G12-like properties may provide a solution for viruses with SU proteins protected by densely packed glycans, although it is not certain how frequently the immune system can produce this type of Abs. In addition, escape mutants of 2G12 can be found in both nature (Parren et al., 1998; Richman et al., 2003) and in experimental conditions (Poignard et al., 1999), and the major circulating clade C HIV viruses do not bear 2G12 epitopes and are largely resistant to 2G12 (Binley et al., 2004; Bures et al., 2002).

2.4. Decoys of SU proteins

The term "decoy" has been used in different ways to describe different mechanisms viruses may use to evade NAbs. It has been used to describe SU proteins released or secreted from infected cells that may function to "mop up" some of the NAbs to reduce virus neutralization. In Ebola virus, the SU glycoprotein (GP) is a homotrimer and is produced as transmembrane and soluble forms by transcriptional editing (Sanchez et al., 1996). Cells infected by Ebola virus secrete large amounts of soluble GPs that are maintained as homodimers by disulfide bonds (Sanchez et al., 1998). The "decoy" activity of soluble GPs is demonstrated by its ability to inhibit NAbs against pseudotype virus displaying GP (Ito et al., 2001). Poxvirus also produces a secreted form of the SU protein B5R (Martinez-Pomares et al., 1993) and the spikes of HIV-1 is shed from the virus and infected cells as soluble monomeric gp120 (reviewed by Poignard et al., 2001). Whether these soluble proteins can act as decoys is not known.

"Decoy" effect has also been used to explain the poor induction of NAbs in some viruses. In porcine reproductive and respiratory syndrome virus (PRRSV), it has been suggested that the presence of an immunodominant but non-neutralizing epitope on the SU protein GP5 might act as a "decoy" to delay the immune system from eliciting NAbs against the neutralizing epitope on the same protein (Ostrowski et al., 2002).

2.5. Original antigenic sin

Original antigenic sin (OAS) was first proposed by Francis, Davenport and Hennessy in 1953 to describe a phenomenon that the Ab response in an individual after a second viral infection reacts more strongly to the viral variant originally infected the individual (Davenport et al., 1953; Haaheim, 2003). OAS has been observed in both humans (Fazekas de St and Webster, 1966a; Haaheim, 2003; Webster et al., 1976) and animals (Fazekas de St and Webster, 1966b; Webster, 1966), and also in other infection models such as Dengue (Halstead et al., 1983), Japanese encephalitis, West Nile (Inouye et al., 1984), and enteroviruses (Tsuchiya et al., 2000).

OAS presents a challenging issue in the design of vaccines for viruses exist as diverse variants in the population (Anderson et al., 2001; Muller, 2004). For instance, immunization with an influenza vaccine with a particular set of SU proteins (HA and NA) and subsequent infection with a heterologous strain will boost mainly Ab responses against the vaccine strain. It is suggested that the boosting of memory B cells against the first virus may interfere with the activation of naïve B cells to the second virus by an undefined mechanism, resulting in a condition that specific NAb responses against the second virus will not be as great comparing to a naïve host that has never encountered the first virus. Viruses may take advantage of OAS to avoid a strong NAb response against the SU proteins when infecting an immune host.

2.6. Spacing and antigenicity of SU proteins

Ags with a highly organized, repetitive and densely packed structural pattern are better immunogens because they can cluster B cell receptors (BCRs) and stimulate B cells strongly. Through the studies of haptens, bacteria and virus Ags, it is shown that Ags with a minimal of 10 to 30 repeating determinants spacing 8 to 10 nm at a high enough concentration can stimulate B cells directly without help from T cells by cross-linking BCRs and activating Ab production (reviewed by Zinkernagel et al., 2001). This type of rigid paracrystalline structure is expected in many non-enveloped viruses such as poliovirus and polyomavirus (Fehr et al., 1998; Hogle et al., 1985; Liddington et al., 1991; Szomolanyi-Tsuda et al., 1998; Szomolanyi-Tsuda and Welsh, 1996) but can also be found in enveloped viruses such as VSV (Bachmann et al., 1993; Thomas et al., 1985). For the VSV-mouse infection model, part of the Ab repertoire against VSV is encoded in the murine germline therefore infection by VSV elicits a rapid protective Ab response in mice within 2-4 days post-infection (Zinkernagel, 1997). The protective NAbs are mainly IgM and are T-cell independent (TI) (Bachmann and Zinkernagel, 1996; Fehr et al., 1998; Leist et al., 1987; Pinschewer et al., 2004). The rapidly generated IgM can be functional in neutralizing viruses and protective *in vivo* (Bachmann and Zinkernagel, 1996; Szomolanyi-Tsuda and Welsh, 1996) and the protection is diminished in the absence of complement (Ochsenbein et al., 1999). Similar TI-Ab responses can be artificially elicited against LCMV, a virus that normally only induces late protective Abs, in mice transgenic with the IgH variable region of a NAb against LCMV (Hangartner et al., 2003). TI-Ab responses were also noted in polyomavirus (Szomolanyi-Tsuda et al., 1998). Injection of soluble VP1, virus-like particles (VLPs) and live virus infection all resulted in IgM production although 10-fold higher titres were found for VLPs and infection.

To escape NAbs, viruses have developed strategies to reduce Ab responses, such as heavy glycosylation of the SU protein (see above). Viruses that do not display their SU proteins or neutralizing epitopes in a dense regular array (such as HIV) will reducet clustering of BCRs and renders the SU proteins less immunogenic. Indeed NAb responses are not produced in many enveloped viruses (*e.g.* HIV, HCV) until weeks or months after infection (Ariyoshi et al., 1992; Chien et al., 1992; Homsy et al., 1990; Moore et al., 1994; Planz et al., 1996; 2003). The late induction of NAbs certainly provides a window for these viruses to establish infection and produce many quasispecies for evasion from NAbs arising later in infection.

2.7. Multiple virus attachment proteins and virus forms

Viruses, particularly RNA viruses, have a very limited genome size (< 10 kb, *e.g.* picornavirus, HIV, flavivirus etc) therefore usually encode a single protein for cell attachment in order to conserve genetic materials. In some virus families such as *Herpesviridae* and *Poxviridae* which have a large genome (>100 kb), this constraint becomes less significant. Viruses in these families can afford more genetic resource to machineries to improve the fitness of the viruses. Virus entry is one of the functions which these viruses dedicate multiple genes, *e.g.* to broaden tropism and enhance the kinetics of attachment for successful infection. Another reason for these viruses to encode multiple attachment proteins may be due to the fact that these viruses are less able to mutate to produce escape mutants because of the nature of their DNA genomes. Therefore multiple virus attachment proteins may be important for their survival if one of the attachment proteins is blocked.

In herpes simpex virus (HSV), 11 proteins are identified to be exposed on the virus envelope and at least 10 of them are glycosylated (Roizman and Knipe, 2001). The glycoproteins gB, gC and gD are implicated in cell attachment while gB, gD and the heterodimer gH/gL are required for cell fusion (reviewed by Spear, 2004). As discussed below, the heterodimer gE/gI, act as a viral-encoded Fc receptor and facilitates cell-to-cell virus spread (Dingwell and Johnson, 1998) (see below). The generally accepted model for HSV entry is that HSV initially binds to the abundant cell surface glycosaminoglycans (GAGs) followed by interactions of gD with cellular receptors to trigger membrane fusion (Spear, 2004). Both gB and gC can bind to GAGs to enhance infectivity, although gC is dispensable for growth in tissue culture (Herold et al., 1991). The cellular receptors include herpes virus entry mediator (HVEM), herpesvirus Ig-like receptor (nectin-1 and -2) and 3-*O*-sulfated heparan sulfates. The glycoprotein gD contains the major neutralizing determinants of HSV (Balachandran et al., 1982; Cohen et al., 1984; 1972; Dix et al., 1981; Muggeridge et al., 1990).

Poxviruses also encode multiple proteins for entry (Moss, 2001). In vaccinia virus (VACV), two different forms of infectious particles, intracellular mature virus (IMV) and extracellular enveloped virus (EEV), are produced during an infection. IMV represents the majority of infectivity during an infection but is only released when the cells lyse. Early in infection a small fraction of IMV is wrapped by intracellular membranes and disseminated from the cells as EEV by a process homologous to exocytosis without causing cell lysis (Smith and Law, 2004; Smith et al., 2002). Therefore EEV is an IMV particle covered with an additional layer of host cellular membrane in which host and EEV-specific proteins are incorporated. Since IMV particles are robust and resistant to physical factors such as heat and dryness it is thought that IMV is used for transmission between hosts while EEV is important for spreading within an infected host. The cellular receptors for poxvirus are unknown but these two forms of virus appear to bind to different receptors and enter cells by different mechanisms (Vanderplasschen et al., 1998a; Vanderplasschen and Smith, 1997). The IMV membrane proteins A27L, D8L and H3L are implicated for binding to cells via GAGs (Hsiao et al., 1998; Hsiao et al., 1999; Lin et al., 2000) while proteins A27L, A28L and L1R are important for penetration (Ichihashi and Oie, 1996; Rodriguez et al., 1987; Senkevich et al., 2004). Potent NAbs targeting A27L, H3L and L1R have been identified (Czerny and Mahnel, 1990; Gordon et al., 1991; Ichihashi and Oie, 1996; Rodriguez et al., 1985; Wolffe et al., 1995; Zinoviev et al., 1994) but NAbs against IMV *per se* are insufficient for protection in animal models (Boulter and Appleyard, 1973) and unlikely to be effective in preventing smallpox (Kaplan, 1962; Kaplan et al., 1965) because the immunodominant IMV SU proteins are masked by the EEV membrane. More recent studies showed unequivocally that a better protection requires NAbs against the SU proteins of both forms of virus (Fogg et al., 2004; Hooper et al., 2003; Hooper et al., 2004; Law et al., 2005). In EEV, the SU glycoprotein B5R is both a neutralizing and protective determinant (Bell et al., 2004; Galmiche et al., 1999; Law and Smith, 2001) while Abs to A33R protein do not neutralize but do protect (Galmiche et al., 1999; Hooper et al., 2000).

2.8. Antigenic changes in SU proteins as a universal strategy for evading host Ab responses in other pathogens

Changes in viral SU proteins to evade Ab responses during a persistent infection are in principle very similar to that those found in parasites and are worth considering briefly here. The protozoans, *Plasmodium falciparum* and *Trypanosoma brucei*, the causative agents of malaria and sleeping sickness, respectively, have developed a wide range of strategies to evade host immunity and allow prolonged infection (reviewed by Zambrano-Villa et al., 2002). Chronic infection is important to provide sufficient time for these parasites to complete their life cycle and subsequently transmit to a new host by insect vectors. Inside the mammalian hosts, one of the strategies is frequent changes in exposed SU antigens (antigenic variation) to escape protective Abs. The exposed antigens can either be on the protozoans, or on infected red blood cells (I-RBC) for intracellular *Plasmodium*, and are encoded by large families of non-allelic genes.

T. brucei covers its entire cell surface with approximately 5 million dimers of a single molecule called variable surface glycoprotein (VSG). This molecule changes repeatedly to ensure a fraction of the trypanosome population is antigenically diverse enough to escape Ab-mediated killing during a chronic infection (reviewed by Pays et al., 2004). It is estimated that a large repertoire consisting of tens of hundreds of VSG genes are present in the trypososome genome, although many of them are pseudogenes (reviewed by McCulloch, 2004). The process of antigenic variation depends on transcriptional switching and DNA recombination of the VSG loci frequently found near the telomeres, but it is unclear how only a single VSG gene is active in one trypanosome at a time. Since the VSG variants are so different in primary sequences, particularly at the N termini, infected hosts are unable to produce Abs that are broad and potent enough to eliminate all trypanosomes. Consequently, repopulation of new parasitic variants results in the successive surges of parasitemia in the hosts.

In *P. falciparum* (the *Plasmodium* species responsible for nearly all malaria-specific mortality in humans), the most studied variant antigen is the *P. falciparum* erythrocyte membrane protein 1 (PfEMP-1) (reviewed by Kyes et al., 2001). PfEMP-1 is exposed on the I-RBC surface during the merozoite stage of malaria infection when the parasites reside and replicate inside I-RBC. It is a virulence factor that promotes cytoadhesion of I-RBC to vascular endothelium and delays clearance of I-RBC by the spleen. Abs to I-RBC interfere with cytoadherence and ameliorate disease (Bull et al., 1998; Marsh et al., 1989). As for VSG of trypanosomes, PfEMP-1 is encoded by a family of non-allelic genes (~50), the *var* gene family, which is present on different chromosomes of the parasite. The *var* genes are extremely divergent in sequence and size. Escape of Abs in the blood stage of malaria is associated with transcriptional switching of variant PfEMP-1 (Hisaeda et al., 2005; Kyes et al., 2001). Antigenic variation in malaria may be regulated by host immune responses (Recker et al., 2004). Using mathematical models, Recker and co-workers proposed that Ab responses against malaria variant antigens can be classified into long-lasting variant-specific Abs and short-lived partially cross-reactive Abs. The latter target minor epitopes on a variant antigen such as PfEMP-1 and may help synchronise variant protein expression to avoid rapid exhaustion of the variant repertoire, hence prolonging the survival of the parasites under Ab surveillance.

3. Evasion of humoral responses by direct cell-to-cell virus dissemination

Viruses released from infected cells are immediately vulnerable to humoral immune effectors and infection can be terminated at any stages during virus release, diffusion, attachment and entry to target cells. Viruses transmitted directly from cell to cell can reduce the efficacy of humoral effectors. The evasion of humoral responses by direct cell-to-cell spread has been reported for alpha-herpesviruses, VACV and retroviruses.

For alpha-herpesviruses including HSV, varicellar-zoster virus (VZV) and pseudorabies virus (PRV), the majority of virus infectivity is cell associated and NAbs are relatively ineffective in preventing virus spread between cells in culture and limiting progression of diseases. Virus mutants with defects in cell-to-cell spread are highly attenuated in animal models (reviewed by Johnson and

Huber, 2002). Dingwell et al. propose that spreading of HSV-1 is directed across cell junctions between epithelial cells and neurons by the gE-gI complex and this pathway is resistant to NAbs (Dingwell et al., 1994; Dingwell and Johnson, 1998). The gE-gI complex accumulates at sites of cell-cell contact which possibly interacts with junctional components to mediate virus transmission (Dingwell and Johnson, 1998). The deletion of gE or gI proteins does not reduce virus replication and entry kinetics but does reduce virus cell-to-cell spread significantly (Dingwell et al., 1994). The addition of NAbs in the culture medium further inhibits the yields of the mutants but not the wild-type virus showing that direct cell-to-cell dissemination can be an efficient way to evade NAbs. The expression of VZV gE in polarized epithelial cells alters the organization of F-actin and accelerats the formation of tight junctions between cells, by which the direct cell-to-cell spread of the virus may be enhanced (Mo et al., 2000).

In poxviruses, the majority of virus infectivity is also cell-associated and virus particles are actively delivered into neighboring cells by polymerizing actin bundles (actin tails) (reviewed by Smith and Law, 2004). Virus-tipped actin tails are generated by a mechanism mimicking the receptor tyrosine kinase signaling pathways that control actin polymerization at the plasma membrane (Frischknecht et al., 1999). VACV is transported from the cytoplasm to the cell surface on microtubules and the virus neutralizing determinant B5R may provide the switching signal to activate Src kinase at the plasma membrane to phosphorylate another virus protein A36R which in turn elicits the cascade of actin nucleation and switches virus transport from using microtubules to actin (Newsome et al., 2004). The actin tail-mediated cell-cell transmission is highly efficient and resistant to NAbs (Law et al., 2002). Mutants that fail to induce actin tails produce small plaques in cell culture and are attenuated in animal models.

For retroviruses that infect immune cells such as human T cell leukemia virus type 1 (HTLV-1) and HIV-1, virus spread is facilitated by the formation of "virological synapses" between infected and uninfected cells (Ganesh et al., 2004; Igakura et al., 2003; Jolly et al., 2004; McDonald et al., 2003). The virological synapse resembles the immunological synapse that describes the interface between T cells and Ag-presenting cells (APCs) in many ways (reviewed by Piguet and Sattentau, 2004). Infected cells form a stable adhesion junction with uninfected cells and release viral materials across the synapse by cellular machinery. This mode of spreading is relatively NAb-resistant (Gupta et al., 1989) and can occur between different cell populations, such as from dendritic cells (DCs) to T cells, and from T cells to T cells, in HIV-1. In the case of DCs, HIV-1 can hide in these professional Ag-presenting cells from humoral responses by binding to a cell surface lectin (*e.g.* DC-SIGN) and being internalized but not degraded inside the cells by an unknown mechanism (Kwon et al., 2002). "Infected" DCs then travel from the site of infection to lymphoid tissues and act as a "Trojan horse" to transmit the virus directly to CD4+ T cells. Virological synapses allow the virus to transmit efficiently to uninfected cells in the presence of strong humoral responses and to use the secondary lymphoid tissues as a factory for virus production.

4. Viral subversion of the complement system

The complement system has a number of anti-viral activities (reviewed by Cooper, 1998). Complement can enhance virus neutralization by binding to Ab-virus complexes and sterically interfering with virus entry, by inducing lysis of the membrane of enveloped viruses, and by promoting clearance of virus through complement receptors. Complement can also promote the lysis and/or phagocytosis of virally infected cells. In the presence of antigen-associated Abs (most notably for IgM, IgG1 and IgG3 in humans), complement can be readily activated through the classical pathway by the binding of C1q to the Fc region of Abs. In addition, several viruses (Sindbis, Newcastle disease, rubella, HTLV, HIV and other retroviruses of avian, rodent and feline origin) can activate the classical pathway directly in the absence of immune Abs (Cooper and Nemerow, 1986). The activation can be due to the direct binding of C1q as occurs with the gp120 and gp41 of HIV (Süsal et al., 1994; Thielens et al., 1993), or binding of mannose-binding lectin (MBL) to the carbohydrate moeities of influenza virus (Anders et al., 1990) and HIV-1 and HIV-2 (Haurum et al., 1993). On the other hand, some viruses (Sindbis virus, VSV, Epstein-Barr virus (EBV) and influenza virus) are direct activators of the alternative pathway (Cooper and Nemerow, 1986).

Viruses have developed a wide range of mechanisms to evade destruction by complement. The glycoprotein (g)C of HSV-1 and HSV-2 binds C3, C3b and C3c and protects the virus and infected cells from complement attack (Friedman et al., 1984; Friedman et al., 1996; Kostavasili et al., 1997; McNearney et al., 1987) (and see below). Herpesvirus saimiri (HVS) encodes counterparts of CD46 and CD59 and expression of these virus proteins on cells increases the resistance of these cells to complement cytolysis (Albrecht et al., 1992; Bramley et al., 1997). Vaccinia virus encodes a viral homologue of host complement control protein (CCP), VCP, which protects infected cells from complement activated through the alternative pathway (Kotwal and Moss, 1988; Okada et al., 1987; Sahu et al., 1998; Wakamiya et al., 1989). VCP binds human C3b and C4b, accelerates the decay of C3 convertases of both classical and alternative pathways and serves as a cofactor of factor I for the cleavage of both C3b and C4b (Kotwal et al., 1990; McKenzie et al., 1992; Sahu et al., 1998). In contrast to factor H and CR1 which displayed cofactor activity in factor I-mediated cleavage at 3 sites of C3b, VCP-assisted factor I cleavage of C3b was primarily between Arg^{1281}-Ser^{1282} in a fluid phase assay (Sahu et al., 1998). The infiltration of immune cells induced by the complement cleavage products C3a, C4a and C5a is also suppressed as a result of the inhibition of complement activation by VCP (Kotwal et al., 1998; Miller et al., 1997).

In addition to virus-encoded proteins that inhibit complement activation, viruses also exploit host membrane and soluble CCPs for complement evasion. Host CCPs such as CD46, CD55 and CD59 have been found to be incorporated onto the membrane envelopes of several viruses, including SIV, HIV, HTLV-1, cytomegalovirus (CMV) and VACV, and protect these viruses from complement lysis (Marschang et al., 1995; Montefiori et al., 1994; Saifuddin et al., 1997; 1995; Spear et al., 1995; Spiller et al., 1997; Vanderplasschen et al., 1998b). HIV-1 gp120 and gp41 of HIV bind factor H in addition to C1q, C3b, C4b and properdin (Stoiber et al., 1995) and only sera devoid of factor H lysed free virus and virus-infected cells effectively (Stoiber et al., 1996). Human CMV up-regulates the expression of host cell surface CCPs (CD46 and CD55), rendering the infected cells more resistance to complement lysis (Spiller et al., 1996).

5. Fc receptors (FcRs) and complement receptors (C'-Rs) expressed by viruses as potential evasion mechanisms

Various microorganisms have been shown to express receptors for IgGs and complement components of humans and other species. The ability to bind IgGs could serve as a potential immune evasion mechanism. Pathogens expressing FcR molecules include various herpesviruses like HSV type 1, 2, VZV, EBV and CMV, bacteria like *Staphylococcus aureus* (Forsgren and Sjoquist, 1966), streptococci of groups A, C, and G (Kronvall, 1973), and *Haemophilus somnus* (Yarnall et al., 1988), pathogenic trypanosomatidae like *Leishmania amazonensis* and *Trypanosoma cruzi*, but not nonpathogenic ones (De Miranda-Santos and Campos-Neto, 1995; 1981), and parasites like *Schistosoma mansoni* (Torpier et al., 1979). This convergence of functions from unrelated microorganisms suggest an evolutionary advantage in binding host IgG, although evidence that Fc receptors constitute a bona fide virulence factor is still incomplete and pathogen FcRs appear to be generally multifunctional.

In the virus realm, probably the best characterized example of pathogen-encoded FcR is the complex of HSV gE and gI that act as a receptor for the IgG of certain species (*e.g.*, human and rabbit) but not others (*e.g.*, murine IgGs) (Johansson et al., 1986). The existence of a moiety capable of binding IgG on HSV-infected cells was first observed by Watkins in 1964 (Watkins, 1964). Subsequent work by Bauke and Spear demonstrated the involvement of gE in the putative viral IgG receptor (Baucke and Spear, 1979) and Johnson et al. identified the dimeric nature of HSV-encoded FcR (Johnson et al., 1988). Work by various other groups, including major contributions from Friedman and associates, tested the hypothesis that HSV FcR has a role as a virulence factor (Adler et al., 1978; Dowler and Veltri, 1984; Frank and Friedman, 1989; Hidaka et al., 1991; Lubinski et al., 1998; 1999; 2002; Nagashunmugam et al., 1998; Van Vliet et al., 1992).

HSV-encoded FcR has been shown to limit IgG effectiveness both *in vitro* and *in vivo* (Lubinski et al., 1998; 1999; 2002; Nagashunmugam et al., 1998). The primary mechanism of this activity has been suggested to be interference with host effector functions such as complement and ADCC by engaging the Fc of anti-HSV antibodies (Adler et al., 1978; Frank and Friedman, 1989; Lubinski et al., 1998; Van Vliet et al., 1992). However, it is not clear how this occurs in the presence of a vast excess of non-immune serum IgG. It also appears that HSV FcR provides some protection from complement-independent neutralization and this has been related, at least in part, to the coating of viral particles with non-immune IgG (Dowler and Veltri, 1984).

Friedman, Lubinski and associates examined the combined effects of gC and gE in immune evasion on HSV pathogenesis (Lubinski et al., 2002). To this aim they used a mutant allele of gE, NS-gE339, with a four amino acid insertion after gE position 339, which eliminates the ability of the gE/gI complex to bind the Fc domain of human IgG (Nagashunmugam et al., 1998) and a mutant allele of gC, NS-gC, with a deletion of 93 amino acids, which eliminates the ability of gC to bind C3b (Lubinski et al., 1999). Elimination of both FcR and complement binding activities increased susceptibility of HSV-1 to antibody and complement neutralization *in vitro* and reduced virulence *in vivo* as measured by mortality and clinical scores, as well as virus yields (Lubinski et al., 2002). In the murine flank model, gE or gC single-mutant HSV produced lower disease scores in passively immunized mice than the wild type virus and, predictably, gC/gE double-mutant virus was the least virulent (Lubinski et al., 2002). Consistent with the hypothesis that gE and gC are virulence factors for HSV, in C3 knockout mice, disease scores of gC/gE mutant virus were similar to disease scores of wild type virus, indicating that complement evasion may contribute to viral pathogenesis (Lubinski et al., 2002). Lubinski, Friedman and associates also showed that passive immunization with gC monoclonal antibodies prior to HSV infection was protective if the gC antibodies blocked C3b binding (Judson et al., 2003). To circumvent HSV FcR-mediated escape mechanisms, Lubinski, Friedman and associates also used gE-derived peptides as immunogens to induce antibody to HSV FcR and inhibit IgG binding to the FcR (Lin et al., 2004). Consistent with this approach, Sanna et al. (2000) observed that a weak neutralizing antibody to gE strongly synergizes in neutralization of free virus with a good neutralizer to gD.

As discussed above, beside contributing to the HSV FcR, a role for gE in cell-to-cell spread has also been identified (Johnson et al., 2001; Weeks et al., 1997). The gE mutant virus used for the aforementioned experiments was comparable to wild-type virus in cell-to-cell spread in epidermal keratinocytes (Weeks et al., 1997). However, it caused less zosteriform spread disease than wild-type virus, suggesting that the mutant virus may be defective in epidermal-neuronal spread, neuronal transport, or transneuronal spread (Nagashunmugam et al., 1998). Therefore, different gE domains may mediate epidermal and neuronal spread. Enquist and associates working with PRV, a porcine alpha-herpesviruses related to HSV, demonstrated that PRV gE has distinct domains that control spread and neurovirulence (Tirabassi and Enquist, 1999; Tirabassi et al., 1997).

Located at the gC amino terminus is a region that blocks the interaction of complement C5 and properidine (P) with C3b (Kostavasili et al., 1997; Lubinski et al., 1999). Therefore, it is possible that a mutant virus lacking this additional immune evasion domain could be even more attenuated than the gC/gE mutant virus used in the studies by Friedman, Lubinski and associates (Lubinski et al., 1999; 2002). The function of HSV gC goes beyond binding complement. In fact, gC has also been shown to mediate viral attachment to cell surface heparan sulfate (Griffiths et al., 1998; Herold et al., 1991) and is required for apical infection of polarized cells (Sears et al., 1991; WuDunn and Spear, 1989). The gC C5-P interacting domain also plays a major role in virus attachment to heparan sulfate (Rux et al., 2002) and therefore the role of complement in reduced virulence by such a mutant could not be tested *in vivo* without affecting virus attachment (Lubinski et al., 1999; 2002). Lastly, gC also binds factor X avidly (Altieri et al., 1991; Etingin et al., 1990). While unrelated in primary sequence, gC bindsfactor X using a similar ligand recognition domain as the leukocyte integrin Mac1, which also, like gC, binds C3b (Altieri et al., 1991). These observations suggest that gC could be a mediator of vascular pathology during HSV infection (Altieri et al., 1991; Etingin et al., 1990).

These results support the notion that HSV gE/gI and gC act in part as immune escape mechanisms against host complement and IgG. However, as for bacterial FcRs, HSV FcR and complement receptors are multi-functional proteins that affect virulence through additional mechanisms related to virus invasion and spread. While HSV FcR is the best-studied example, HSV is not the only herpesvirus to express FcRs. In fact, as indicated above, FcR are also expressed by other alpha-herpesviruses like PRV, VZV and bovine herpesvirus 1, to name a few (Litwin et al., 1990; Whitbeck et al., 1996), as well as beta-herpesviruses like CMV (Furukawa et al., 1975) and human herpervirus 6 (HHV6) (Farber and Wutzler, 1991).

6. Immunosuppressive activity of viruses

Viruses have also evolved means to affect Ab responses and production. Measles virus (MV) induces cellular and humoral immunosuppression that contributes to secondary infections and mortality (Hirsch et al., 1984; Whittle et al., 1973). MV immunosuppressive actions are in part due to its ability to infect cells of the immune system and induce cell cycle arrest of infected T and B lymphocytes (McChesney et al., 1987; Naniche and Oldstone, 2000). MV infection of unstimulated B lymphocytes also suppresses differentiation into immunoglobulin-secreting cells (McChesney et al., 1987). Additionally, MV nucleocapsid protein (N) has been shown to bind to inhibitory FcRII isoforms on human and murine B lymphocytes *in vitro,* which could explain, at least in part, MV interference with Ab production (Ravanel et al., 1997). A mAb capable of blocking the interaction between murine FcRII and immune complexes also inhibited nucleocapsid protein binding to murine FcRII indicating that immune complexes and nucleocapsid protein share the same binding site on FcRII (Ravanel et al., 1997). An apparent paradox is that, despite the apparently crucial role for nucleocapsid protein in immunosuppression, high amounts of Abs to the protein are produced during the acute phase of measles infection (Graves et al., 1984).

7. Concluding remarks

Viruses have adopted a wide range of strategies to subvert humoral immunity. The selection pressures on different viruses to adopt evasion strategies vary considerably and lead to great differences in the repertoires of different viruses. We emphasized the very different pressures acting upon acute (hit and run) and chronic (hit and stay) viruses. The chronic viruses, in particular, have often acquired many evasion strategies. We also emphasized fundamental differences between DNA and RNA viruses with respect to evasion. RNA viruses, with inherent genetic instability, rely in many instances on sequence variation to avoid Ab responses. DNA viruses, with more sophisticated but stable genomes, rely on establishing latency and/or have more surface proteins and virus-encoded immunomodulators that can provide evasion mechanisms. A recurring theme is the importance of the nature of the virus SU glycoprotein(s) in evasion. For many viruses, these proteins have acquired a bewildering array of features associated with evasion.

Viral evasion from Ab responses has profound implications for vaccine design. In so far as the general logic of vaccination is to try and mimic natural infection without the morbidity, then the evasion mechanisms operating in natural infection will be felt in vaccination. Thus an SU protein that minimizes NAb responses in the course of natural infection is likely to do so when presented as a vaccine designed to mimic natural infection. Since chronic RNA viruses typically have evolved the most effective Ab evasion mechanisms, one might anticipate they will present the greatest difficulties for vaccine design. This appears to be the case. Indeed, we suggest that successful vaccine design for viruses most evolved to avoid Ab responses will require understanding the, probably minor, fraction of the response that is effective against the virus and building upon that understanding to design a new generation of vaccines.

Acknowledgements

ML is supported in part by a scholarship (grant # PF-77497) from the Elizabeth Glaser Pediatric AIDS Foundation. DRB and PPS are supported by grants from the National Institutes of Health and DRB by a grant from the International AIDS Vaccine Initiative.

References

Adler, R., Glorioso, J. C., Cossman, J., and Levine, M. (1978). Possible role of Fc receptors on cells infected and transformed by herpesvirus: escape from immune cytolysis. Infect. Immun. *21*, 442-447.

Albrecht, J. C., Nicholas, J., Cameron, K. R., Newman, C., Fleckenstein, B., and Honess, R. W. (1992). Herpesvirus saimiri has a gene specifying a homologue of the cellular membrane glycoprotein CD59. Virology *190*, 527-530.

Altieri, D. C., Etingin, O. R., Fair, D. S., Brunck, T. K., Geltosky, J. E., Hajjar, D. P., and Edgington, T. S. (1991). Structurally homologous ligand binding of integrin Mac-1 and viral glycoprotein C receptors. Science *254*, 1200-1202.

Anders, E. M., Hartley, C. A., and Jackson, D. C. (1990). Bovine and mouse serum beta inhibitors of influenza A viruses are mannose-binding lectins. Proc. Natl. Acad. Sci. USA. *87*, 4485-4489.

Anderson, D. E., Carlos, M. P., Nguyen, L., and Torres, J. V. (2001). Overcoming original (antigenic) sin. Clin. Immunol. *101*, 152-157.

Anderson, J. P., Daifuku, R., and Loeb, L. A. (2004). Viral error catastrophe by mutagenic nucleosides. Annu. Rev. Microbiol. *58*, 183-205.

Ariyoshi, K., Harwood, E., Chiengsong-Popov, R., and Weber, J. (1992). Is clearance of HIV-1 viraemia at seroconversion mediated by neutralising antibodies? Lancet *340*, 1257-1258.

Arnold, J. J., and Cameron, C. E. (1999). Poliovirus RNA-dependent RNA polymerase (3Dpol) is sufficient for template switching *in vitro*. J. Biol. Chem. *274*, 2706-2716.

Bachmann, M. F., Hunziker, L., Zinkernagel, R. M., Storni, T., and Kopf, M. (2004). Maintenance of memory CTL responses by T helper cells and CD40-CD40 ligand: antibodies provide the key. Eur. J. Immunol. *34*, 317-326.

Bachmann, M. F., Rohrer, U. H., Kundig, T. M., Burki, K., Hengartner, H., and Zinkernagel, R. M. (1993). The influence of antigen organization on B cell responsiveness. Science *262*, 1448-1451.

Bachmann, M. F., and Zinkernagel, R. M. (1996). The influence of virus structure on antibody responses and virus serotype formation. Immunol. Today *17*, 553-558.

Back, N. K., Smit, L., De Jong, J. J., Keulen, W., Schutten, M., Goudsmit, J., and Tersmette, M. (1994). An N-glycan within the human immunodeficiency virus type 1 gp120 V3 loop affects virus neutralization. Virology *199*, 431-438.

Balachandran, N., Bacchetti, S., and Rawls, W. E. (1982). Protection against lethal challenge of BALB/c mice by passive transfer of monoclonal antibodies to five glycoproteins of herpes simplex virus type 2. Infect. Immun. *37*, 1132-1137.

Baucke, R. B., and Spear, P. G. (1979). Membrane proteins specified by herpes simplex viruses. V. Identification of an Fc-binding glycoprotein. J. Virol. *32*, 779-789.

Baumgarth, N. (2000). A two-phase model of B-cell activation. Immunol. Rev. *176*, 171-180.

Bell, E., Shamim, M., Whitbeck, J. C., Sfyroera, G., Lambris, J. D., and Isaacs, S. N. (2004). Antibodies against the extracellular enveloped virus B5R protein are mainly responsible for the EEV neutralizing capacity of vaccinia immune globulin. Virology *325*, 425-431.

Binley, J. M., Wrin, T., Korber, B., Zwick, M. B., Wang, M., Chappey, C., Stiegler, G., Kunert, R., Zolla-Pazner, S., Katinger, H., et al. (2004). Comprehensive cross-clade neutralization analysis of a panel of anti-human immunodeficiency virus type 1 monoclonal antibodies. J. Virol. *78*, 13232-13252.

Bizebard, T., Barbey-Martin, C., Fleury, D., Gigant, B., Barrere, B., Skehel, J. J., and Knossow, M. (2001). Structural studies on viral escape from antibody neutralization. Curr. Top Microbiol. Immunol. *260*, 55-64.

Bizebard, T., Gigant, B., Rigolet, P., Rasmussen, B., Diat, O., Bosecke, P., Wharton, S. A., Skehel, J. J., and Knossow, M. (1995). Structure of influenza virus haemagglutinin complexed with a neutralizing antibody. Nature *376*, 92-94.

Boulter, E. A., and Appleyard, G. (1973). Differences between extracellular and intracellular forms of poxvirus and their implications. Prog Med. Virol. *16*, 86-108.

Bramley, J. C., Davies, A., and Lachmann, P. J. (1997). Herpesvirus saimiri CD59--baculovirus expression and characterisation of complement inhibitory activity. Biochem. Soc. Trans *25*, 354S.

Bull, P. C., Lowe, B. S., Kortok, M., Molyneux, C. S., Newbold, C. I., and Marsh, K. (1998). Parasite antigens on the infected red cell surface are targets for naturally acquired immunity to malaria. Nat. Med. *4*, 358-360.

Bures, R., Morris, L., Williamson, C., Ramjee, G., Deers, M., Fiscus, S. A., Abdool-Karim, S., and Montefiori, D. C. (2002). Regional clustering of shared neutralization determinants on primary isolates of clade C human immunodeficiency virus type 1 from South Africa. J. Virol. *76*, 2233-2244.

Burton, D. R. (2002). Antibodies, viruses and vaccines. Nat. Rev. Immunol. *2*, 706-713.

Calarese, D. A., Scanlan, C. N., Zwick, M. B., Deechongkit, S., Mimura, Y., Kunert, R., Zhu, P., Wormald, M. R., Stanfield, R. L., Roux, K. H., et al. (2003). Antibody domain exchange is an immunological solution to carbohydrate cluster recognition. Science *300*, 2065-2071.

Caton, A. J., Brownlee, G. G., Yewdell, J. W., and Gerhard, W. (1982). The antigenic structure of the influenza virus A/PR/8/34 hemagglutinin (H1 subtype). Cell *31*, 417-427.

Chackerian, B., Rudensey, L. M., and Overbaugh, J. (1997). Specific N-linked and O-linked glycosylation modifications in the envelope V1 domain of simian immunodeficiency virus variants that evolve in the host alter recognition by neutralizing antibodies. J. Virol. *71*, 7719-7727.

Cheng, B., Vogan, E.M., Gong, H, Skehel, J.J., Wiley, D.C., and Harrison, S.C. (2005). Structure of an unliganded simian immunodeficiency virus gp120 core. Nature *433*, 834-841.

Cheng-Mayer, C., Brown, A., Harouse, J., Luciw, P. A., and Mayer, A. J. (1999). Selection for neutralization resistance of the simian/human immunodeficiency virus SHIVSF33A variant *in vivo* by virtue of sequence changes in the extracellular envelope glycoprotein that modify N-linked glycosylation. J. Virol. *73*, 5294-5300.

Chien, D. Y., Choo, Q. L., Tabrizi, A., Kuo, C., McFarland, J., Berger, K., Lee, C., Shuster, J. R., Nguyen, T., Moyer, D. L., and et al. (1992). Diagnosis of hepatitis C virus (HCV) infection using an immunodominant chimeric polyprotein to capture circulating antibodies: reevaluation of the role of HCV in liver disease. Proc. Natl. Acad. Sci. USA. *89*, 10011-10015.

Coffin, J. M. (1995). HIV population dynamics *in vivo*: implications for genetic variation, pathogenesis, and therapy. Science *267*, 483-489.

Cohen, G. H., Dietzschold, B., Ponce de Leon, M., Long, D., Golub, E., Varrichio, A., Pereira, L., and Eisenberg, R. J. (1984). Localization and synthesis of an antigenic determinant of herpes simplex virus glycoprotein D that stimulates the production of neutralizing antibody. J. Virol. *49*, 102-108.

Cohen, G. H., Ponce de Leon, M., and Nichols, C. (1972). Isolation of a herpes simplex virus-specific antigenic fraction which stimulates the production of neutralizing antibody. J. Virol. *10*, 1021-1030.

Colman, P. M. (1994). Influenza virus neuraminidase: structure, antibodies, and inhibitors. Protein Sci. *3*, 1687-1696.

Colman, P. M., Varghese, J. N., and Laver, W. G. (1983). Structure of the catalytic and antigenic sites in influenza virus neuraminidase. Nature *303*, 41-44.

Cooper, N. R. (1998). Complement-dependent virus neutralization. In The complement system, K. Rother, G. O. Till, and G. M. Hänsch, eds. (New York, Springer), pp. 302-309.

Cooper, N. R., and Nemerow, G. R. (1986). Complement dependent mechanisms of virus neutralization. In Immunobiology of the complement system : an introduction for research and clinical medicine, G. D. Ross, ed. (New York, Academic Press), pp. 139-162.

Czerny, C. P., and Mahnel, H. (1990). Structural and functional analysis of orthopoxvirus epitopes with neutralizing monoclonal antibodies. J. Gen. Virol. *71*, 2341-2352.

Davenport, F. M., Hennessy, A. V., and Francis, T., Jr. (1953). Epidemiologic and immunologic significance of age distribution of antibody to antigenic variants of influenza virus. J. Exp. Med. *98*, 641-656.

De Miranda-Santos, I. K., and Campos-Neto, A. (1995). Immune evasion by *Trypanosomatidae*: normal aggregated immunoglobulin protects against lysis by the alternative complement pathway. Braz J. Med. Biol. Res. *28*, 585-589.

De Miranda-Santos, I. K., and Compos-Neto, A. (1981). Receptor for immunoglobulin Fc on pathogenic but not on nonpathogenic protozoa of the Trypanosomatidae. J. Exp. Med. *154*, 1732-1742.

Dingwell, K. S., Brunetti, C. R., Hendricks, R. L., Tang, Q., Tang, M., Rainbow, A. J., and Johnson, D. C. (1994). Herpes simplex virus glycoproteins E and I facilitate cell-to-cell spread *in vivo* and across junctions of cultured cells. J. Virol. *68*, 834-845.

Dingwell, K. S., and Johnson, D. C. (1998). The herpes simplex virus gE-gI complex facilitates cell-to-cell spread and binds to components of cell junctions. J. Virol. *72*, 8933-8942.

Dix, R. D., Pereira, L., and Baringer, J. R. (1981). Use of monoclonal antibody directed against herpes simplex virus glycoproteins to protect mice against acute virus-induced neurological disease. Infect. Immun. *34*, 192-199.

Dowler, K. W., and Veltri, R. W. (1984). In vitro neutralization of HSV-2: inhibition by binding of normal IgG and purified Fc to virion Fc receptor (FcR). J. Med. Virol *13*, 251-259.

Etingin, O. R., Silverstein, R. L., Friedman, H. M., and Hajjar, D. P. (1990). Viral activation of the coagulation cascade: molecular interactions at the surface of infected endothelial cells. Cell *61*, 657-662.

Farber, I., and Wutzler, P. (1991). FC receptors induced by human herpesvirus-6. Acta Virol. *35*, 400.

Farci, P., Shimoda, A., Wong, D., Cabezon, T., De Gioannis, D., Strazzera, A., Shimizu, Y., Shapiro, M., Alter, H. J., and Purcell, R. H. (1996). Prevention of hepatitis C virus infection in chimpanzees by hyperimmune serum against the hypervariable region 1 of the envelope 2 protein. Proc. Natl. Acad. Sci. USA. *93*, 15394-15399.

Fazekas de St, G., and Webster, R. G. (1966a). Disquisitions of Original Antigenic Sin. I. Evidence in man. J. Exp. Med. *124*, 331-345.

Fazekas de St, G., and Webster, R. G. (1966b). Disquisitions on Original Antigenic Sin. II. Proof in lower creatures. J. Exp. Med. *124*, 347-361.

Fehr, T., Naim, H. Y., Bachmann, M. F., Ochsenbein, A. F., Spielhofer, P., Bucher, E., Hengartner, H., Billeter, M. A., and Zinkernagel, R. M. (1998). T-cell independent IgM and enduring protective IgG antibodies induced by chimeric measles viruses. Nat. Med. *4*, 945-948.

Fleury, D., Wharton, S. A., Skehel, J. J., Knossow, M., and Bizebard, T. (1998). Antigen distortion allows influenza virus to escape neutralization. Nat. Struct Biol. *5*, 119-123.

Fogg, C., Lustig, S., Whitbeck, J. C., Eisenberg, R. J., Cohen, G. H., and Moss, B. (2004). Protective immunity to vaccinia virus induced by vaccination with multiple recombinant outer membrane proteins of intracellular and extracellular virions. J. Virol. *78*, 10230-10237.

Forsgren, A., and Sjoquist, J. (1966). "Protein A" from *S. aureus*. I. Pseudo-immune reaction with human gamma-globulin. J. Immunol. *97*, 822-827.

Fournillier, A., Wychowski, C., Boucreux, D., Baumert, T. F., Meunier, J. C., Jacobs, D., Muguet, S., Depla, E., and Inchauspe, G. (2001). Induction of hepatitis C virus E1 envelope protein-specific immune response can be enhanced by mutation of N-glycosylation sites. J. Virol. *75*, 12088-12097.

Frank, I., and Friedman, H. M. (1989). A novel function of the herpes simplex virus type 1 Fc receptor: participation in bipolar bridging of antiviral immunoglobulin G. J. Virol. *63*, 4479-4488.

Friedman, H. M., Cohen, G. H., Eisenberg, R. J., Seidel, C. A., and Cines, D. B. (1984). Glycoprotein C of herpes simplex virus 1 acts as a receptor for the C3b complement component on infected cells. Nature *309*, 633-635.

Friedman, H. M., Wang, L., Fishman, N. O., Lambris, J. D., Eisenberg, R. J., Cohen, G. H., and Lubinski, J. (1996). Immune evasion properties of herpes simplex virus type 1 glycoprotein gC. J. Virol. *70*, 4253-4260.

Frischknecht, F., Moreau, V., Röttger, S., Gonfloni, S., Rechmann, I., Superti-Furga, G., and Way, M. (1999). Actin-based motility of vaccinia virus mimics receptor tyrosine kinase signalling. Nature *401*, 926-929.

Furukawa, T., Hornberger, E., Sakuma, S., and Plotkin, S. A. (1975). Demonstration of immunoglobulin G receptors induced by human cytomegalovirus. J. Clin. Microbiol. *2*, 332-336.

Galmiche, M. C., Goenaga, J., Wittek, R., and Rindisbacher, L. (1999). Neutralizing and protective antibodies directed against vaccinia virus envelope antigens. Virology *254*, 71-80.

Ganesh, L., Leung, K., Lore, K., Levin, R., Panet, A., Schwartz, O., Koup, R. A., and Nabel, G. J. (2004). Infection of specific dendritic cells by CCR5-tropic human immunodeficiency virus type 1 promotes cell-mediated transmission of virus resistant to broadly neutralizing antibodies. J. Virol. *78*, 11980-11987.

Gerhard, W. (2001). The role of the antibody response in influenza virus infection. Curr. Top Microbiol. Immunol. *260*, 171-190.

Gordon, J., Mohandas, A., Wilton, S., and Dales, S. (1991). A prominent antigenic surface polypeptide involved in the biogenesis and function of the vaccinia virus envelope. Virology *181*, 671-686.

Govorkova, E. A., Rehg, J. E., Krauss, S., Yen, H. L., Guan, Y., Peiris, M., Nguyen, T. D., Hanh, T. H., Puthavathana, P., Long, H. T., et al. (2005). Lethality to ferrets of H5N1 influenza viruses isolated from humans and poultry in 2004. J. Virol. *79*, 2191-2198.

Graves, M., Griffin, D. E., Johnson, R. T., Hirsch, R. L., de Soriano, I. L., Roedenbeck, S., and Vaisberg, A. (1984). Development of antibody to measles virus polypeptides during complicated and uncomplicated measles virus infections. J. Virol. *49*, 409-412.

Griffiths, A., Renfrey, S., and Minson, T. (1998). Glycoprotein C-deficient mutants of two strains of herpes simplex virus type 1 exhibit unaltered adsorption characteristics on polarized or non-polarized cells. J. Gen. Virol. *79 (Pt 4)*, 807-812.

Gupta, P., Balachandran, R., Ho, M., Enrico, A., and Rinaldo, C. (1989). Cell-to-cell transmission of human immunodeficiency virus type 1 in the presence of azidothymidine and neutralizing antibody. J. Virol. *63*, 2361-2365.

Haaheim, L. R. (2003). Original antigenic sin. A confounding issue? Dev. Biol. (Basel) *115*, 49-53.

Hahn, Y. S. (2003). Subversion of immune responses by hepatitis C virus: immunomodulatory strategies beyond evasion? Curr. Opin. Immunol. *15*, 443-449.

Halstead, S. B., Rojanasuphot, S., and Sangkawibha, N. (1983). Original antigenic sin in dengue. Amer. J. Trop Med. Hyg *32*, 154-156.

Hangartner, L., Senn, B. M., Ledermann, B., Kalinke, U., Seiler, P., Bucher, E., Zellweger, R. M., Fink, K., Odermatt, B., Burki, K., et al. (2003). Antiviral immune responses in gene-targeted mice expressing the immunoglobulin heavy chain of virus-neutralizing antibodies. Proc. Natl. Acad. Sci. USA. *100*, 12883-12888.

Hartley, O., Klasse, P. J., Sattentau, Q. J., and Moore, J. P. (2005). V3: HIV's Switch-Hitter. AIDS Res. Hum. Retroviruses *21*, 171-189.

Haurum, J. S., Thiel, S., Jones, I. M., Fischer, P. B., Laursen, S. B., and Jensenius, J. C. (1993). Complement activation upon binding of mannan-binding protein to HIV envelope glycoproteins. Aids *7*, 1307-1313.

Herold, B. C., WuDunn, D., Soltys, N., and Spear, P. G. (1991). Glycoprotein C of herpes simplex virus type 1 plays a principal role in the adsorption of virus to cells and in infectivity. J. Virol. *65*, 1090-1098.

Heyer, W. D., Ehmsen, K. T., and Solinger, J. A. (2003). Holliday junctions in the eukaryotic nucleus: resolution in sight? Trends Biochem. Sci. *28*, 548-557.

Hidaka, Y., Sakai, Y., Toh, Y., and Mori, R. (1991). Glycoprotein C of herpes simplex virus type 1 is essential for the virus to evade antibody-independent complement-mediated virus inactivation and lysis of virus-infected cells. J. Gen. Virol. *72 (Pt 4)*, 915-921.

Hilleman, M. R. (2004). Strategies and mechanisms for host and pathogen survival in acute and persistent viral infections. Proc. Natl. Acad. Sci. USA. *101 Suppl 2*, 14560-14566.

Hirsch, R. L., Griffin, D. E., Johnson, R. T., Cooper, S. J., Lindo de Soriano, I., Roedenbeck, S., and Vaisberg, A. (1984). Cellular immune responses during complicated and uncomplicated measles virus infections of man. Clin. Immunol. Immunopathol *31*, 1-12.

Hisaeda, H., Yasutomo, K., and Himeno, K. (2005). Malaria: immune evasion by parasites. Int. J. Biochem. Cell Biol. *37*, 700-706.

Ho, D. D., Neumann, A. U., Perelson, A. S., Chen, W., Leonard, J. M., and Markowitz, M. (1995). Rapid turnover of plasma virions and CD4 lymphocytes in HIV-1 infection. Nature *373*, 123-126.

Hogle, J. M., Chow, M., and Filman, D. J. (1985). Three-dimensional structure of poliovirus at 2.9 A resolution. Science *229*, 1358-1365.

Homsy, J., Meyer, M., and Levy, J. A. (1990). Serum enhancement of human immunodeficiency virus (HIV) infection correlates with disease in HIV-infected individuals. J. Virol. *64*, 1437-1440.

Hooper, J. W., Custer, D. M., Schmaljohn, C. S., and Schmaljohn, A. L. (2000). DNA vaccination with vaccinia virus L1R and A33R genes protects mice against a lethal poxvirus challenge. Virology *266*, 329-339.

Hooper, J. W., Custer, D. M., and Thompson, E. (2003). Four-gene-combination DNA vaccine protects mice against a lethal vaccinia virus challenge and elicits appropriate antibody responses in nonhuman primates. Virology *306*, 181-195.

Hooper, J. W., Thompson, E., Wilhelmsen, C., Zimmerman, M., Ichou, M. A., Steffen, S. E., Schmaljohn, C. S., Schmaljohn, A. L., and Jahrling, P. B. (2004). Smallpox DNA vaccine protects nonhuman primates against lethal monkeypox. J. Virol. *78*, 4433-4443.

Hsiao, J. C., Chung, C. S., and Chang, W. (1998). Cell surface proteoglycans are necessary for A27L protein-mediated cell fusion: identification of the N-terminal region of A27L protein as the glycosaminoglycan-binding domain. J. Virol. *72*, 8374-8379.

Hsiao, J. C., Chung, C. S., and Chang, W. (1999). Vaccinia virus envelope D8L protein binds to cell surface chondroitin sulfate and mediates the adsorption of intracellular mature virions to cells. J. Virol. *73*, 8750-8761.

Ichihashi, Y., and Oie, M. (1996). Neutralizing epitope on penetration protein of vaccinia virus. Virology *220*, 491-494.

Igakura, T., Stinchcombe, J. C., Goon, P. K., Taylor, G. P., Weber, J. N., Griffiths, G. M., Tanaka, Y., Osame, M., and Bangham, C. R. (2003). Spread of HTLV-I between lymphocytes by virus-induced polarization of the cytoskeleton. Science *299*, 1713-1716.

Inouye, S., Matsuno, S., and Tsurukubo, Y. (1984). "Original antigenic sin" phenomenon in experimental flavivirus infections of guinea pigs: studies by enzyme-linked immunosorbent assay. Microbiol. Immunol. *28*, 569-574.

Ito, H., Watanabe, S., Takada, A., and Kawaoka, Y. (2001). Ebola virus glycoprotein: proteolytic processing, acylation, cell tropism, and detection of neutralizing antibodies. J. Virol. *75*, 1576-1580.

Janin, J., and Chothia, C. (1990). The structure of protein-protein recognition sites. J. Biol. Chem. *265*, 16027-16030.

Johansson, P. J., Schroder, A. K., Nardella, F. A., Mannik, M., and Christensen, P. (1986). Interaction between herpes simplex type 1-induced Fc receptor and human and rabbit immunoglobulin G (IgG) domains. Immunology *58*, 251-255.

Johnson, D. C., Frame, M. C., Ligas, M. W., Cross, A. M., and Stow, N. D. (1988). Herpes simplex virus immunoglobulin G Fc receptor activity depends on a complex of two viral glycoproteins, gE and gI. J. Virol. *62*, 1347-1354.

Johnson, D. C., and Huber, M. T. (2002). Directed egress of animal viruses promotes cell-to-cell spread. J. Virol. *76*, 1-8.

Johnson, D. C., Webb, M., Wisner, T. W., and Brunetti, C. (2001). Herpes simplex virus gE/gI sorts nascent virions to epithelial cell junctions, promoting virus spread. J. Virol. *75*, 821-833.

Johnson, W. E., and Desrosiers, R. C. (2002). Viral persistance: HIV's strategies of immune system evasion. Annu. Rev. Med. *53*, 499-518.

Jolly, C., Kashefi, K., Hollinshead, M., and Sattentau, Q. J. (2004). HIV-1 cell to cell transfer across an Env-induced, actin-dependent synapse. J. Exp. Med. *199*, 283-293.

Judson, K. A., Lubinski, J. M., Jiang, M., Chang, Y., Eisenberg, R. J., Cohen, G. H., and Friedman, H. M. (2003). Blocking immune evasion as a novel approach for prevention and treatment of herpes simplex virus infection. J. Virol. *77*, 12639-12645.

Kaplan, C. (1962). A non-infectious smallpox vaccine. Lancet *2*, 1027-1028.

Kaplan, C., Benson, P. F., and Butler, N. R. (1965). Immunogenicity of ultraviolet irradiated, non-infectious, vaccinia virus in infants and young children. Lancet *1*, 573-574.

Kirkegaard, K., and Baltimore, D. (1986). The mechanism of RNA recombination in poliovirus. Cell *47*, 433-443.

Kostavasili, I., Sahu, A., Friedman, H. M., Eisenberg, R. J., Cohen, G. H., and Lambris, J. D. (1997). Mechanism of complement inactivation by glycoprotein C of herpes simplex virus. J. Immunol. *158*, 1763-1771.

Kotwal, G. J., Isaacs, S. N., McKenzie, R., Frank, M. M., and Moss, B. (1990). Inhibition of the complement cascade by the major secretory protein of vaccinia virus. Science *250*, 827-830.

Kotwal, G. J., Miller, C. G., and Justus, D. E. (1998). The inflammation modulatory protein (IMP) of cowpox virus drastically diminishes the tissue damage by down-regulating cellular infiltration resulting from complement activation. Mol. Cell Biochem. *185*, 39-46.

Kotwal, G. J., and Moss, B. (1988). Vaccinia virus encodes a secretory polypeptide structurally related to complement control proteins. Nature *335*, 176-178.

Kronvall, G. (1973). A surface component in group A, C, and G streptococci with non-immune reactivity for immunoglobulin G. J. Immunol. *111*, 1401-1406.

Kwon, D. S., Gregorio, G., Bitton, N., Hendrickson, W. A., and Littman, D. R. (2002). DC-SIGN-mediated internalization of HIV is required for trans-enhancement of T cell infection. Immunity *16*, 135-144.

Kwong, P. D., Wyatt, R., Robinson, J., Sweet, R. W., Sodroski, J., and Hendrickson, W. A. (1998). Structure of an HIV gp120 envelope glycoprotein in complex with the CD4 receptor and a neutralizing human antibody. Nature *393*, 648-659.

Kyes, S., Horrocks, P., and Newbold, C. (2001). Antigenic variation at the infected red cell surface in malaria. Annu. Rev. Microbiol. *55*, 673-707.

Labrijn, A. F., Poignard, P., Raja, A., Zwick, M. B., Delgado, K., Franti, M., Binley, J., Vivona, V., Grundner, C., Huang, C. C., et al. (2003). Access of antibody molecules to the conserved coreceptor binding site on glycoprotein gp120 is sterically restricted on primary human immunodeficiency virus type 1. J. Virol. *77*, 10557-10565.

Law, M., Hollinshead, R., and Smith, G. L. (2002). Antibody-sensitive and antibody-resistant cell-to-cell spread of vaccinia virus: role of the A33R protein in antibody-resistant spread. J. Gen. Virol. *83*, 209-222.

Law, M., Putz, M. M., and Smith, G. L. (2005). An investigation of the therapeutic value of vaccinia-immune IgG in a mouse pneumonia model. J. Gen. Virol. *86*, 991-1000.

Law, M., and Smith, G. L. (2001). Antibody neutralization of the extracellular enveloped form of vaccinia virus. Virology *280*, 132-142.

Lawrence, M. C., and Colman, P. M. (1993). Shape complementarity at protein/protein interfaces. J. Mol. Biol. *234*, 946-950.

Lee, W. R., Syu, W. J., Du, B., Matsuda, M., Tan, S., Wolf, A., Essex, M., and Lee, T. H. (1992). Nonrandom distribution of gp120 N-linked glycosylation sites important for infectivity of human immunodeficiency virus type 1. Proc. Natl. Acad. Sci. USA. *89*, 2213-2217.

Leist, T. P., Cobbold, S. P., Waldmann, H., Aguet, M., and Zinkernagel, R. M. (1987). Functional analysis of T lymphocyte subsets in antiviral host defense. J. Immunol. *138*, 2278-2281.

Leonard, C. K., Spellman, M. W., Riddle, L., Harris, R. J., Thomas, J. N., and Gregory, T. J. (1990). Assignment of intrachain disulfide bonds and characterization of potential glycosylation sites of the type 1 recombinant human immunodeficiency virus envelope glycoprotein (gp120) expressed in Chinese hamster ovary cells. J. Biol. Chem. *265*, 10373-10382.

Li, K. S., Guan, Y., Wang, J., Smith, G. J., Xu, K. M., Duan, L., Rahardjo, A. P., Puthavathana, P., Buranathai, C., Nguyen, T. D., et al. (2004). Genesis of a highly pathogenic and potentially pandemic H5N1 influenza virus in eastern Asia. Nature *430*, 209-213.

Li, Y., Luo, L., Rasool, N., and Kang, C. Y. (1993). Glycosylation is necessary for the correct folding of human immunodeficiency virus gp120 in CD4 binding. J. Virol. *67*, 584-588.

Liddington, R. C., Yan, Y., Moulai, J., Sahli, R., Benjamin, T. L., and Harrison, S. C. (1991). Structure of simian virus 40 at 3.8-A resolution. Nature *354*, 278-284.

Liem, N. T., and Lim, W. (2005). Lack of H5N1 avian influenza transmission to hospital employees, Hanoi, 2004. Emerg. Infect. Dis. *11*, 210-215.

Lin, C. L., Chung, C. S., Heine, H. G., and Chang, W. (2000). Vaccinia virus envelope H3L protein binds to cell surface heparan sulfate and is important for intracellular mature virion morphogenesis and virus infection *in vitro* and *in vivo*. J. Virol. *74*, 3353-3365.

Lin, X., Lubinski, J. M., and Friedman, H. M. (2004). Immunization strategies to block the herpes simplex virus type 1 immunoglobulin G Fc receptor. J. Virol. *78*, 2562-2571.

Litwin, V., Sandor, M., and Grose, C. (1990). Cell surface expression of the varicella-zoster virus glycoproteins and Fc receptor. Virology *178*, 263-272.

Losman, B., Bolmstedt, A., Schonning, K., Bjorndal, A., Westin, C., Fenyo, E. M., and Olofsson, S. (2001). Protection of neutralization epitopes in the V3 loop of oligomeric human immunodeficiency virus type 1 glycoprotein 120 by N-linked oligosaccharides in the V1 region. AIDS Res. Hum. Retroviruses *17*, 1067-1076.

Lu, X., Mehta, A., Dwek, R., Butters, T., and Block, T. (1995). Evidence that N-linked glycosylation is necessary for hepatitis B virus secretion. Virology *213*, 660-665.

Lubinski, J., Nagashunmugam, T., and Friedman, H. M. (1998). Viral interference with antibody and complement. Semin Cell Dev. Biol. *9*, 329-337.

Lubinski, J., Wang, L., Mastellos, D., Sahu, A., Lambris, J. D., and Friedman, H. M. (1999). In vivo role of complement-interacting domains of herpes simplex virus type 1 glycoprotein gC. J. Exp. Med. *190*, 1637-1646.

Lubinski, J. M., Jiang, M., Hook, L., Chang, Y., Sarver, C., Mastellos, D., Lambris, J. D., Cohen, G. H., Eisenberg, R. J., and Friedman, H. M. (2002). Herpes simplex virus type 1 evades the effects of antibody and complement *in vivo*. J. Virol. *76*, 9232-9241.

Ly, A., and Stamatatos, L. (2000). V2 loop glycosylation of the human immunodeficiency virus type 1 SF162 envelope facilitates interaction of this protein with CD4 and CCR5 receptors and protects the virus from neutralization by anti-V3 loop and anti-CD4 binding site antibodies. J. Virol. *74*, 6769-6776.

Mansky, L. M., and Temin, H. M. (1995). Lower *in vivo* mutation rate of human immunodeficiency virus type 1 than that predicted from the fidelity of purified reverse transcriptase. J. Virol. *69*, 5087-5094.

Marschang, P., Sodroski, J., Wurzner, R., and Dierich, M. P. (1995). Decay-accelerating factor (CD55) protects human immunodeficiency virus type 1 from inactivation by human complement. Eur. J. Immunol. *25*, 285-290.

Marsh, K., Otoo, L., Hayes, R. J., Carson, D. C., and Greenwood, B. M. (1989). Antibodies to blood stage antigens of *Plasmodium falciparum* in rural Gambians and their relation to protection against infection. Trans R Soc. Trop Med. Hyg *83*, 293-303.

Martinez-Pomares, L., Stern, R. J., and Moyer, R. W. (1993). The ps/hr gene (B5R open reading frame homolog) of rabbitpox virus controls pock colour, is a component of extracellular enveloped virus, and is secreted into the medium. J. Virol. *67*, 5450-5462.

Masuda, T., Matsushita, S., Kuroda, M. J., Kannagi, M., Takatsuki, K., and Harada, S. (1990). Generation of neutralization-resistant HIV-1 *in vitro* due to amino acid interchanges of third hypervariable env region. J. Immunol. *145*, 3240-3246.

McChesney, M. B., Kehrl, J. H., Valsamakis, A., Fauci, A. S., and Oldstone, M. B. (1987). Measles virus infection of B lymphocytes permits cellular activation but blocks progression through the cell cycle. J. Virol. *61*, 3441-3447.

McCoy, A. J., Chandana Epa, V., and Colman, P. M. (1997). Electrostatic complementarity at protein/protein interfaces. J. Mol. Biol. *268*, 570-584.

McCulloch, R. (2004). Antigenic variation in African trypanosomes: monitoring progress. Trends Parasitol *20*, 117-121.

McDonald, D., Wu, L., Bohks, S. M., KewalRamani, V. N., Unutmaz, D., and Hope, T. J. (2003). Recruitment of HIV and its receptors to dendritic cell-T cell junctions. Science *300*, 1295-1297.

McKenzie, R., Kotwal, G. J., Moss, B., Hammer, C. H., and Frank, M. M. (1992). Regulation of complement activity by vaccinia virus complement-control protein. J. Infect. Dis. *166*, 1245-1250.

McMichael, A. J., and Phillips, R. E. (1997). Escape of human immunodeficiency virus from immune control. Annu. Rev. Immunol. *15*, 271-296.

McNearney, T. A., Odell, C., Holers, V. M., Spear, P. G., and Atkinson, J. P. (1987). Herpes simplex virus glycoproteins gC-1 and gC-2 bind to the third component of complement and provide protection against complement- mediated neutralization of viral infectivity. J. Exp. Med. *166*, 1525-1535.

Meunier, J. C., Fournillier, A., Choukhi, A., Cahour, A., Cocquerel, L., Dubuisson, J., and Wychowski, C. (1999). Analysis of the glycosylation sites of hepatitis C virus (HCV) glycoprotein E1 and the influence of E1 glycans on the formation of the HCV glycoprotein complex. J. Gen. Virol. *80 (Pt 4)*, 887-896.

Mikkelsen, J. G., and Pedersen, F. S. (2000). Genetic reassortment and patch repair by recombination in retroviruses. J. Biomed Sci. *7*, 77-99.

Miller, C. G., Shchelkunov, S. N., and Kotwal, G. J. (1997). The cowpox virus-encoded homolog of the vaccinia virus complement control protein is an inflammation modulatory protein. Virology *229*, 126-133.

Mo, C., Schneeberger, E. E., and Arvin, A. M. (2000). Glycoprotein E of varicella-zoster virus enhances cell-cell contact in polarized epithelial cells. J. Virol. *74*, 11377-11387.

Montefiori, D. C., Cornell, R. J., Zhou, J. Y., Zhou, J. T., Hirsch, V. M., and Johnson, P. R. (1994). Complement control proteins, CD46, CD55, and CD59, as common surface constituents of human and simian immunodeficiency viruses and possible targets for vaccine protection. Virology *205*, 82-92.

Moore, J. P., Cao, Y., Ho, D. D., and Koup, R. A. (1994). Development of the anti-gp120 antibody response during seroconversion to human immunodeficiency virus type 1. J. Virol. *68*, 5142-5155.

Moss, B. (2001). Poxviridae: the viruses and their replication. In Virology, B. N. Fields, D. M. Knipe, P. M. Howley, R. M. Chanock, J. Melnick, T. P. Monath, B. Roizman, and S. E. Straus, eds. (Philadelphia, Lippincott-Raven Publishers), pp. 2849-2883.

Mozdzanowska, K., Maiese, K., and Gerhard, W. (2000). Th cell-deficient mice control influenza virus infection more effectively than Th- and B cell-deficient mice: evidence for a Th-independent contribution by B cells to virus clearance. J. Immunol. *164*, 2635-2643.

Muggeridge, M. I., Roberts, S. R., Isola, V. J., Cohen, G. H., and Eisenberg, R. J. (1990). Herpes simplex virus glycoproteins. In Immunochemistry of Viruses: The Basis for Serodiagnosis and Vaccines, M. H. van Regenmortel, and A. R. Neurath, eds. (New York, Elsevier), pp. 459-481.

Muller, S. (2004). Avoiding deceptive imprinting of the immune response to HIV-1 infection in vaccine development. Int. Rev. Immunol. *23*, 423-436.

Nagashunmugam, T., Lubinski, J., Wang, L., Goldstein, L. T., Weeks, B. S., Sundaresan, P., Kang, E. H., Dubin, G., and Friedman, H. M. (1998). *In vivo* immune evasion mediated by the herpes simplex virus type 1 immunoglobulin G Fc receptor. J. Virol. *72*, 5351-5359.

Naniche, D., and Oldstone, M. B. (2000). Generalized immunosuppression: how viruses undermine the immune response. Cell Mol. Life Sci. *57*, 1399-1407.

Neirynck, S., Deroo, T., Saelens, X., Vanlandschoot, P., Jou, W. M., and Fiers, W. (1999). A universal influenza A vaccine based on the extracellular domain of the M2 protein. Nat. Med. *5*, 1157-1163.

Neumann, A. U., Lam, N. P., Dahari, H., Gretch, D. R., Wiley, T. E., Layden, T. J., and Perelson, A. S. (1998). Hepatitis C viral dynamics *in vivo* and the antiviral efficacy of interferon-alpha therapy. Science *282*, 103-107.

Newsome, T. P., Scaplehorn, N., and Way, M. (2004). SRC mediates a switch from microtubule- to actin-based motility of vaccinia virus. Science *306*, 124-129.

Nuss, J. M., Whitaker, P. B., and Air, G. M. (1993). Identification of critical contact residues in the NC41 epitope of a subtype N9 influenza virus neuraminidase. Proteins *15*, 121-132.

Ochsenbein, A. F., Pinschewer, D. D., Odermatt, B., Carroll, M. C., Hengartner, H., and Zinkernagel, R. M. (1999). Protective T cell-independent antiviral antibody responses are dependent on complement. J. Exp. Med. *190*, 1165-1174.

Ochsenbein, A. F., and Zinkernagel, R. M. (2000). Natural antibodies and complement link innate and acquired immunity. Immunol. Today *21*, 624-630.

Ohgimoto, S., Shioda, T., Mori, K., Nakayama, E. E., Hu, H., and Nagai, Y. (1998). Location-specific, unequal contribution of the N. glycans in simian immunodeficiency virus gp120 to viral infectivity and removal of multiple glycans without disturbing infectivity. J. Virol. *72*, 8365-8370.

Ohno, H., Aguilar, R. C., Fournier, M. C., Hennecke, S., Cosson, P., and Bonifacino, J. S. (1997). Interaction of endocytic signals from the HIV-1 envelope glycoprotein complex with members of the adaptor medium chain family. Virology *238*, 305-315.

Okada, H., Wakamiya, N., Okada, N., and Kato, S. (1987). Sensitization of human tumor cells to homologous complement by vaccinia virus treatment. Cancer Immunol. Immunother *25*, 7-9.

O'Neil, P. K., Sun, G., Yu, H., Ron, Y., Dougherty, J. P., and Preston, B. D. (2002). Mutational analysis of HIV-1 long terminal repeats to explore the relative contribution of reverse transcriptase and RNA polymerase II to viral mutagenesis. J. Biol. Chem. *277*, 38053-38061.

Ostrowski, M., Galeota, J. A., Jar, A. M., Platt, K. B., Osorio, F. A., and Lopez, O. J. (2002). Identification of neutralizing and nonneutralizing epitopes in the porcine reproductive and respiratory syndrome virus GP5 ectodomain. J. Virol. *76*, 4241-4250.

Pal, R., Hoke, G. M., and Sarngadharan, M. G. (1989). Role of oligosaccharides in the processing and maturation of envelope glycoproteins of human immunodeficiency virus type 1. Proc. Natl. Acad. Sci. USA. *86*, 3384-3388.

Paradiso, P. R., and Lindberg, A. A. (1996). Glycoconjugate vaccines: future combinations. Dev. Biol. Stand *87*, 269-275.

Parren, P. W., Moore, J. P., Burton, D. R., and Sattentau, Q. J. (1999). The neutralizing antibody response to HIV-1: viral evasion and escape from humoral immunity. Aids *13 Suppl A*, S137-162.

Parren, P. W., Wang, M., Trkola, A., Binley, J. M., Purtscher, M., Katinger, H., Moore, J. P., and Burton, D. R. (1998). Antibody neutralization-resistant primary isolates of human immunodeficiency virus type 1. J. Virol. *72*, 10270-10274.

Pays, E., Vanhamme, L., and Perez-Morga, D. (2004). Antigenic variation in *Trypanosoma brucei*: facts, challenges and mysteries. Curr. Opin. Microbiol. *7*, 369-374.

Perelson, A. S., Neumann, A. U., Markowitz, M., Leonard, J. M., and Ho, D. D. (1996). HIV-1 dynamics *in vivo*: virion clearance rate, infected cell life-span, and viral generation time. Science *271*, 1582-1586.

Piguet, V., and Sattentau, Q. (2004). Dangerous liaisons at the virological synapse. J. Clin. Invest *114*, 605-610.

Pikora, C. A. (2004). Glycosylation of the ENV spike of primate immunodeficiency viruses and antibody neutralization. Curr. HIV Res. *2*, 243-254.

Pinschewer, D. D., Perez, M., Jeetendra, E., Bachi, T., Horvath, E., Hengartner, H., Whitt, M. A., de la Torre, J. C., and Zinkernagel, R. M. (2004). Kinetics of protective antibodies are determined by the viral surface antigen. J. Clin. Invest *114*, 988-993.

Planz, O., Seiler, P., Hengartner, H., and Zinkernagel, R. M. (1996). Specific cytotoxic T cells eliminate B cells producing virus-neutralizing antibodies [corrected]. Nature *382*, 726-729.

Planz, O., Seiler, P., Hengartner, H., and Zinkernagel, R. M. (2003). Specific cytotoxic T cells eliminate cells producing neutralizing antibodies. Nature *426*, 474.

Poignard, P., Sabbe, R., Picchio, G. R., Wang, M., Gulizia, R. J., Katinger, H., Parren, P. W., Mosier, D. E., and Burton, D. R. (1999). Neutralizing antibodies have limited effects on the control of established HIV-1 infection *in vivo*. Immunity *10*, 431-438.

Poignard, P., Saphire, E. O., Parren, P. W., and Burton, D. R. (2001). gp120: Biologic aspects of structural features. Annu. Rev. Immunol. *19*, 253-274.

Ramratnam, B., Bonhoeffer, S., Binley, J., Hurley, A., Zhang, L., Mittler, J. E., Markowitz, M., Moore, J. P., Perelson, A. S., and Ho, D. D. (1999). Rapid production and clearance of HIV-1 and hepatitis C virus assessed by large volume plasma apheresis. Lancet *354*, 1782-1785.

Ravanel, K., Castelle, C., Defrance, T., Wild, T. F., Charron, D., Lotteau, V., and Rabourdin-Combe, C. (1997). Measles virus nucleocapsid protein binds to FcgammaRII and inhibits human B cell antibody production. J. Exp. Med. *186*, 269-278.

Recker, M., Nee, S., Bull, P. C., Kinyanjui, S., Marsh, K., Newbold, C., and Gupta, S. (2004). Transient cross-reactive immune responses can orchestrate antigenic variation in malaria. Nature *429*, 555-558.

Richman, D. D., Wrin, T., Little, S. J., and Petropoulos, C. J. (2003). Rapid evolution of the neutralizing antibody response to HIV type 1 infection. Proc. Natl. Acad. Sci. USA. *100*, 4144-4149.

Rizzuto, C., and Sodroski, J. (2000). Fine definition of a conserved CCR5-binding region on the human immunodeficiency virus type 1 glycoprotein 120. AIDS Res. Hum. Retroviruses *16*, 741-749.

Rizzuto, C. D., Wyatt, R., Hernandez-Ramos, N., Sun, Y., Kwong, P. D., Hendrickson, W. A., and Sodroski, J. (1998). A conserved HIV gp120 glycoprotein structure involved in chemokine receptor binding. Science *280*, 1949-1953.

Robertson, D. L., Sharp, P. M., McCutchan, F. E., and Hahn, B. H. (1995). Recombination in HIV-1. Nature *374*, 124-126.

Rodriguez, J. F., Janeczko, R., and Esteban, M. (1985). Isolation and characterization of neutralizing monoclonal antibodies to vaccinia virus. J. Virol. *56*, 482-488.

Rodriguez, J. F., Paez, E., and Esteban, M. (1987). A 14,000-Mr envelope protein of vaccinia virus is involved in cell fusion and forms covalently linked trimers. J. Virol. *61*, 395-404.

Rohm, C., Zhou, N., Suss, J., Mackenzie, J., and Webster, R. G. (1996). Characterization of a novel influenza hemagglutinin, H15: criteria for determination of influenza A subtypes. Virology *217*, 508-516.

Roizman, B., and Knipe, D. M. (2001). Herpes simplex viruses and their replication. In Virology, B. N. Fields, D. M. Knipe, P. M. Howley, R. M. Chanock, J. Melnick, T. P. Monath, B. Roizman, and S. E. Straus, eds. (Philadelphia, Lippincott-Raven Publishers), pp. 2399-2459.

Rossmann, M. G. (1989). The canyon hypothesis. Viral Immunol. *2*, 143-161.

Rossmann, M. G., Arnold, E., Erickson, J. W., Frankenberger, E. A., Griffith, J. P., Hecht, H. J., Johnson, J. E., Kamer, G., Luo, M., Mosser, A. G., and et al. (1985). Structure of a human common cold virus and functional relationship to other picornaviruses. Nature *317*, 145-153.

Rux, A. H., Lou, H., Lambris, J. D., Friedman, H. M., Eisenberg, R. J., and Cohen, G. H. (2002). Kinetic analysis of glycoprotein C of herpes simplex virus types 1 and 2 binding to heparin, heparan sulfate, and complement component C3b. Virology *294*, 324-332.

Sahu, A., Isaacs, S. N., Soulika, A. M., and Lambris, J. D. (1998). Interaction of vaccinia virus complement control protein with human complement proteins: factor I-mediated degradation of C3b to iC3b1 inactivates the alternative complement pathway. J. Immunol. *160*, 5596-5604.

Saifuddin, M., Hedayati, T., Atkinson, J. P., Holguin, M. H., Parker, C. J., and Spear, G. T. (1997). Human immunodeficiency virus type 1 incorporates both glycosyl phosphatidylinositol-anchored CD55 and CD59 and integral membrane CD46 at levels that protect from complement-mediated destruction. J. Gen. Virol. *78*, 1907-1911.

Saifuddin, M., Parker, C. J., Peeples, M. E., Gorny, M. K., Zolla-Pazner, S., Ghassemi, M., Rooney, I. A., Atkinson, J. P., and Spear, G. T. (1995). Role of virion-associated glyco sylphosphatidylinositol-linked proteins CD55 and CD59 in complement resistance of cell line-derived and primary isolates of HIV-1. J. Exp. Med. *182*, 501-509.

Sanchez, A., Trappier, S. G., Mahy, B. W., Peters, C. J., and Nichol, S. T. (1996). The virion glycoproteins of Ebola viruses are encoded in two reading frames and are expressed through transcriptional editing. Proc. Natl. Acad. Sci. USA. *93*, 3602-3607.

Sanchez, A., Yang, Z. Y., Xu, L., Nabel, G. J., Crews, T., and Peters, C. J. (1998). Biochemical analysis of the secreted and virion glycoproteins of Ebola virus. J. Virol. *72*, 6442-6447.

Sanders, R. W., Venturi, M., Schiffner, L., Kalyanaraman, R., Katinger, H., Lloyd, K. O., Kwong, P. D., and Moore, J. P. (2002). The mannose-dependent epitope for neutralizing antibody 2G12 on human immunodeficiency virus type 1 glycoprotein gp120. J. Virol. *76*, 7293-7305.

Sanna, P. P., Ramiro-Ibanez, F., and De Logu, A. (2000). Synergistic interactions of antibodies in rate of virus neutralization. Virology *270*, 386-396.

Scanlan, C. N., Pantophlet, R., Wormald, M. R., Ollmann Saphire, E., Stanfield, R., Wilson, I. A., Katinger, H., Dwek, R. A., Rudd, P. M., and Burton, D. R. (2002). The broadly neutralizing anti-human immunodeficiency virus type 1 antibody 2G12 recognizes a cluster of alpha1-->2 mannose residues on the outer face of gp120. J. Virol. *76*, 7306-7321.

Scholtissek, C., Koennecke, I., and Rott, R. (1978). Host range recombinants of fowl plague (influenza A) virus. Virology *91*, 79-85.

Schonning, K., Jansson, B., Olofsson, S., and Hansen, J. E. (1996). Rapid selection for an N-linked oligosaccharide by monoclonal antibodies directed against the V3 loop of human immunodeficiency virus type 1. J. Gen. Virol. *77 (Pt 4)*, 753-758.

Sears, A. E., McGwire, B. S., and Roizman, B. (1991). Infection of polarized MDCK cells with herpes simplex virus 1: two asymmetrically distributed cell receptors interact with different viral proteins. Proc. Natl. Acad. Sci. USA. *88*, 5087-5091.

Senkevich, T. G., Ward, B. M., and Moss, B. (2004). Vaccinia virus entry into cells is dependent on a virion surface protein encoded by the A28L gene. J. Virol. *78*, 2357-2366.

Shortridge, K. F., Zhou, N. N., Guan, Y., Gao, P., Ito, T., Kawaoka, Y., Kodihalli, S., Krauss, S., Markwell, D., Murti, K. G., et al. (1998). Characterization of avian H5N1 influenza viruses from poultry in Hong Kong. Virology *252*, 331-342.

Sims, L. D., Ellis, T. M., Liu, K. K., Dyrting, K., Wong, H., Peiris, M., Guan, Y., and Shortridge, K. F. (2003). Avian influenza in Hong Kong 1997-2002. Avian Dis. *47*, 832-838.

Skehel, J. J., Stevens, D. J., Daniels, R. S., Douglas, A. R., Knossow, M., Wilson, I. A., and Wiley, D. C. (1984). A carbohydrate side chain on hemagglutinins of Hong Kong influenza viruses inhibits recognition by a monoclonal antibody. Proc. Natl. Acad. Sci. USA. *81*, 1779-1783.

Skehel, J. J., and Wiley, D. C. (2000). Receptor binding and membrane fusion in virus entry: the influenza hemagglutinin. Annu. Rev. Biochem. *69*, 531-569.

Smith, G. L., and Law, M. (2004). The exit of Vaccinia virus from infected cells. Virus Res. *106*, 189-197.

Smith, G. L., Vanderplasschen, A., and Law, M. (2002). The formation and function of extracellular enveloped vaccinia virus. J. Gen. Virol. *83*, 2915-2931.

Smith, T. J. (2001). Antibody interactions with rhinovirus: lessons for mechanisms of neutralization and the role of immunity in viral evolution. Curr. Top Microbiol. Immunol. *260*, 1-28.

Smith, T. J., Chase, E. S., Schmidt, T. J., Olson, N. H., and Baker, T. S. (1996). Neutralizing antibody to human rhinovirus 14 penetrates the receptor-binding canyon. Nature *383*, 350-354.

Spear, G. T., Lurain, N. S., Parker, C. J., Ghassemi, M., Payne, G. H., and Saifuddin, M. (1995). Host cell-derived complement control proteins CD55 and CD59 are incorporated into the virions of two unrelated enveloped viruses. Human T cell leukemia/lymphoma virus type I (HTLV-I) and human cytomegalovirus (HCMV). J. Immunol. *155*, 4376-4381.

Spear, P. G. (2004). Herpes simplex virus: receptors and ligands for cell entry. Cell Microbiol. *6*, 401-410.

Spiller, O. B., Hanna, S. M., Devine, D. V., and Tufaro, F. (1997). Neutralization of cytomegalovirus virions: the role of complement. J. Infect. Dis. *176*, 339-347.

Spiller, O. B., Morgan, B. P., Tufaro, F., and Devine, D. V. (1996). Altered expression of host-encoded complement regulators on human cytomegalovirus-infected cells. Eur. J. Immunol. *26*, 1532-1538.

Stanfield, R. L., Fieser, T. M., Lerner, R. A., and Wilson, I. A. (1990). Crystal structures of an antibody to a peptide and its complex with peptide antigen at 2.8 A. Science *248*, 712-719.

Starcich, B. R., Hahn, B. H., Shaw, G. M., McNeely, P. D., Modrow, S., Wolf, H., Parks, E. S., Parks, W. P., Josephs, S. F., Gallo, R. C., and et al. (1986). Identification and characterization of conserved and variable regions in the envelope gene of HTLV-III/LAV, the retrovirus of AIDS. Cell *45*, 637-648.

Steinhauer, D. A. (1999). Role of hemagglutinin cleavage for the pathogenicity of influenza virus. Virology *258*, 1-20.

Stoiber, H., Pinter, C., Siccardi, A. G., Clivio, A., and Dierich, M. P. (1996). Efficient destruction of human immunodeficiency virus in human serum by inhibiting the protective action of complement factor H and decay accelerating factor (DAF, CD55). J. Exp. Med. *183*, 307-310.

Stoiber, H., Schneider, R., Janatova, J., and Dierich, M. P. (1995). Human complement proteins C3b, C4b, factor H and properdin react with specific sites in gp120 and gp41, the envelope proteins of HIV-1. Immunobiology *193*, 98-113.

Szomolanyi-Tsuda, E., Le, Q. P., Garcea, R. L., and Welsh, R. M. (1998). T-Cell-independent immunoglobulin G responses *in vivo* are elicited by live-virus infection but not by immunization with viral proteins or virus-like particles. J. Virol. *72*, 6665-6670.

Szomolanyi-Tsuda, E., and Welsh, R. M. (1996). T cell-independent antibody-mediated clearance of polyoma virus in T cell-deficient mice. J. Exp. Med. *183*, 403-411.

Süsal, C., Kirschfink, M., Kröpelin, M., Daniel, V., and Opelz, G. (1994). Complement activation by recombinant HIV-1 glycoprotein gp120. J. Immunol. *152*, 6028-6034.

Thielens, N. M., Bally, I. M., Ebenbichler, C. F., Dierich, M. P., and Arlaud, G. J. (1993). Further characterization of the interaction between the C1q subcomponent of human C1 and the transmembrane envelope glycoprotein gp41 of HIV-1. J. Immunol. *151*, 6583-6592.

Thomas, D., Newcomb, W. W., Brown, J. C., Wall, J. S., Hainfeld, J. F., Trus, B. L., and Steven, A. C. (1985). Mass and molecular composition of vesicular stomatitis virus: a scanning transmission electron microscopy analysis. J. Virol. *54*, 598-607.

Tirabassi, R. S., and Enquist, L. W. (1999). Mutation of the YXXL endocytosis motif in the cytoplasmic tail of pseudorabies virus gE. J. Virol. *73*, 2717-2728.

Tirabassi, R. S., Townley, R. A., Eldridge, M. G., and Enquist, L. W. (1997). Characterization of pseudorabies virus mutants expressing carboxy-terminal truncations of gE: evidence for envelope incorporation, virulence, and neurotropism domains. J. Virol. *71*, 6455-6464.

Topham, D. J., and Doherty, P. C. (1998). Clearance of an influenza A virus by CD4+ T cells is inefficient in the absence of B cells. J. Virol. *72*, 882-885.

Torpier, G., Capron, A., and Ouaissi, M. A. (1979). Receptor for IgG(Fc) and human beta2-microglobulin on S. mansoni schistosomula. Nature *278*, 447-449.

Tortorella, D., Gewurz, B. E., Furman, M. H., Schust, D. J., and Ploegh, H. L. (2000). Viral subversion of the immune system. Annu. Rev. Immunol. *18*, 861-926.

Tran, T. H., Nguyen, T. L., Nguyen, T. D., Luong, T. S., Pham, P. M., Nguyen, V. C., Pham, T. S., Vo, C. D., Le, T. Q., Ngo, T. T., et al. (2004). Avian influenza A (H5N1) in 10 patients in Vietnam. N. Engl. J. Med. *350*, 1179-1188.

Trkola, A., Purtscher, M., Muster, T., Ballaun, C., Buchacher, A., Sullivan, N., Srinivasan, K., Sodroski, J., Moore, J. P., and Katinger, H. (1996). Human monoclonal antibody 2G12 defines a distinctive neutralization epitope on the gp120 glycoprotein of human immunodeficiency virus type 1. J. Virol. *70*, 1100-1108.

Tsuchiya, H., Furukawa, M., Matsui, M., Katsuki, K., and Inouye, S. (2000). 'Original antigenic sin' phenomenon in neutralizing antibody responses in children with enterovirus meningitis. J. Clin. Virol. *19*, 205-207.

van Doorn, L. J., Capriles, I., Maertens, G., DeLeys, R., Murray, K., Kos, T., Schellekens, H., and Quint, W. (1995). Sequence evolution of the hypervariable region in the putative envelope region E2/NS1 of hepatitis C virus is correlated with specific humoral immune responses. J. Virol. *69*, 773-778.

Van Vliet, K. E., De Graaf-Miltenburg, L. A., Verhoef, J., and Van Strijp, J. A. (1992). Direct evidence for antibody bipolar bridging on herpes simplex virus-infected cells. Immunology *77*, 109-115.

Vanderplasschen, A., Hollinshead, M., and Smith, G. L. (1998a). Intracellular and extracellular vaccinia virions enter cells by different mechanisms. J. Gen. Virol. *79*, 877-887.

Vanderplasschen, A., Mathew, E., Hollinshead, M., Sim, R. B., and Smith, G. L. (1998b). Extracellular enveloped vaccinia virus is resistant to complement because of incorporation of host complement control proteins into its envelope. Proc. Natl. Acad. Sci. USA. *95*, 7544-7549.

Vanderplasschen, A., and Smith, G. L. (1997). A novel virus binding assay using confocal microscopy: demonstration that the intracellular and extracellular vaccinia virions bind to different cellular receptors. J. Virol. *71*, 4032-4041.

Wakamiya, N., Okada, N., Wang, Y. L., Ito, T., Ueda, S., Kato, S., and Okada, H. (1989). Tumor cells treated with vaccinia virus can activate the alternative pathway of mouse complement. Jpn J. Cancer Res. *80*, 765-770.

Ward, B. M., and Moss, B. (2000). Golgi network targeting and plasma membrane internalization signals in vaccinia virus B5R envelope protein. J. Virol. *74*, 3771-3780.

Watkins, J. F. (1964). Adsorption of Sensitized Sheep Erythrocytes to Hela Cells Infected with Herpes Simplex Virus. Nature *202*, 1364-1365.

Webster, R. G. (1966). Original antigenic sin in ferrets: the response to sequential infections with influenza viruses. J. Immunol. *97*, 177-183.

Webster, R. G., Kasel, J. A., Couch, R. B., and Laver, W. G. (1976). Influenza virus subunit vaccines. II. Immunogenicity and original antigenic sin in humans. J. Infect. Dis. *134*, 48-58.

Weeks, B. S., Sundaresan, P., Nagashunmugam, T., Kang, E., and Friedman, H. M. (1997). The herpes simplex virus-1 glycoprotein E (gE) mediates IgG binding and cell-to-cell spread through distinct gE domains. Biochem. Biophys Res. Commun. *235*, 31-35.

Wei, X., Decker, J. M., Wang, S., Hui, H., Kappes, J. C., Wu, X., Salazar-Gonzalez, J. F., Salazar, M. G., Kilby, J. M., Saag, M. S., et al. (2003). Antibody neutralization and escape by HIV-1. Nature *422*, 307-312.

Wei, X., Ghosh, S. K., Taylor, M. E., Johnson, V. A., Emini, E. A., Deutsch, P., Lifson, J. D., Bonhoeffer, S., Nowak, M. A., Hahn, B. H., and et al. (1995). Viral dynamics in human immunodeficiency virus type 1 infection. Nature *373*, 117-122.

Weis, W., Brown, J. H., Cusack, S., Paulson, J. C., Skehel, J. J., and Wiley, D. C. (1988). Structure of the influenza virus haemagglutinin complexed with its receptor, sialic acid. Nature *333*, 426-431.

Whitbeck, J. C., Knapp, A. C., Enquist, L. W., Lawrence, W. C., and Bello, L. J. (1996). Synthesis, processing, and oligomerization of bovine herpesvirus 1 gE and gI membrane proteins. J. Virol. *70*, 7878-7884.

Whittle, H. C., Bradley-Moore, A., Fleming, A., and Greenwood, B. M. (1973). Effects of measles on the immune response of Nigerian children. Arch. Dis. Child *48*, 753-756.

Whitton, J. L., and Oldstone, M. B. A. (2001). The immune responses to virual infections. In Virology, B. N. Fields, D. M. Knipe, P. M. Howley, R. M. Chanock, J. Melnick, T. P. Monath, B. Roizman, and S. E. Straus, eds. (Philadelphia, Lippincott-Raven Publishers), pp. 285-320.

Wiley, D. C., and Skehel, J. J. (1987). The structure and function of the hemagglutinin membrane glycoprotein of influenza virus. Annu. Rev. Biochem. *56*, 365-394.

Wiley, D. C., Wilson, I. A., and Skehel, J. J. (1981). Structural identification of the antibody-binding sites of Hong Kong influenza haemagglutinin and their involvement in antigenic variation. Nature *289*, 373-378.

Wilson, I. A., and Cox, N. J. (1990). Structural basis of immune recognition of influenza virus hemagglutinin. Annu. Rev. Immunol. *8*, 737-771.

Wilson, I. A., Skehel, J. J., and Wiley, D. C. (1981). Structure of the haemagglutinin membrane glycoprotein of influenza virus at 3 A resolution. Nature *289*, 366-373.

Wolffe, E. J., Vijaya, S., and Moss, B. (1995). A myristylated membrane protein encoded by the vaccinia virus L1R open reading frame is the target of potent neutralizing monoclonal antibodies. Virology *211*, 53-63.

Wright, K. E., Salvato, M. S., and Buchmeier, M. J. (1989). Neutralizing epitopes of lymphocytic choriomeningitis virus are conformational and require both glycosylation and disulfide bonds for expression. Virology *171*, 417-426.

Wright, P. F., and Webster, R. G. (2001). Orthomyxoviruses. In Virology, B. N. Fields, D. M. Knipe, P. M. Howley, R. M. Chanock, J. Melnick, T. P. Monath, B. Roizman, and S. E. Straus, eds. (Philadelphia, Lippincott-Raven Publishers), pp. 1533-1579.

WuDunn, D., and Spear, P. G. (1989). Initial interaction of herpes simplex virus with cells is binding to heparan sulfate. J. Virol. *63*, 52-58.

Wyatt, R., Kwong, P. D., Desjardins, E., Sweet, R. W., Robinson, J., Hendrickson, W. A., and Sodroski, J. G. (1998). The antigenic structure of the HIV gp120 envelope glycoprotein. Nature *393*, 705-711.

Wyatt, R., and Sodroski, J. (1998). The HIV-1 envelope glycoproteins: fusogens, antigens, and immunogens. Science *280*, 1884-1888.

Wyss, S., Berlioz-Torrent, C., Boge, M., Blot, G., Honing, S., Benarous, R., and Thali, M. (2001). The highly conserved C-terminal dileucine motif in the cytosolic domain of the human immunodeficiency virus type 1 envelope glycoprotein is critical for its association with the AP-1 clathrin adaptor [correction of adapter]. J. Virol. *75*, 2982-2992.

Xu, X., Subbarao, Cox, N. J., and Guo, Y. (1999). Genetic characterization of the pathogenic influenza A/Goose/Guangdong/1/96 (H5N1) virus: similarity of its hemagglutinin gene to those of H5N1 viruses from the 1997 outbreaks in Hong Kong. Virology *261*, 15-19.

Yarnall, M., Gogolewski, R. P., and Corbeil, L. B. (1988). Characterization of two Haemophilus somnus Fc receptors. J. Gen. Microbiol. *134*, 1993-1999.

Ye, Y., Si, Z. H., Moore, J. P., and Sodroski, J. (2000). Association of structural changes in the V2 and V3 loops of the gp120 envelope glycoprotein with acquisition of neutralization resistance in a simian-human immunodeficiency virus passaged *in vivo*. J. Virol. *74*, 11955-11962.

Yoshiyama, H., Mo, H., Moore, J. P., and Ho, D. D. (1994). Characterization of mutants of human immunodeficiency virus type 1 that have escaped neutralization by a monoclonal antibody to the gp120 V2 loop. J. Virol. *68*, 974-978.

Zambrano-Villa, S., Rosales-Borjas, D., Carrero, J. C., and Ortiz-Ortiz, L. (2002). How protozoan parasites evade the immune response. Trends Parasitol *18*, 272-278.

Zhang, M., Gaschen, B., Blay, W., Foley, B., Haigwood, N., Kuiken, C., and Korber, B. (2004). Tracking global patterns of N-linked glycosylation site variation in highly variable viral glycoproteins: HIV, SIV, and HCV envelopes and influenza hemagglutinin. Glycobiology *14*, 1229-1246.

Zinkernagel, R. M. (1997). Felix Hoppe-Seyler Lecture 1997. Protective antibody responses against viruses. Biol. Chem. *378*, 725-729.

Zinkernagel, R. M. (2003). On natural and artificial vaccinations. Annu. Rev. Immunol. *21*, 515-546.

Zinkernagel, R. M., LaMarre, A., Ciurea, A., Hunziker, L., Ochsenbein, A. F., McCoy, K. D., Fehr, T., Bachmann, M. F., Kalinke, U., and Hengartner, H. (2001). Neutralizing antiviral antibody responses. Adv. Immunol. *79*, 1-53.

Zinoviev, V. V., Tchikaev, N. A., Chertov, O., and Malygin, E. G. (1994). Identification of the gene encoding vaccinia virus immunodominant protein p35. Gene *147*, 209-214.

Zlateva, K. T., Lemey, P., Vandamme, A. M., and Van Ranst, M. (2004). Molecular evolution and circulation patterns of human respiratory syncytial virus subgroup a: positively selected sites in the attachment g glycoprotein. J. Virol. *78*, 4675-4683.

Zwick, M. B., Saphire, E. O., and Burton, D. R. (2004). gp41: HIV's shy protein. Nat. Med. *10*, 133-134.

8

Viruses and Dendritic Cells: A Prominent Mechanism for Subverting the Immune Response

Elina I. Zuniga, Kurt H. Edelmann and Michael B.A. Oldstone*

Abstract

Viruses have evolved multiple strategies to counteract host immune responses. The essential role played by DCs in host defense to pathogens makes them an ideal choice for viruses to suppress the immune response and maximize their chances of survival, replication and transmission. Indeed many viruses that cause diseases and major health problems are able to interfere with the ability of DCs to prime an efficient and effective anti-viral immune response. By analogy to human viral infections, lymphocytic choriomeningitis virus (LCMV) infection in its natural host (the mouse) provides a convenient and powerful model system to study how viruses influence the DC compartment, avoid immunological surveillance and cause a persistent infection.

In this chapter we briefly review the DC fundamental role during immune response, the several strategies of human viruses to subvert DC function and finally we describe in more detail the DC-virus interaction and its consequences in the LCMV murine model system

1. Introduction

Host defense against microbial infections relies on complementary response between innate and adaptive immune system. Dendritic cells (DCs) are key elements for protection of the host as they are responsible for bridging these two arms of immunity. These cells are a widely distributed, migratory group of bone marrow-derived leukocytes that are specialized for the up-take, transport, processing and presentation of antigens to T cells. At an "immature" stage of development, DCs act

*For correspondence email ezuniga@scripps.edu

as sentinels in peripheral tissues, continuously sampling the antigenic environment. These cells are able to detect components of invading pathogens which trigger their activation/maturation. Specifically, pathogen associated molecular patters (PAMPs), as these components have been termed, range from lipoproteins to proteins to nucleic acids carried by potential invaders. These PAMPs are recognized by evolutionary conserved "pattern recognition receptors" (PRRs) on DCs. Among PRRs, the toll-like receptors (TLRs) have emerged as critical players in determining DC imprinting on the ensuing immune response. TLR triggering has pleitropic effects on DCs, promoting survival, chemokine secretion, expression of chemokine receptors, migration, cytoskeletal and shape changes or endocytic remodelling (Reis e Sousa, 2004). After interacting with these pathogen signatures, activated DCs change their chemokine receptors and initiate migration to lymph nodes (LNs) and spleen. The pathogen's antigens are processed and presented as peptides associated with major histocompatibility complex (MHC) molecules. During transit the cells undergo a program of maturation events that take them from being poorly immunogenic to being the most potent antigen presenting cells (APC) for naïve T cells. Indeed, migrating DCs up-regulate co-stimulatory molecules (CD80, CD86, ICAM-1) that are required for effective interactions with T cells. In the LNs and spleen, the now mature DCs, with up-regulated co-stimulatory molecules (CD80, CD86, ICAM-1, etc) provide the necessary signals to activate any T cells with a receptor that is specific for the foreign peptide-MHC complex. DCs are considered the most powerful APC, not only because they are more potent than B cells and macrophages to activate naïve T cells, but also because they have the capacity to migrate into the T cells areas of the secondary lymphoid organs, where the T cell response is initiated. In addition with the secretion of specific cytokines and chemokines, activated DCs either directly or indirectly play a crucial role in the recruitment and or/activation of other effector cells, including NK and NKT, macrophages and B cells (reviewed in Itano and Jenkins, 2003; Liu, 2001; Shortman and Liu, 2002).

Interestingly, DCs are a heterogeneous cell population composed of different subtypes which present unique and overlapping functions. As many as six different subsets of DCs occupy the lymph nodes (reviewed in Shortman and Liu, 2002). Three major defined populations of DCs have been recognized in the mouse spleen: Lymphoid DCs ($CD11c^+CD8\alpha^+CD11b^-$), myeloid DCs ($CD11c^+CD11b^+CD8\alpha^-$), and plasmacytoid DCs ($CD11c^+Ly6C^+B220^+$). DC subpopulations differ not only in surface phenotype but also in functional potential and localization. In this regard, it was observed that myeloid and lymphoid DCs are potent activators of naïve T lymphocytes, with the latter being better for cross-presentation of exogenous antigens. In contrast, plasmacytoid DCs, are poor activators of T cells, even after stimulation *in vitro*. They have been proposed to play a more protagonist role during innate immunity by secreting specific cytokines and chemokines, such as type 1 interferons (IFN), and activation of a broad range of effector cells, such as natural killer (NK) cells. Thus, the heterogeneity inherent to DC populations significantly influences the flavors of immune responses to different pathogens which are subsequently amplified by cross-talked between the various subsets. (Dalod et al., 2003; Kelsall et al., 2002)

2. DC as common targets of many virus infections

Viruses have evolved multiple strategies to counteract host immune responses. The essential role played by DCs in host defense to pathogens makes them an ideal choice for viruses to suppress the immune response and maximize their chances of survival, replication and transmission. Indeed many viruses that cause diseases and major health problems are able to interfere with the ability of DCs to prime an efficient and effective anti-viral immune response.

The chain of events leading to proper immune function of DCs can be broken down into several discrete steps: Ag detection, DC maturation, migration, Ag-presentation to lymphocytes, and effector cytokine elicitation. Remarkably, viruses have developed different mechanisms to subvert each of these stages of DC biology (Table 1). Indeed, with the greater understanding of DC immune pathways comes the discovery of novel viral immune-evasion strategies. As mentioned above, TLRs function in DCs as an early sensor against pathogens, therefore impairment in this pathway would likely confer a selective advantage to the infectious agent. To this end, Vaccinia virus (VV) was the first reported to block TLR signaling and the subsequent maturation of DCs. Specifically, two proteins of VV suppress TIR domain dependent intracellular signaling, including IL-1 and TLR-4 mediated activation of the transcription factor NFkB (Bowie et al., 2000).

Migration of DCs, which carry antigen (obtained via infection or phagocytosis) to the lymphoid compartments, is crucial step in initiating the adaptive immune response. Two viruses that have developed mechanisms to prevent migration of infected DCs to lymphoid organs are Herpes simplex virus (HSV) 1 and Human cytomegalovirus (HCMV) (Moutaftsi et al., 2002; Salio et al., 1999). In both cases there is an inhibition of complete DC maturation and subsequent expression of chemokine homing receptors. This strategy forces the immune system to rely on cross-presentation of virus antigens by uninfected bystander DCs in order to initiate immunity (Arrode and Davrinche, 2003). In addition, HCMV takes the strategy of inhibiting DC migration one step further by inhibiting APCs from arriving at a site of infection by producing homologs of chemokines that interfere with host proinflammatory chemokine gradients (Moutaftsi et al., 2004).

An effective immune-evasion strategy used by viruses on DCs is the impairment of Ag-presentation and T cell stimulation functions resulting in either prevention of or non-productive activation of Ag-specific T cells. The ability to disrupt Ag presentation has evolved in many different virus species including adenovirus and Human immunodeficiency virus (HIV); however herpesviruses represent the epitome of evolution in blocking host cell Ag-presentation. Some mechanisms utilized by herpesviruses to disrupt Ag-presentation pathway include blocking peptide transport to the endoplasmic reticulum (ER) through interference with TAP proteins (HSV ICP47, HCMV US6), transport of particular MHC class I (MHC-I) heavy chains from the ER to the cytosol where they are destroyed (HCMV US11, HCMV US2), retention of specific MHC-I heavy chains in the ER (HCMV US3, Murine CMV –MCMV- m152), and disruption of T cell recognition of MHC-I on the cell surface (MCMV m04). That viruses have independently evolved numerous mechanisms to disrupt Ag presentation indicates the effectiveness and importance of this strategy to the survival of a variety of viruses with different infectious lifecycles.

Table 1. Viral mechanisms to subvert each of the stages of DC biology

Virus	Migr.	Ag pres.	Impaired T allostimulation	IL-12 impairment	Maturation	Costim	DC apoptosis
HSV-2	+	impaired	+	+		No effect	+
CMV	+	impaired	+			inhibits	
Dengue					promotes		
Measles			+		promotes		
HIV				impaired			
HCV				impaired			
LCMV		impaired	+		inhibits		
HSV-1		impaired	+	+			
ebola			+	+		inhibits	
lassa			+	+		inhibits	

Maturation of DCs, as mentioned above, results in up-regulation of co-stimulatory molecules and expression of cytokines that make DCs excellent stimulators of naïve T cells. Viruses that can impair T cell stimulation by preventing the up-regulation of co-stimulatory molecules include Ebola, Lassa fever virus (LFV), HSV-1, and HIV. Additionally, a number of viruses, Hepatitis C virus (HCV), HIV, Measles virus (MV) and Dengue virus (DV), are also able to inhibit IL-12 production. HCV does this through the action of its core and nonstructural protein 3 (NS3) which induces production of IL-10 (Dolganiuc et al., 2003). DV, on the other hand, is able to inhibit IL-12 production through an IL-10 independent mechanism (Libraty et al., 2001). Inhibiting IL-12 can be adventitious for pathogens since lower levels of IL-12 can contribute to T cell skewing towards Th2 responses instead of the more effective TH1 responses resulting in the blunting of the CD8 T cell response towards the pathogen.

The fact that not all viruses are able to block DC maturation does not necessarily represent a failure of the pathogen or a win for the host. A good example of this is observed in MV infection which exploits the ability of DCs to mature and migrate to lymphoid organs in response to infection (Dubois et al., 2001). MV benefits greatly by having infected DCs home to lymphoid compartments where the infected cells are able to actively suppress T cell proliferation (mediated through T cell contact with surface viral glycoproteins) and also facilitate virus spread to more lymphoid cells. Therefore, the full understanding of the virus-host relationship requires not only studying the active mechanisms that viruses use to disable the immune system but also asking what benefits a virus gains by leaving a particular immune function.

3. Lessons learned from Lymphocytic Choriomeningitis Virus

By analogy to human viral infections, lymphocytic choriomeningitis virus (LCMV) infection in its natural host (the mouse) provides a convenient and powerful model system to study how viruses influence the DC compartment, avoid immunological surveillance and cause a persistent infection. In this section we describe in more detail DC-virus interaction and its consequences in the LCMV murine model system.

3.1. LCMV structure and cellular receptor

LCMV is the prototype member of the family Arenaviridae, which includes important human pathogens such as Lassa fever, Machupo, Sabia and Junin viruses. Its genome consists of two negative stranded RNA segments with an ambisense coding strategy. The long (L) RNA fragment encodes the viral RNA-dependent RNA polymerase and a small RING finger protein, called Z, while the short (S) RNA fragment encodes the three major structural proteins: the nucleoprotein (NP) and a precursor polypeptide which is cleaved into two products, the peripheral glycoprotein GP1 and the transmembrane glycoprotein GP2 (reviewed in Oldstone, 2002a; 2002b).

The exposed GP1 of LCMV interacts with α-dystroglycan (α-DG), which was identified as a common receptor for LCMV and several arenaviruses that are pathogenic in humans, including Lassa Fever virus (LFV). DG, initially encoded as a single protein, undergoes posttranslational cleavage to form α– and β–DG. The amino terminal part of DG, α–DG, interacts with components of the extracellular matrix and is non-covalently associated with β-DG, the transmembrane component of DG, thereby providing a link between the cell exterior and the cytoskeleton. The DG complex is expressed in most organs at different levels, typically in cell types that adjust to basement membranes. At those sites, DG plays a fundamental role in cell assembly and organization of basement membranes. Interestingly, among the cells of the immune system α-DG is primarily (>99% of total) expressed on DCs, although its specific function on these cells is still unknown.

α-DG undergoes high affinity interactions with the extracellular matrix (ECM) proteins laminin-1, laminin-2, agrin and perlecan. Interestingly, LCMV and ECM proteins, such as laminin-1, utilize an overlapping binding site on α-DG and may compete for binding to their receptor. Thus, co-expression of ECM components with DG may influence the susceptibility of a specific cell for LCMV infection and represent an important determinant for virus tissue tropism *in vivo*. Therefore, the final infectivity of a particular LCMV isolate on a DG+ cell type will be determined by a) its ability to successfully compete with ECM proteins to bind the receptor and b) the expression of ECM molecules in the surrounding environment (reviewed in Kunz et al., 2002).

3.2. LCMV and immunosuppressive variants

Infection of adult mice with the parental LCMV Armstrong 53b (ARM) virus results in an acute infection in which virus is cleared within 7 to 10 days. Virus clearance is mediated by CD8+ antiviral cytotoxic T lymphocytes (CTL) (Byrne and Oldstone, 1984). This protective response proceeds mostly by lysing virally infected cells through a perforin mediated pathway. Studies infecting neonatal mice with LCMV-ARM identified the emergence of viral variants that present an immunosuppressive phenotype (Ahmed and Oldstone, 1988). Clone 13 (Cl13) is the prototype of these variants which, in contrast to the parental virus, ARM, fails to generate LCMV specific CTL responses when injected into immunocompetent adult mice and caused a persistent infection. Remarkably, Cl13 not only evades immune surveillance, but can also induce a generalized immunosuppression as evidenced by an impaired primary CTL response to other RNA and DNA viruses, as well as an inability to generate antibodies when immunized with soluble and particulate foreign antigens (Tishon et al., 1993).

The persistent and immunosuppressive phenotype of Cl13 has been associated with its particular tropism (Borrow et al., 1995). The initial pattern of infection within the spleen of mice infected with either LCMV ARM or Cl13 is similar with both viruses almost exclusively infecting cells of the marginal zone one day post-infection. However, by 3 days post-inoculation, LCMV Cl13 preferentially replicates in the white pulp of the spleen and infects DCs, with more than 80% of splenic DCs infected during the first 20 days after infection (Sevilla et al., 2000). Recent studies showed that Cl13 infects all DC subset present in spleen, including the recently described murine plasmacytoid DCs (unpublished results). In contrast, LCMV-ARM localizes primarily to the red pulp of the spleen, with almost a complete absence in the inner white pulp, and mainly infect F4/80-positive macrophages and few (<10%) DCs. The differential ability of these variants to infect DCs was explained by *in vitro* studies showing that high affinity binding LCMV variants like Cl13 compete successfully with laminin-1 and laminin-2 for α-DG binding. On the other hand, laminin was able to block binding of the low-affinity binding variant ARM to α-DG (Kunz et al., 2001). Interestingly, Cl13 genetically differs from ARM by only 5 nucleotides, of which only two result in residue changes in open reading frames (Salvato et al., 1991). One of the amino acid (aa) substitutions is in position 260 of GP-1, and the other one is in position 1079 of the virus polymerase protein. The aa substitution in the virus polymerase results in enhanced viral replication and may contributes in part to virus persistence (Matloubian et al., 1993). However, previous studies in our laboratory (Dockter et al., 1996; Salvato et al., 1991) mapped the molecular basis of persistence and immunosuppression to the aa change in GP-1 at position 260 from a bulky aromatic aa (Phe) in ARM to a small aliphatic aa (Leu) in Cl13. This single aa change gives Cl13 the advantage to bind more vigorously than ARM to α-DG, and thus more effectively infect DCs (Cao et al., 1998). These findings were extended to other isolates of LCMV demonstrating that variants that bind α-DG with high affinity are associated with virus replication in the white pulp, specifically in DCs, and cause persistent infection and generalized immunosuppression (Sevilla et al., 2000; Smelt et al., 2001). These findings point out to the key role that virus tropism plays in pathogenesis, as two genetically very closely related virus isolates which have differential tropism within the spleen have widely distinct biological outcomes. In addition, the expression of α-DG associated ECM in lymphoid tissues such as spleen may be an important determinant for the differential anatomic distribution of LCMV variants with different receptor binding affinities and the ability to cause different course of disease.

3.3. Mechanisms underlying LCMV subversion of DCs

As a consequence of virus infection of DCs, LCMV Cl13 model underscores the central role of DCs, the most powerful antigen presenting cells, in mediating viral persistence and immunosuppression. Indeed, splenocytes from Cl13-infected mice are not efficient at stimulating a primary mixed lymphocyte reaction, whereas splenocytes from ARM infected mice were as effective as those from uninfected mice at all timepoints after infection (Borrow et al., 1995; Sevilla et al., 2003). These data demonstrate that there is a functional deficit in antigen presenting cells in LCMV Cl13 infected mice.

Remarkably, Cl13 evolved multiple strategies for subverting DCs, including their immunopathological elimination, inhibition of their maturation and blockade of their development from hematopoietic bone marrow progenitors (Figure 1).

3.3.1. Immunopathological elimination of DCs

Immunohistochemical staining of spleenic sections using macrophage and DC-specific antibodies revealed a loss of interdigitating DCs in the central periarterial sheaths in mice infected with LCMV Cl13. In contrast, LCMV-ARM infected mice show normal levels of DCs even when spleens display activation induced changes in their architecture associated with an ongoing immune response (Borrow et al., 1995). Of note, staining with macrophage specific antibodies revealed that there was no significant depletion of the cells expressing these markers in mice infected with

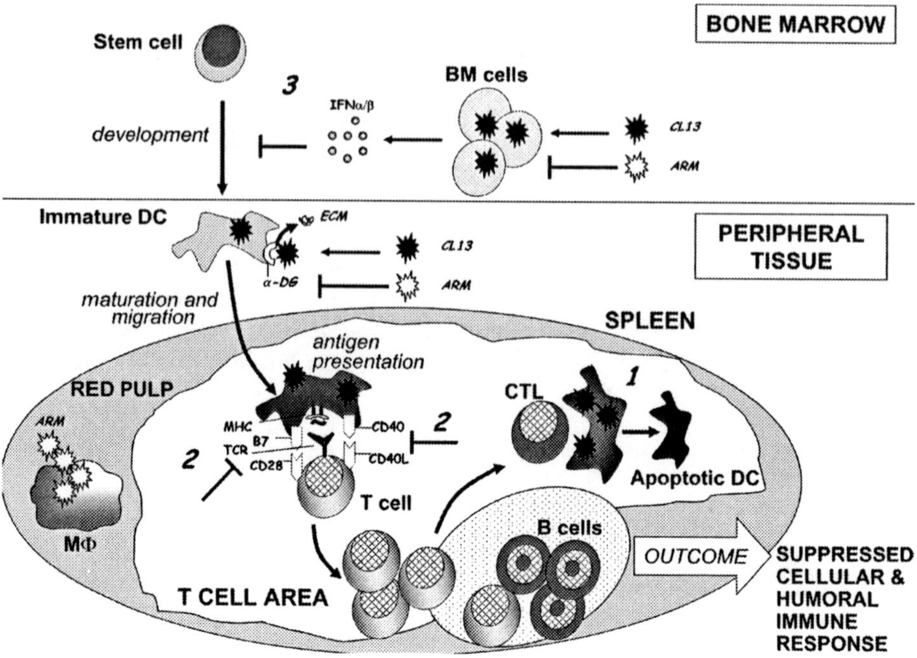

Figure 1. LCMVCl13 differs from its parental virus LCMV ARM by two amino acids. Genetic mapping indicate that the one amino acid difference in viral glycoprotein (GP) at amino acid gp 260 (Cl13 Leu, ARM Phe) is responsible for Cl13 binding at high affinity to its receptor α-dystroglycan (α-DG) and its entry into DCs. Thus in contrast to ARM which localize mainly in the red pulp of the spleen, LCMV Cl13 strategically infects DCs within the white pulp and displays multiple strategies to affect these crucial immune cells at different levels: 1) immunopathological destruction of DCs 2) impairment of DC maturation 3) blockade of DC development from BM progenitors. Subversion of DCs by LCMV-Cl13 causes a general immunosuppression and gives the virus a window of opportunity to chronically establish in the host.

any of LCMV variants. Further experiments addressed the mechanism by which DCs are depleted following LCMV Cl13 infection. By *in vivo* depletion of CD8+ T cells using a subset-specific monoclonal antibody prior to infection of mice with LCMV, it was demonstrated that DC loss during Cl13 infection was dependent on CD8+ T cells. As described above, the immunosuppressive Cl13 LCMV variant establishes a persistent infection because the CTL response generated to it is insufficient to achieve viral clearance (Ahmed et al., 1984). Indeed, 7 days post-infection when the CTL response to LCMV ARM peaks, only low levels of LCMV-specific CTL activity are detected in Cl13 infected mice (Ahmed et al., 1984; Borrow et al., 1995). However, when the LCMV-specific CTL response mounted at earlier time points was measured, Cl13 infected mice were found to make a strong anti-viral CTL response. It is likely that DCs are depleted from periarterial sheaths because of lysis mediated by antiviral CD8+ CTL activated at early times post-infection. In turn, this depletion of DCs would compromise the maintenance of CD8 T cell response, providing a window of time for the virus to replicate and chronically establish in the host

3.3.2. Impairment of DC maturation

Apart from the immunopathological destruction of DCs in Cl13 infected mice, recent findings of Sevilla et al. (2003) revealed an additional mechanism by which the virus alters the luxury function of these powerful antigen presenting cells without affecting its viability. A similar finding in which the virus maintains the normal anatomical architecture of the infected cell, but impairs its differentiated function was observed previously in the endocrine and nervous system (reviewed in Oldstone, 1993). Within the immune system, the infection of DCs by Cl13 leads to impairment of their maturation and a concomitant reduction in the expression of MHC-I, MHC-II, CD40, CD80 and CD86 molecules. The maximum reduction of MHC and costimulatory molecules occurred between days 7 and 15 after infection with immunosuppressive Cl13 variant. At the peak of Cl13 infection, expression of MHC-II and CD80 molecules was decreased to near-undetectable levels relative to DCs from ARM infected mice. By day 120 after infection, when Cl13-infected mice have cleared the infection, the expression of CD40 and MHC-I molecules recovered completely, whereas a significant reduction remained for MHC-II, CD86 and CD80 molecules. However, by day 360 after infection the expression of all molecules recovered completely. A similar pattern of changes was observed in both CD8α^+ and CD8α^- DCs, although less pronounced in the latter ones. Supporting Cl13 impairment of DC maturation, it was shown that Cl13 blocks the expression of MHC-II molecules in myeloid DCs derived from bone marrow (BM) cells cultured in presence of Granulocyte-macrophage colony stimulating factor (GM-CSF). Specifically, myeloid DCs generated from uninfected BM progenitors contained abundant intracellular MHC-II molecules, primarily in the perinuclear region. At the same time post-culture, Cl13-infected myeloid DCs showed a significant reduction in the intensity of MHC-II molecules, indicating an association between virus replication and a decreased expression of MHC-II molecules. Furthermore, after stimulation of uninfected DCs with bacterial lypopolysaccharide (LPS), the level of MHC-II substantially increased in the cytoplasm, and MHC molecules migrated to the plasma membrane. In contrast, Cl13-infected CD11c$^+$ DCs showed barely detectable levels of MHC-II molecules in the intracellular compartment and on the plasma membrane.

Since antigen presenting and co-stimulatory molecules on DCs play an essential role for T cell activation, the lack of their expression on DCs likely contribute to the defective antigen presenting capacity observed in LCMV Cl13 infected mice.

3.3.3. Blockade of DC development

Finally as part of the multiple strategies used by Cl13 to disable DC guard, this immunosuppressive virus affects BM hematopoietic progenitor cells by blocking their development into DCs (Sevilla et al., 2003). Fms-like tyrosine kinase 3 ligand (Flt3-L) is well known to induce expansion of DC precursors and trigger DC development. Indeed, uninfected mice treated with Flt3L showed a 15-fold increase in the number of CD11c+ DCs, which were predominantly CD8α$^+$. Similarly, DCs from ARM infected mice treated with Flt3L showed 20-fold expansion. However, DCs harvested from Cl13 infected mice were refractory to the stimulatatory effect that Flt3L normally has on hematopoietic progenitors of both the myeloid and lymphoid lineage. This defective DC development observed in Cl13 infected mice was associated with the fact that following Cl13 infection, but not ARM infection, a great proportion of BM cells are infected by the virus making it likely that Cl13 interference with DC development is related to its capacity to infect BM cells. Particularly, about 20% of myeloid and plasmacytoid DCs in the BM are infected (unpublished observations). Similar to DCs, expansion of NK cells, which is also triggered by Flt3L injection was also blocked in Cl13 infected mice, suggesting that Cl13 infection alters early progenitor cells that are up-stream of the division between NK and DC developmental pathways. Accordingly, we observed that mice infected with LCMV-Cl13 display a marked reduction of the undifferentiated lineage negative flt3 positive DC progenitors (unpublished observations).

LCMV Cl13 interference with DC development is further supported by its ability to abort DC development in BM cells cultured in the presence of GM-CSF. The decline in the percentage of CD11c+ cells in GM-CSF cultures in the presence of Cl13 was not observed in cultures infected with LCMV-ARM and was associated with the number of DCs infected. However, addition of Cl13 virus at late time points of GM-CSF cultures fails to disrupt DC development, further suggesting that Cl13 impact must occur at an early stage of development to affect the generation of CD11c+DCs.

Type I IFN, including IFN-α and IFN-β are critical for both innate and adaptive immunity and they also play a major role in protection against virus infection (Le Bon and Tough, 2002). In addition, the positive impact of type I IFN in DC maturation and crosspresentation has been very well established. However, recent findings in Cl13 model system indicate that the virus can subvert the known antiviral effect of type I IFNs and use it to benefit its survival. Particularly, experiments using IFNα/β KO mice infected with Cl13 showed that impairment of DC differentiation from stem cells is dependent on these cytokines, indicating that INF-α/β has a critical role in blocking DC generation from undifferentiated BM cells. This is supported by the fact that, the transient reduction in BM cellularity and the concomitant pancitopenia observed after LCMV infection do not occur in the absence of INF-α/β receptor (Binder et al., 1997). However, while this transient aplasia in BM as well as high serum levels of INF-α/β seem to be a common factor during infection with several strains and variants of LCMV, the impairment of stem cell differentiation to DCs occurs during Cl13 but not ARM infection. This suggests that impairment of stem cell maturation to DCs is a selective advantage acquired by immunosuppressive variants of LCMV in order to persist in the host. Because mDCs were reported to secrete IFN α/β during LCMV Cl13 but not ARM infection (Diebold et al., 2003), conceivably mDCs may be used by Cl13 to achieve higher levels of type I IFN in the BM and block DC development from stem cells. Disruption of DC development complements other evasive mechanisms used by Cl13 to escape the immune response and establish a persistent infection. Moreover, since BM is the primary site for hematopoiesis in adult mice, it is possible that

Cl13 blocks the development of other immune cells derived from these stem cells, which further enhances the generalized immunosuppression observed during this infection

4. Concluding remarks

For a virus or any infectious agent to persist, it must utilize strategies to suppress the host's immune response. As exemplified in more detail by LCMV Cl13 infection, targeting of DCs represent a common maneuver of many viruses to avoid host surveillance and establish chronic or persistent infections. A major challenge in biomedical research is to counterattack virus subversion modulating their effects on DCs for the benefit of the host and the eradication of the pathogen. The better understanding of the mechanisms used by viruses to subvert DC performance during immune response will provide new tools for designing DC-based immunotherapeutic strategies for treatment and/or prevention of viral infections.

Acknowledgments

This is publication No. 16871-NP from Division of Virology, Department of Neuropharmacology, The Scripps Research Institute (TSRI). These studies were supported by grants from USPHS grant AI45927 and AI09484 and training grant NS041219 (KHE). EIZ was supported by Pew Foundation Latin American Fellowship and Fundacion Antorchas.

References

Ahmed, R., and Oldstone, M. B. (1988). Organ-specific selection of viral variants during chronic infection. J. Exp. Med. *167*, 1719-1724.

Ahmed, R., Salmi, A., Butler, L. D., Chiller, J. M., and Oldstone, M. B. (1984). Selection of genetic variants of lymphocytic choriomeningitis virus in spleens of persistently infected mice. Role in suppression of cytotoxic T lymphocyte response and viral persistence. J. Exp. Med. *160*, 521-540.

Arrode, G., and Davrinche, C. (2003). Dendritic cells and HCMV cross-presentation. Curr. Top Microbiol. Immunol. *276*, 277-294.

Binder, D., Fehr, J., Hengartner, H., and Zinkernagel, R. M. (1997). Virus-induced transient bone marrow aplasia: major role of interferon-alpha/beta during acute infection with the noncytopathic lymphocytic choriomeningitis virus. J. Exp. Med. *185*, 517-530.

Borrow, P., Evans, C. F., and Oldstone, M. B. (1995). Virus-induced immunosuppression: immune system-mediated destruction of virus-infected dendritic cells results in generalized immune suppression. J. Virol. *69*, 1059-1070.

Bowie, A., Kiss-Toth, E., Symons, J. A., Smith, G. L., Dower, S. K., and O'Neill, L. A. (2000). A46R and A52R from vaccinia virus are antagonists of host IL-1 and toll-like receptor signaling. Proc. Natl. Acad. Sci. USA. *97*, 10162-10167.

Byrne, J. A., and Oldstone, M. B. (1984). Biology of cloned cytotoxic T lymphocytes specific for lymphocytic choriomeningitis virus: clearance of virus *in vivo*. J. Virol. *51*, 682-686.

Cao, W., Henry, M. D., Borrow, P., Yamada, H., Elder, J. H., Ravkov, E. V., Nichol, S. T., Compans, R. W., Campbell, K. P., and Oldstone, M. B. (1998). Identification of alpha-dystroglycan as a receptor for lymphocytic choriomeningitis virus and Lassa fever virus. Science *282*, 2079-2081.

Dalod, M., Hamilton, T., Salomon, R., Salazar-Mather, T. P., Henry, S. C., Hamilton, J. D., and Biron, C. A. (2003). Dendritic cell responses to early murine cytomegalovirus infection: subset functional specialization and differential regulation by interferon alpha/beta. J. Exp. Med. *197*, 885-898.

Diebold, S. S., Montoya, M., Unger, H., Alexopoulou, L., Roy, P., Haswell, L. E., Al-Shamkhani, A., Flavell, R., Borrow, P., and Reis e Sousa, C. (2003). Viral infection switches non-plasmacytoid dendritic cells into high interferon producers. Nature *424*, 324-328.

Dockter, J., Evans, C. F., Tishon, A., and Oldstone, M. B. (1996). Competitive selection *in vivo* by a cell for one variant over another: implications for RNA virus quasispecies *in vivo*. J. Virol. *70*, 1799-1803.

Dolganiuc, A., Kodys, K., Kopasz, A., Marshall, C., Do, T., Romics, L., Jr., Mandrekar, P., Zapp, M., and Szabo, G. (2003). Hepatitis C virus core and nonstructural protein 3 proteins induce pro- and anti-inflammatory cytokines and inhibit dendritic cell differentiation. J. Immunol. *170*, 5615-5624.

Dubois, B., Lamy, P. J., Chemin, K., Lachaux, A., and Kaiserlian, D. (2001). Measles virus exploits dendritic cells to suppress CD4+ T-cell proliferation via expression of surface viral glycoproteins independently of T-cell trans-infection. Cell Immunol. *214*, 173-183.

Itano, A. A., and Jenkins, M. K. (2003). Antigen presentation to naive CD4 T cells in the lymph node. Nat. Immunol. *4*, 733-739.

Kelsall, B. L., Biron, C. A., Sharma, O., and Kaye, P. M. (2002). Dendritic cells at the host-pathogen interface. Nat. Immunol. *3*, 699-702.

Kunz, S., Borrow, P., and Oldstone, M. B. (2002). Receptor structure, binding, and cell entry of arenaviruses. Curr. Top Microbiol. Immunol. *262*, 111-137.

Kunz, S., Sevilla, N., McGavern, D. B., Campbell, K. P., and Oldstone, M. B. (2001). Molecular analysis of the interaction of LCMV with its cellular receptor [alpha]-dystroglycan. J. Cell Biol. *155*, 301-310.

Le Bon, A., and Tough, D. F. (2002). Links between innate and adaptive immunity via type I interferon. Curr. Opin. Immunol. *14*, 432-436.

Libraty, D. H., Pichyangkul, S., Ajariyakhajorn, C., Endy, T. P., and Ennis, F. A. (2001). Human dendritic cells are activated by dengue virus infection: enhancement by gamma interferon and implications for disease pathogenesis. J. Virol. *75*, 3501-3508.

Liu, Y. J. (2001). Dendritic cell subsets and lineages, and their functions in innate and adaptive immunity. Cell *106*, 259-262.

Matloubian, M., Kolhekar, S. R., Somasundaram, T., and Ahmed, R. (1993). Molecular determinants of macrophage tropism and viral persistence: importance of single amino acid changes in the polymerase and glycoprotein of lymphocytic choriomeningitis virus. J. Virol. *67*, 7340-7349.

Moutaftsi, M., Brennan, P., Spector, S. A., and Tabi, Z. (2004). Impaired lymphoid chemokine-mediated migration due to a block on the chemokine receptor switch in human cytomegalovirus-infected dendritic cells. J. Virol. *78*, 3046-3054.

Moutaftsi, M., Mehl, A. M., Borysiewicz, L. K., and Tabi, Z. (2002). Human cytomegalovirus inhibits maturation and impairs function of monocyte-derived dendritic cells. Blood *99*, 2913-2921.

Oldstone, M. B. (1993). Rous-Whipple Award Lecture. Viruses and diseases of the twenty-first century. Amer. J. Pathol. *143*, 1241-1249.

Oldstone, M. B. (2002a). Arenaviruses. II. The molecular pathogenesis of arenavirus infections. Introduction. Curr. Top Microbiol. Immunol. *263*, V-XII.

Oldstone, M. B. (2002b). Biology and pathogenesis of lymphocytic choriomeningitis virus infection. Curr. Top Microbiol. Immunol. *263*, 83-117.

Reis e Sousa, C. (2004). Toll-like receptors and dendritic cells: for whom the bug tolls. Semin Immunol. *16*, 27-34.

Salio, M., Cella, M., Suter, M., and Lanzavecchia, A. (1999). Inhibition of dendritic cell maturation by herpes simplex virus. Eur. J. Immunol. *29*, 3245-3253.

Salvato, M., Borrow, P., Shimomaye, E., and Oldstone, M. B. (1991). Molecular basis of viral persistence: a single amino acid change in the glycoprotein of lymphocytic choriomeningitis virus is associated with suppression of the antiviral cytotoxic T-lymphocyte response and establishment of persistence. J. Virol. *65*, 1863-1869.

Sevilla, N., Kunz, S., Holz, A., Lewicki, H., Homann, D., Yamada, H., Campbell, K. P., de La Torre, J. C., and Oldstone, M. B. (2000). Immunosuppression and resultant viral persistence by specific viral targeting of dendritic cells. J. Exp. Med. *192*, 1249-1260.

Sevilla, N., Kunz, S., McGavern, D., and Oldstone, M. B. (2003). Infection of dendritic cells by lymphocytic choriomeningitis virus. Curr. Top Microbiol. Immunol. *276*, 125-144.

Shortman, K., and Liu, Y. J. (2002). Mouse and human dendritic cell subtypes. Nat. Rev. Immunol. *2*, 151-161.

Smelt, S. C., Borrow, P., Kunz, S., Cao, W., Tishon, A., Lewicki, H., Campbell, K. P., and Oldstone, M. B. (2001). Differences in affinity of binding of lymphocytic choriomeningitis virus strains to the cellular receptor alpha-dystroglycan correlate with viral tropism and disease kinetics. J. Virol. *75*, 448-457.

Tishon, A., Borrow, P., Evans, C., and Oldstone, M. B. (1993). Virus-induced immunosuppression. 1. Age at infection relates to a selective or generalized defect. Virology *195*, 397-405.

9

Subversion of the Immune System by Measles Virus: A Model for the Intricate Interplay Between a Virus and the Human Immune System

Denis Gerlier, Hélène Valentin, David Laine, Chantal Rabourdin-Combe, and Christine Servet-Delprat*

Abstract

Despite the development of a successful live attenuated vaccine, measles remains one of the major infectious diseases with high mortality in developing countries and continuing outbreaks in other nations. A major physiopathological feature of measles infection is the concomitance of the induction of a long lasting efficient anti-viral immune response and a profound immunosuppression with impairment of humoral and cellular immune response to unrelated antigens. The analytical analysis of natural infection in human and experimental disease in monkeys allied to the discovery of many cellular proteins recruited by viral proteins has led to partly uncover many molecular and cellular mechanisms which govern the measles virus interplay with the human immune system. Indeed, this model of immune subversion by a pathogen has been thoroughly studied for many years and has led to some seminal findings dealing with the regulation of the immune system.

1. Introduction

Measles virus (MV[§]) is responsible for an acute childhood disease that, in 2002, has infected almost 40 million infants and caused 0.8 million deaths (Stein et al., 2003). Natural infection by this virus is restricted to human, although transmission to some monkey species can occur. It is characterized by a profound cellular immunodepression favouring opportunistic secondary infections, but

[§]See the full list of abbrevations in Appendix 1 at the end of this chapter

*For correspondence email Denis.Gerlier@univ-lyon1.fr

also the induction of efficient life long humoral and cellular immunity against MV (Auwaerter et al., 1999; Griffin et al., 1994; Hirsch et al., 1984). The immunodepression is characterized by ablation of delayed-type hypersensitivity, severe lymphopenia, impaired *in vitro* lymphocyte proliferation and a shift towards a Th2 response (Auwaerter et al., 1999; Griffin et al., 1994). While immunization with an attenuated MV strain provides efficient protection, the use of inactivated MV vaccine in the early sixties resulted in the induction of an exacerbated atypical measles disease after secondary natural infection (Fulginiti et al., 1967). MV infection thus represents a challenging model to understand how a virus can interplay with the immune system.

2. Natural measles diseases

2.1. Physiopathology

MV is extremely infectious with high rate of secondary transmission, reaching 400 secondary cases from a single infected child. The childhood infection is characterized by a maculopapular rash, dry cough, coryza, conjunctivitis and photophobia. Measles is a severe disease which was clinically distinguished from small pox after the XVIIth century. Severity is increased below the age of one year and after the age of 12 years. While usually mild in the developed world, the infection has higher mortality (~5%) and morbidity (~30%) rate in developing countries, and the causes for this striking differences are yet unclear (Griffin, 2001; Rall, 2003; Rima, 2000).

MV infects and probably replicates in epithelial cells of the respiratory tract, which are likely to be the site of virus release for interhuman spread of the disease. Infection of local trafficking macrophages and probably dendritic cells (DCs) disseminates the virus into the draining lymph nodes (Figure 1a). Then, a systemic infection occurs with cell-associated viraemia (infected monocytes and lymphocytes) and a very limited and transient plasma viraemia (Esolen et al., 1993). All lymphoid organs including the thymus, spleen, lymph nodes, appendix and tonsils are severely infected. Infection of dermal endothelial cells associated to MV specific cell mediated immunity results in the characteristic rash (Burnet, 1968; Markowitz et al., 1988). Among the other target organs are the lungs with bronchopneumonia and the gut with gastroenteritis. Infected tissues show characteristic multinucleated Warthin-Finkeldey cells (WFC) (Tajima and Kudow, 1976) and local inflammation. The viraemia occurs within the 10-14 days latent period to culminate with 2-3 days of rash and the appearance of the clinical signs. The virus is cleared after about 20 days. Many infections by other opportunistic pathogens can be observed after the rash and they usually are severe and life-threatening.

About 1 out of 1000 cases of measles suffers from often fatal encephalitis, characterized by a lack of virus in the brain and likely to be autoimmune in nature. Two fatal persistent infections of the brain by MV are subacute sclerosis pan-encephalitis (SSPE) and measles inclusion body encephalitis (MIBE). These diseases are rare with usually ~10 cases per million population below the age of 20

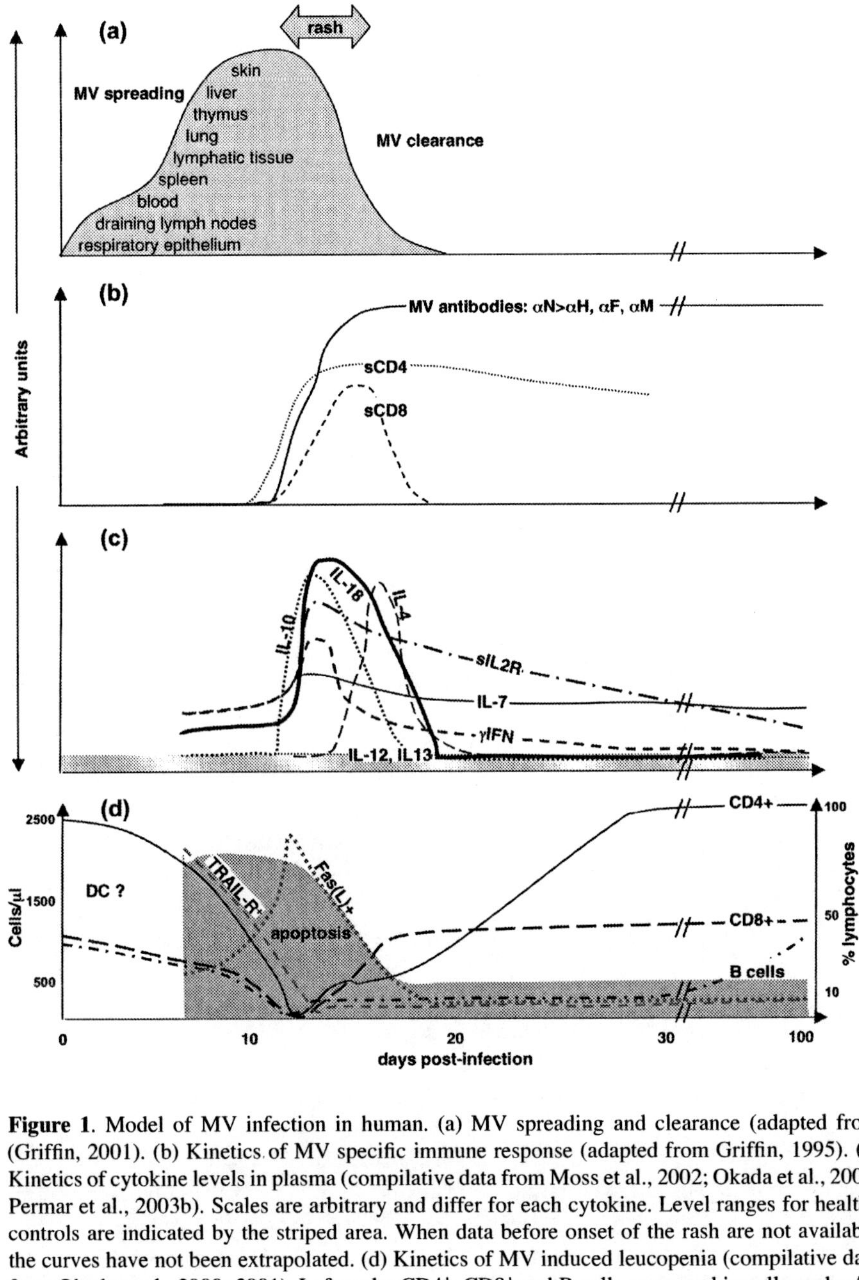

Figure 1. Model of MV infection in human. (a) MV spreading and clearance (adapted from (Griffin, 2001). (b) Kinetics of MV specific immune response (adapted from Griffin, 1995). (c) Kinetics of cytokine levels in plasma (compilative data from Moss et al., 2002; Okada et al., 2001; Permar et al., 2003b). Scales are arbitrary and differ for each cytokine. Level ranges for healthy controls are indicated by the striped area. When data before onset of the rash are not available, the curves have not been extrapolated. (d) Kinetics of MV induced leucopenia (compilative data from Okada et al., 2000; 2001). Left scale: $CD4^+$, $CD8^+$ and B cells expressed in cell number/µl. Right scale: % of lymphocytes expressing Fas/FasL, TRAIL-Rs and primed to *in vitro* induced aptotosis (shaded curve).

years (Bojinova et al., 2004; Detels et al., 1973; Jabbour et al., 1972) but can be 10 times more frequent (Takasu et al., 2003), occurs about ~7 years after the primary infection, are linked to virus defective for budding, and are disappearing with the progress of the vaccination.

2.2. MV induces a strong and lifelong immunity

Measles is a one shot infectious disease, because it induces a strong and lifelong immunity. The earliest IgM antibodies directed against the nucleoprotein (N) and phosphoprotein (P) are detected at the onset of the rash (figure 1b). A switch to IgG subclasses, IgA and IgE antibodies occurs thereafter. Antibodies against glycoproteins hemagglutinin (H) and fusion (F) proteins with neutralising activities are detected 2-4 days after the onset of the rash. During the acute phase, MHC-I restricted $CD8^+$ cytolytic T lymphocytes (CTL) and, later on, MHC-II restricted $CD4^+$ cells are detected in the blood. Accordingly soluble CD8 and CD4 proteins are shed into the plasma, transiently for the former and lasting over 30 days for the latter (Figure 1b). CTL specifically recognize the five major proteins, N, P, M, F and H. A typical change in $CD4^+$ T cell response occurs with a switch from Th1 to Th2 response (see Griffin, 1995 for review).

The acute infection is characterized by a specific pattern of cytokine plasma levels (Figure 1c) (Moss et al., 2002; Okada et al., 2001). IFNγ and IL-18 increase few days before the onset of the rash and peak during the rash. IL-18 returns to normal levels about ten days after the onset of the rash and IFNγ slowly declines over more than 30 days. There is an acute and transient increase of IL-2 (Griffin et al., 1990) and IL-10, a Th2 cell-derived cytokine suppressing cytokine production by Th1, followed by a peak of IL-4 in agreement with the Th1 to Th2 switch. Soluble IL-2 receptor (sIL2R), a hallmark of activated T cells, is shed acutely on the onset of the rash to decrease slowly over more than 100 days after the rash. IL-7 levels are sustained for over 100 days, suggesting an active lymphocyte homeostasis. The plasma level of the Th1 inducer IL-12 remains within normal values, although close to the lower limit, in agreement with a quick Th1 to Th2 switch. Likewise, the IL-13 level remains unchanged throughout the infection (Moss et al., 2002; Okada et al., 2001).

Measles immunity is acquired for life and there are no reports of secondary infection. Whereas this lifelong immunity relies mainly on the presence of neutralising antibodies, the initial control of MV infection requires an intact cellular immunity as shown by the severe disease in immunocompromised children but not in agammaglobulinemia patients (see Griffin, 2001 for review). A central role of the $CD8^+$ T cells is observed in the control and clearance of MV infection in the closely related rhesus macaque model (Permar et al., 2003a). MV-specific CD8 and CD4 T cells is long lasting over 34 years after vaccination, the latter declining slowly over 21 years (Naniche et al., 2004).

2.3. MV induces a profound immuno-suppression

In 1908, von Pirquet made a key observation: in patients suffering from measles disease, the tuberculin reaction disappeared and this lasted for several weeks after recovery (Von Pirquet, 1908). Indeed, MV induces a severe depression of cellular immunity which accounts for the frequent additional infections with opportunistic pathogens such as bacteria. Although the underlying mechanisms are yet to be uncovered, many biological features of this immunosuppression have been characterised. At the onset of rash, there is a severe lymphopenia affecting CD8, CD4 and B lymphocytes (Figure 1d) (Okada et al., 2000) whereas the % of (activated) natural killer (NK) cells sharply rises to compensate the lymphopenia. Typically, $CD4^+$ level lower than 200 cells/μl can be observed in young children with measles rash. While $CD8^+$ T lymphocytes go back to normal level within one week, normalisation of CD4 levels takes 2-3 weeks and that of B cells around 100 days. Paradoxically, only a few cells (<0.01 %) get infected. Instead, just before the onset of the rash and during the rash, most of the lymphocytes look pre-apoptotic with high TNF-related apoptosis-inducing ligand receptor (TRAIL-R) expression just before the rash and high Fas Ligand (FasL) expression during the rash. Accordingly, a peak of shed soluble Fas and FasL appears with similar kinetics in the plasma (not shown in the figure). And when they are cultured *in vitro,* these cells die from apoptosis with DNA fragmentation. Thus, the profound lymphopenia results, at least in part, from the extended death of uninfected blood cells caused by apoptosis (Okada et al., 2000). In agreement with the more severe disease in young infants (<1 year) and adults (> 15 years), the immunosuppression is more profound and lasts for a longer time.

When stimulated *in vitro* using polyclonal activators, peripheral blood cells from MV infected patients display reduced proliferation (Whittle et al., 1978), poor IL-12 secretion and recovery to normal values takes more than 32 weeks (Atabani et al., 2001).

2.4. Animal models of MV infection

The animal model, closest to the human measles disease, is the macaque, the infection of which reproduces most if not all of the clinical and biological features of the measles disease. This model is very useful for analyseing *in vivo* both MV induced immunity and immunosuppression (McChesney et al., 1997; Tajima and Kudow, 1976; van Binnendijk et al., 1995; Zhu et al., 1997). The cotton rat (Sigmodon hispidus) inoculated with MV shares many clinical and immunologic features of the human infection (Niewiesk et al., 1997; Pfeuffer et al., 2003). However, none of the proteins from this animal host has been identified so far as a partner of MV proteins. Apart from the intracranial inoculation of suckling mice with adapted MV strains, mice are poorly sensitive to MV infection even when they express human CD46 or CD150 receptors, and generalized MV infection *in vivo* is observed only in mice devoid of IFNα/β receptor (IFNAR) ((Mrkic et al., 2000; Mrkic et al., 1998) and (Manchester and Rall, 2001) for review). CD46 transgenic mice are however useful for investigating some of the mechanisms of the MV induced immunodepression. Indeed, this can be mimicked in a human CD46-transgenic mouse model inoculated with MV and *Listeria monocytogenes.* MV-infected mice are more susceptible to infection with Listeria than the control mice. MV-infected mice have significantly decreased numbers of splenic macrophages and neutrophils and show substantial defects in IFNγ production by $CD4^+$ T cells (Slifka et al., 2003).

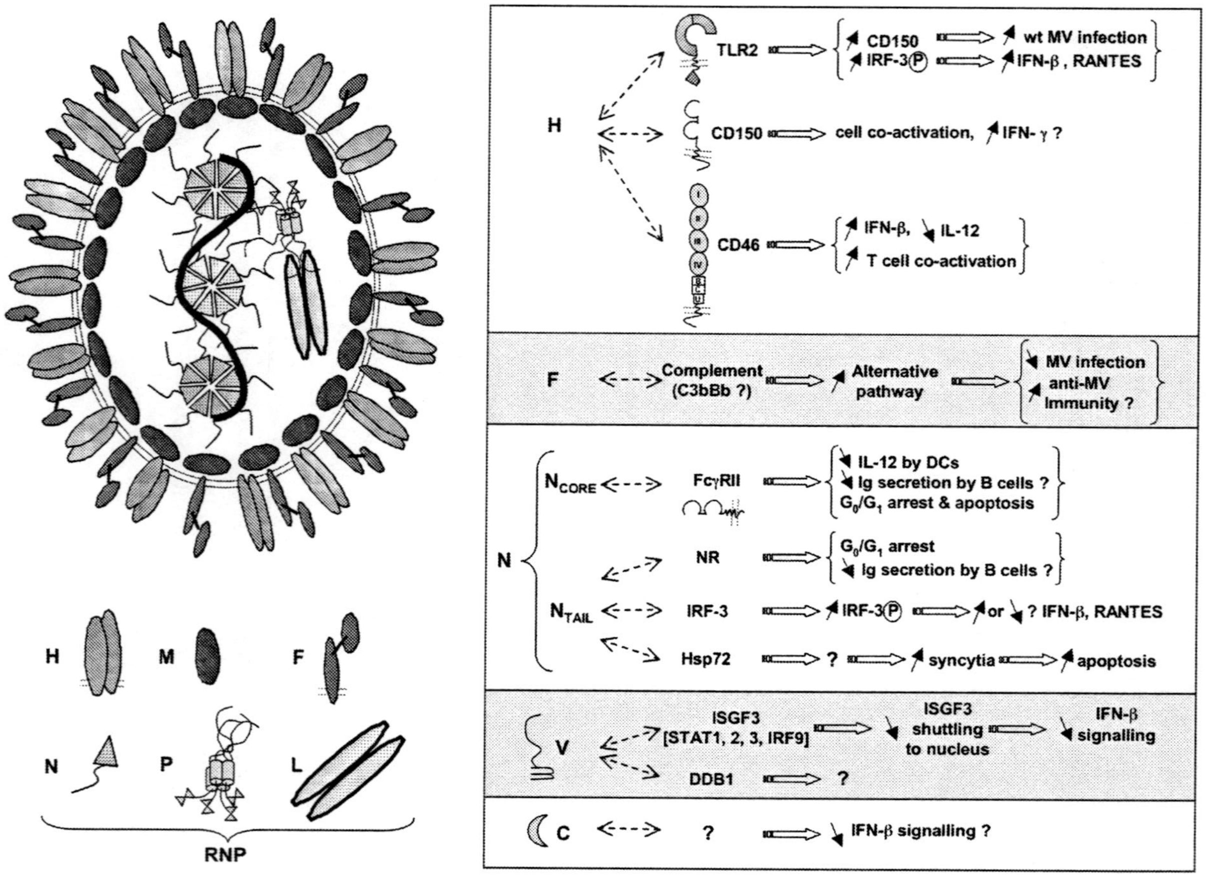

Figure 2. Schematic view of MV, viral proteins and their cellular partners. Structural (N, P, M, F, H and L) proteins and non structural V and C proteins are schematically represented. The nucleocapsid, inside the virion, is in an extended view for simplicity. Natively unfolded C-terminal domain of N and N-terminal domain of P are represented by lines, P is a homotetramer, H a homodimer and F a disulfide bridge heterodimer associating F1 and F2 subunits. FcγRII, IgG Fc receptor; NR, N protein receptor; IFN, Interferon, IRF, Interferon regulatory factor; ISGF, interferon stimulated gene factor; STAT, Signal transducer and activator of transcription; DDB1: Damage-specific DNA binding protein; TLR, Toll-like receptor. Direct interactions are indicated by double dotted arrows. Demonstrated or postulated (question marks) effects on cellular functions are indicated on the right. Effects on the virus cycle are not shown.

3. Molecular and cellular biology of measles virus

MV is an enveloped virus belonging to the *Mononegavirales* order, *Paramyxoviridae* family and *Morbillivirus* genus (Rima, 2000). The replicative strategy, common to all of these viruses, involves a RNA dependent RNA polymerase (RdRp) using a single strand RNA in complex with a nucleoprotein as a template.

3.1. Virus structure and genome organization

MV is a pleiomorphic virus with a diameter of 100-200 nm. A single particle contains several nucleocapsid (NC) or ribonucleoparticles (RNP) and is therefore polyploid (Rager et al., 2002). Measles NC is helicoidal with 1.2 μm length, 18-21 nm diameter and a central core of 5 nm diameter (Rima, 2000). The genome is a non segmented and negative strand RNA. It comprises 6 genes, nucleoprotein (N), phosphoprotein (P), matrix (M), fusion (F), hemagglutinin (H) and large protein, or polymerase (L) in that order, flanked by two short leader (Le) and trailer (Tr) sequences, which contain the genome and antigenome promoter, respectively. The P gene also codes the V protein, by RNA editing, and C protein by the use of an alternative open reading frame. Thus, V protein shares the same N-terminal sequence than P protein, and C protein has no sequence similarities.

Owing to the precise binding of 6 nucleotides by N protein and the complete and exact cover of the whole genome length (15 894 nt), a single RNP is made of 2649 N proteins to which are bound ~300 P proteins and ~20-30 L proteins. The NC is wrapped into a membrane envelope made of H and F glycoproteins in tight complex. Underneath lies a lattice made of M protein, which links the NC to the cytoplasmic tails of the glycoproteins (Figure 2).

3.2. MV is a monotypic virus with limited variability restricted to H and N sequences

From numerous sequence analyses, distinct genetic lineages of wild-type viruses exist and co-circulate. Of the 6 genes on the viral genome, the H and N genes are the most variable. They contain up to 7% variability at the nucleotide level. The most variable part of the measles genome is the 450 nucleotides coding the N protein COOH-terminus where nucleotide variability rises up to 12% between various wild-type viruses. The variation between MV strains in the H protein coding sequence reaches a maximum of 5% at the nucleotide level and 3% at the amino acid level. This low variation, a remarkably low mutation rate (6×10^{-3} per annum) and phylogenic analysis indicate a lack of antigenic drift in the MV-H gene. Thus, the virus is monotypic. There are 8 clades designated A, B, C, D, E, F, G, H, and 15 genotypes designated A, B1, B2, C1, C2, D1, D2, D3, D4, D5, D6, E, F, G, H (World Health Organization, 1998).

The availability of an extensive sequence database for wild-type MVes has made molecular epidemiological studies possible to identify the source and transmission pathways of the virus and to monitor circulating viral strains.

3.3. Virus replication cycle

3.3.1. Virus entry

To enter its target cell, MV binds to a cellular receptor through the interaction with H glycoprotein. This binding activates MV-F protein, which mediates the fusion of the viral envelope with the plasma cell membrane at neutral pH and allows the intracytoplasmic delivery of the RNP. This neutral pH fusion is also responsible for cell to cell fusion and syncytia formation, a hallmark of MV infection. Indeed, H and F glycoproteins are exported to the cell surface, and can interact with neighbouring cells expressing a MV cellular receptor. MV-H is a type II glycoprotein which associates into disulfide bridged homodimers made of 78 kDa monomers (see Gerlier et al., 1995 for review). The residues 1-151 constitute the "minimal unit" sufficient for H dimerization (Plemper et al., 2000). The tridimentional structure has not been solved, but is modelled upon the Newcastle virus hemagglutinin/neuraminidase protein as a stalk and globular head made of a six-bladed β-sheet propeller (Vongpunsawad et al., 2004). Despite this homology, MV-H neither relies on sialic acid for binding to the cell membrane nor displays significant neuraminidase activity (Langedijk et al., 1997).

3.3.1.1. At least two cellular receptors (CD46 and CD150) bind a single viral attachment glycoprotein

To date, two cellular receptors able to bind to H protein and mediate MV entry have been identified so far, the ubiquitously expressed membrane cofactor protein, MCP or CD46 and the signalling lymphocyte activation molecule, SLAM or CD150, expressed only on activated immune cells (Figure 2). Wild type and vaccine strains show variable usage of these two receptors. The use of a third cellular receptor by MV is postulated.

a) CD46 as ubiquitous cellular receptor

Mouse and rodent cells are refractory to MV entry. The expression of recombinant human CD46 enables murine cells to bind to MV, to fuse with cells expressing H and F and renders murine B lymphocytes and other rodent cells permissive for MV replication into infectious progeny (Dorig et al., 1993; Naniche et al., 1993a). Furthermore, anti-CD46 antibodies inhibit MV entry. CD46 is a type I glycoprotein expressed in all human cells except erythrocytes. In some simian species, CD46 is also expressed on red blood cells in agreement with the description of H activities of MV reported in 1960 (see Gerlier et al., 1995 for review).

CD46 (or MCP) belongs to the regulators of complement activation (RCA) gene family. CD46 is a key player which inhibits the spontaneous complement activation on self tissue (Liszewski et al., 1991). The activation of the alternative pathway leads to C3b deposition, and the mannan binding lectin and classical pathway lead to C3b and C4b deposition. CD46 acts as a cofactor of the soluble serine-protease, Factor I, to cleave and inactivate covalently bound cell surface C3b and C4b into C3bi and C4bi (see Liszewski et al., 1991 for review). CD46 acts preferentially on the alternative pathway (Kojima et al., 1993) by inhibiting the amplification loop of the C3b deposition (Devaux et al., 1999). Complement activation relies solely on an enzymatic cascade and a tight regulation is needed to prevent destruction of self-tissue. As a matter of proof, knock-out of the gene coding Crry1, the mouse CD46 functional counterpart, results in abortion with a C3-mediated complement destruction of the fetuses, which is prevented if mice are also null for C3 expression (Xu et al., 2000). The dominant structural units of CD46 are four short consensus repeat (SCR I to IV) modules of 60-64 amino acids that are responsible for complement binding and regulatory functions (Figure 2). Variable RNA splicing generates multiple CD46 isoforms that differ in the heavily glycosylated serine-threonine-proline (STP)-rich regions (designated A, B and C) and cytoplasmic tail (Cyt1, 2, 3 or 4). The mouse CD46 and one human CD46 isoform is expressed on spermatozoids. The regulation of complement activation maps to SCR II, III and IV, SCR I being dispensable (Adams et al., 1991; Liszewski et al., 2000).

MV binding to CD46 involves an interaction of the ectodomain of MV-H (Devaux et al., 1996) with the most membrane-distal SCRs I and II (Iwata et al., 1995; Manchester et al., 1995). Accordingly, all of the common CD46 isoforms can serve as receptors for MV (see Gerlier et al., 1995 for review). From molecular modelling and crystal structure of SCR-I-II CD46 fragment, amino acids involved in MV-H binding define a large, glycan-free surface that extends from the top of SCR1 to the bottom of SCR2 with a pronounced interdomain bend and some flexibility at the domain interface (Casasnovas et al., 1999; Mumenthaler et al., 1997). From the sensitivity to point mutations, the regulation of the alternative pathway maps to the side face, whereas MV-H binding site is localized on the top of the molecule (Christiansen et al., 2000a). On the other hand, two C4b-regulatory domains overlap with MV-H interactive regions on SCR II, indicating a possible interference with the control of the classical pathway (Liszewski et al., 2000). Although dispensable for MV-H binding, CD46 SCR IV contributes to enhanced MV-H binding onto the SCR I and II interface (Christiansen et al., 2000c). The interaction of CD46 with the virus probably involves the receptor multimerization. A long CD46-CD55 hybrid protein, which cannot mediate fusion, acts as a dominant negative receptor and prevent wild type CD46 and H/F mediated cell to cell fusion (Christiansen et al., 2002). Furthermore, a chimeric disulfide-linked homo-octamer fusion protein encompassing the CD46 ectodomain linked to the C-terminal part of the C4b binding protein (C4bp) α chain (sCD46-C4bpα) has a higher avidity for MV-H than soluble monomeric sCD46. sCD46-C4bpα is more efficient than monomeric sCD46 in inhibiting MV binding to CD46, H/F-mediated cell-cell fusion, and in neutralizing MV *in vitro*. Moreover, the octameric sCD46-C4bpα, but not the monomeric sCD46, fully protected CD46 transgenic mice against a lethal intracranial MV challenge (Christiansen et al., 2000b).

MV infection, MV binding or MV-H expression results in down-regulation of CD46 expression from the cell surface (Naniche et al., 1993b; Schneider-Schaulies et al., 1995). MV-H binding to CD46 mediates receptor down-regulation (Krantic et al., 1995; Naniche et al., 1993b) and, thus, can be uncoupled from virus fusion (Christiansen et al., 2000c; Firsching et al., 1999). The cytoplasmic Tyr-X-X-Leu motif of CD46 is essential for its down-regulation by the MV-H protein in persistently infected cells (Hirano et al., 1996; Yant et al., 1997). After MV infection, CD46 down-regulation results mainly from increased internalisation rate (Naniche et al., 1993b; Welstead et al., 2004), although some intracellular retention may occur (Yant et al., 1997). In non lymphoid cells, CD46 constitutively internalized via clathrin-coated pits, traffics to multivesicular bodies, and is recycled to the cell surface. The cross-linking of CD46 by MV, induces, in all cells, including lymphoid cells, pseudopodia that engulf the ligand in a macropinocytosis-like process, and leads to the degradation of cell surface CD46 (Crimeen-Irwin et al., 2003). MV-H affinity to CD46 is not strong enough to allow co-immunoprecipitation of the two proteins (Welstead et al., 2004).

b) CD150 as cellular receptor restricted to cells from the immune system

Transfection with a human CD150 complementary DNA enables non-susceptible cell lines to bind MV, support MV replication and develop cytopathic effects. The distribution of CD150 on various cell lines is consistent with their susceptibility to clinical isolates of MV (Tatsuo et al., 2000). Indeed, lymphocytes (B, T, NK), DCs and monocytes express CD150 after activation (Kruse et al., 2001; Minagawa et al., 2001). CD150 is a receptor shared by several *morbilliviruses* as human, canine, and bovine CD150 act as cellular receptors for MV, CDV, and RPV, respectively Tatsuo and Yanagi, 2002).

CD150 is a self-ligand glycoprotein closely related to the CD2 family. Its function remains largely unclear. It is thought to play an important role in adhesion between T cells and antigen presenting cells (APCs) and acts as co-receptor in T cell receptor (TCR) dependent responses (Veilette, 2004). Intracellular signalling is mediated in T and NK cells by the recruitment of the small SH2 domain-containing adaptor SAP, and by the recruitment of the SAP homologue EAT-2 in B cells, monocytes and DCs. CD150 null mice show a defect in the IL-4 secretion by $CD4^{+}$ T cells and in the stimulation of macrophages by lipopolysaccharides via the Toll like receptor 4 (TLR4) leading to high sensitivity of mice to parasitic infection (Veilette, 2004; Wang et al., 2004). Thus, CD150 appears to play a central role at the interface of acquired and innate immune response.

The ectodomain of CD150 comprises two immunoglobulin (Ig) superfamily domains (V and C2), and the most distal V domain is necessary and sufficient to interact with MV-H protein and allow MV entry (Ono et al., 2001). Similarly to CD46, CD150 is down-regulated upon MV infection (Erlenhoefer et al., 2001) and MV-H expression (Devaux et al., 2004; Welstead et al., 2004). This down regulation is due to both the formation of CD150-MV-H complexes within the ER and to increase internalisation rate of cell surface CD150 facing H protein on neighbouring cells (Welstead et al., 2004).

c) CD46 and CD150 binding sites in MV H protein

The search for a cellular receptor other than CD46, which lead to the identification of CD150, came from the inability of CD46 expressed in rodent cells and in human epithelial cells to mediate efficient MV infection by MV strains collected from patients and isolated on the simian B lymphoid B95a cell line. This cell line expresses CD150 and a truncated simian CD46 lacking SCR I and thus unable to bind to MV-H (Erlenhoefer et al., 2001; Murakami et al., 1998). Up to now, all MV strains are able to use CD150 as a receptor, whereas vaccine strains and many laboratory strains isolated and/or maintained in non lymphoid cell lines of human or simian origin can efficiently use CD46. Interestingly, clinical MV strains isolated by culture on activated peripheral blood cells can also use CD46 (Manchester et al., 2000). Thus, one can distinguish MV strains as $CD46^{nonusing}$ and $CD46^{using}$. The astonishing finding is that those strains may differ by only two amino acids in position 451 and 481 (Lecouturier et al., 1996). From fusion experiments and recombinant virus infection studies, H binding to CD150 and CD46 involves seven and nine keys amino acids located in contiguous area in propeller beta-sheets 5 and 4, respectively (Vongpunsawad et al., 2004). The H binding sites on CD46 and CD150 probably overlaps since some anti-H antibodies compete for H binding to both of the receptors (Santiago et al., 2002). Accordingly in the three-dimensional surface representation of a structural model, the H505-507-533 CD150 binding site forms the basis of a single continuous site that overlaps with the H546-548-549 CD46 binding site (Masse et al., 2004). Furthermore, CD46 and CD150 mediated entry of $CD46^{nonusing}$ MV are equally sensitive to inhibition by the soluble octameric sCD46-C4bpα (Christiansen et al., 2000b). The elongated and fusion defective CD46-CD55 chimeric receptor efficiently competes out CD46 and CD150-mediated cell-cell fusion by $CD46^{using}$ but not by $CD46^{nonusing}$ MV strains (Christiansen et al., 2002). Accordingly, the association rate of H protein from a $CD46^{using}$ MV strain with CD150 is 20-fold lower than with CD46, although its dissociation rate is 5-fold lower (Santiago et al., 2002). Two clusters of CD150-relevant residues exist that are accessible for receptor contact. Several CD46-relevant amino acids may be shielded from direct receptor contacts. It appears likely that certain residues support receptor-specific H-protein conformational changes. Only accumulation and appropriate mutations in H sequence results in CD46-blind and CD150-blind recombinant MV (Vongpunsawad et al., 2004).

d) Various receptor usage by wild type and vaccine strains and the third receptor hypothesis

The difference in cell tropism on human and simian cells between MV strains is largely determined by virus entry, in which the H proteins of respective MV strains play a decisive role. As a consequence, a CD150 expressing cell line is used to grow new MV wild type isolates, and WHO recommends performing virus isolation in B95a, an Epstein-Barr virus-transformed marmoset B cell line, until a suitable alternative cell line is found (World Health Organization, 1998). Indeed, the cell line used for virus isolation and growth is of great importance as changes in receptor usage depend upon the host cell type. In the natural course of infection, epithelial cells are primary targets of MV. MV can infect primary human small airway epithelial cells (SAEC) and induce formation of large syncytia *in vitro*. mRNA specific for CD150 is not detected in SAEC, indicating that wild-type MV infects SAEC and induces syncytia formation via a CD150-independent mechanism (Takeuchi et al., 2003b). Some patient isolates that have not been adapted to tissue culture cell lines use CD46 as a receptor (Kurita-Taniguchi et al., 2000; Manchester et al., 2000). The passage of these primary MV patient isolates into B95a cells, which express a SCR-I truncated CD46 molecule unable to bind to MV-H, select variants that had alternate receptor usage compared to the original isolate (Manchester et al., 2000), and wild-type strains of MV isolated in B95a cells use CD150 as a cellular receptor (Minagawa et al., 2001). Conversely, MV strains isolated on B95a cells and passaged into Vero cells, select for an H gene with two point mutations (481Asn => Tyr and 546Ser=>Gly), which render them able to agglutinate to CD46 expressing African green monkey red blood cells (Li and Qi, 1998).

Wild-type MV interacts with CD150 with high affinity (Santiago et al., 2002) and can be co-immunoprecipitated (Welstead et al., 2004), but probably possesses the ability to interact with CD46 with low affinity since residues 458 and 459 are conserved and involved in the binding to CD46 (Masse et al., 2002). Indeed, when using a very sensitive assay with recombinant MV expressing the fluorescent GFP protein, differences in CD46 and CD150-mediated entry were much more subtle than differences in the ability to mediate cell to cell fusion (Schneider et al., 2002). This selection process for H either able to bind to CD46 or not, observed *in vitro*, probably occurs also *in vivo* since it results from one or two point mutations made by the viral RNA-dependent RNA polymerase. We predict that multiple sequencings of H gene from various tissue samples harvested from the same patient will reveal H gene sequence heterogeneity with both $CD46^{nonusing}$ and $CD46^{using}$ phenotypes. This would be reminiscent to the CXCR4 to CCR5 coreceptor switch observed in the natural HIV infection.

Whether this variability of the H sequence allowing alternate usage of CD150 and/or CD46 suffices for natural MV propagation in humans or MV uses a third receptor remains to be determined. Indeed, CD150 is restricted to cells of the immune system (lymphocytes, DCs, and monocytes), and histopathological studies in humans and experimentally infected monkeys show that MV infects epithelial, endothelial, and neuronal cells which lack CD150 expression and the usage of a third yet to be identified receptor has been postulated (Hashimoto et al., 2002). H from wild type MV interacts with toll-like receptor 2 (TRL2) (Bieback et al., 2002), but whether this protein acts as a *bona fide* cellular receptor is unknown. In conclusion, MV can adapt itself to replicate in various cell types, and one must pay attention to the strain used and the cell host in which it is propagated.

3.3.1.2. Fusion process

The fusion of the MV envelope with the plasma membrane involves a tripartite protein game between MV-H and MV-F glycoproteins on one hand and CD46 or CD150 cellular receptors on the other hand. MV-F is a type I glycoprotein belonging to the class I fusion protein, the prototype of this class being the Influenza hemagglutinin (Smith and Helenius, 2004). It is synthesized as an homotrimer of the 60 kDa F_0 precursor, which is cleaved into the disulfide bridged heterodimer F_1F_2 by furin proteases upon maturation through the trans-Golgi network (TGN). H homotetramers and F homotrimers associate into H/F complexes within the ER (Plemper et al., 2001) and (Gerlier et al., 1995) for review). At the cell surface, the binding of H to opposite cellular receptors induces a conformational change of the H/F complex resulting in the activation of F_1F_2 which allows the exposure of the N-terminal peptide fusion located on the transmembrane F sub-unit. Indeed, the fusion efficiency relies on the complementarities between H and F glycoproteins (Plemper et al., 2002). The structural model for this sub-unit is the 3D structure of the Newcastle disease F protein (Morrison, 2003). According the current model of the viral class I fusion, the fusion peptide interacts with the outer layer of the opposite plasma membrane and a spring-like secondary conformational change brings the two outer layers of the opposite membrane in close contact so as to fuse (hemifusion step). Then, fusion of the two inner layers occurs to create the fusion pore. During the fusion process, H and F ectodomains remain located outside the cell, and do not translocate into the cytosol (Cardoso et al., 1996; van Binnendijk et al., 1992). As demonstrated by the dominant negative effect of elongated cellular receptors unable to mediate fusion, creation of the fusion pore requires a precise scaffolding of multiple adjacent receptors interacting with several H/F complexes (Buchholz et al., 1996; Christiansen et al., 2002). Furthermore, one out of three F_1 protein of each trimers is cleaved into F_1a and F_1b at a juxtamembrane site, and mutation of this cleavage site reduces fusion efficiency and virus replication. It is postulated that selective loss of the membrane anchor in the external subunits of circularly arranged F protein trimers may disengage them from pulling the membrane centrifugally, thereby facilitating fusion pore formation (von Messling et al., 2004).

Several observations point to differences in the fusion between the virus and a cell and between two adjacent cells. After adaptation for growth in Véro cells, a wild type MV strain induces cytopathic effects in B95a cells but not in Véro cells (Kouomou and Wild, 2002). Likewise, infection of the human HeLa cells with the Schwartz strain is not cytopathic (C. Escoffier and D. Gerlier, unpublished observation). Furthermore, syncytia formation, but not MV infection and spreading through an entire cell culture, is prevented by the fusion inhibitor z-D-Phe-L-Phe-Leu tripeptide (D. Gerlier, unpublished data and (Laine et al., 2003)). Likewise, H and F mediated fusion is modulated by the interaction of their cytoplamic tails with M protein, and a ΔM recombinant MV induces huge syncytia (Cathomen et al., 1998a; 1998b).

3.3.2. Transcription and replication

MV RNA genome is encapsidated by MV-N to form a helical nucleocapsid (NC) which is resistant to nuclease digestion. The structure remains intact after release in the cytoplasm (Lamb and Kolakofsky, 2001) and this, rather than naked RNA, serves as a template for both transcription and replication. After the intracellular delivery of the RNP complex, the initial step is the primary sequential transcription of the different genes byRdRp, comprised of the large (L) protein and phosphoprotein (P). The RdRp enters the genome at the single promoter located at the 3′ end and synthesizes 5′ capped and 3′ polyadenylated mRNAs. Following an ill-defined signal, which may be the accumulation of N protein above a critical threshold (Lamb and Kolakofsky, 2001), the polymerase switches to a replicative mode and uses the upstream replication promoter to synthesize a full-length positive-sense RNA strand. The concomitant N encapsidation of this nascent antigenome is postulated to prevent the polymerase from recognizing the intergenic stop and start signals. The antigenomic promoter at the 3′ end of the RNA is then used by the RdRp to synthesize a full-length negative-sense genomic RNA which is also concomitantly encapsidated (Lamb and Kolakofsky, 2001). The concomitant encapsidation during RNA elongation probably explains the resistance to silencing by RNAi and the rarity of viral recombination events observed in the *Mononagavirales* order (Bitko and Barik, 2001; Spann et al., 2003). The entire replication cycle of MV takes place in the cytoplasm where the RNP complexes are likely to associate with cellular cofactors which contribute to RdRp functions (Horikami and Moyer, 1995).

The L protein (> 250 kDa) is active as a homo-oligomer (Smallwood et al., 2002). The protein is inherently unstable when expressed alone and the P protein, which binds to the N-terminal domain, is known to act as a molecular chaperon (Horikami et al., 1994). The 70 kDa P protein contains two L protein binding sites, in the intrinsically disordered N-terminal domain (PNT) (Curran et al., 1995; Curran et al., 1994; Karlin et al., 2003), and in the C-terminal domain (PCT) (Chen et al., 2003; Liston et al., 1995). The homo-oligomeric P protein acts as a bridge between L and the NC template (Curran, 1996; Curran, 1998). Besides interacting with the assembled form of N (N^{NUC}), the P protein also binds to newly synthesized, monomeric form of the 60 kDa N (N°) to form a soluble complex (N°P) (Curran et al., 1995; Huber et al., 1991; Spehner et al., 1997). The N°P complex is the substrate used by the RdRp to initiate the encapsidation of the nascent RNA chain during replication (Horikami et al., 1992). By sequestering N in the N°P complex, P prevents the illegitimate encapsidation of cellular RNAs (Huber et al., 1991; Spehner et al., 1997). PNT binds to the conserved N-terminal domain (N_{CORE}), while PCT binds to the C-terminal domain of N (N_{TAIL}) (Chen et al., 2003). PCT to N_{TAIL} binding maps to the C-terminus X domain and to the residues N477-505 and involves the induced helix folding of the latter on the compact three α-helices bundle of the former (Johansson et al., 2003; Kingston et al., 2004a; Kingston et al., 2004b; Longhi et al., 2003). N_{CORE} contains the self oligomerisation signals and the RNA binding site(s) (Karlin et al., 2002). V and C are two non-structural proteins which down regulate viral RNA synthesis (Reutter et al., 2001; Tober et al., 1998). Abrogation of C or V expression leads to MV gene accumulation at both mRNA and protein levels whereas over expression of C and V result in a decrease of mRNA synthesis. V protein from *Morbillivirus* prevents N°P complex formation, and interferes with the use of N°P by the RdRp during genome replication (Horikami et al., 1996).

The virus replication cycle relies on cellular factors, since RdRp hardly function *in vitro* and requires the presence of cytoplasmic extracts (Horikami and Moyer, 1995). Hsp72 binds to N_{TAIL} and favour RdRp activity (Oglesbee et al., 1993) (Figure 2). Other cellular partners identified so far are the interferon regulatory factor IRF-3 (tenOever et al., 2002), a key mediator in the activation of interferon which binds also to N_{TAIL}, the interferon stimulated gene factor (ISGF3), a signalling complex downstream to the interferon alpha receptor (IFNAR), and the DNA damage binding protein (DDB1), which bind to V protein (Figure 2) (Lin et al., 1998a; Palosaari et al., 2003).

3.3.3. Assembly and budding

MV assembly is thought to occur at the plasma membrane. Cell membranes are not homogenous and cholesterol and (glyco)sphingolipids tend to segregate into microdomains called membrane rafts. Membrane rafts act as a scaffold and/or assembly platform for many biological events such as signal transduction and virus assembly (see Chazal and Gerlier, 2003 for review). MV assembly occurs within membrane rafts (Manie et al., 2000; Vincent et al., 2000). In infected cells, all structural proteins including the mature F_1-F_2 fusion protein and the hemagglutinin (H) are enriched into the raft fraction while the F_0 precursor remain excluded from the raft fraction. Interestingly, F_0 protein is cleaved into F_1-F_2 protein in the trans-Golgi network, where the membrane rafts are formed. MV-F protein is targeted to rafts in the absence of other viral proteins. Because F_0 and H proteins associate in the ER (Plemper et al., 2001), F protein can drag H protein into rafts, whereas H expressed alone is excluded from these microdomains. When the other structural proteins are expressed alone, only the matrix protein (M) exhibits a low but significant association with membrane rafts. The co-expression of H and F proteins together with M protein do not result in a significant increase in the amount of M associated with rafts, although M protein interacts with the cytoplasmic tail of F (Spielhofer et al., 1998). The inability of F protein to drag M protein into rafts is in agreement with the lack of co-transportation of M protein with MV glycoproteins for efficient surface targeting (Riedl et al., 2002). Independently from H and F glycoproteins, the genomic RNA, N, P and L proteins associate into RNPs in the cytosol allowing the scaffolding of M protein which probably targets the M-RNP complex to membrane rafts (Vincent et al., 2000). Thus, it is proposed that membrane rafts allow H and F glycoprotein, on one hand, and M-RNP complex, on the other hand, to meet there and assemble. Indeed, detergent resistant virus-like particles are observed in electron microscopy (Bohn et al., 1986) and the assembly in membrane rafts is functional, since rafts isolated from infected cells are infectious (Manie et al., 2000).

After assembly of MV within membrane rafts, the envelope-RNP complex looks ready for budding. This virus is rather poorly efficient in budding and single particles contain several RNPs (Rager et al., 2002). Virus released into cell free supernatant is partially made of non-raft membranes, with recovery in detergent resistant membrane rafts of H and F glycoproteins but of no other virus structural proteins (Manie et al., 2000). Thus, either particles assembled in membrane rafts are not precursors of budding mature viruses, or after assembly involving the coalescence of several membrane rafts, virus budding through membrane rafts is associated with the capture of adjacent non-raft membranes and simultaneously initiates a shift of the RNPs from raft to non-raft regions. Two observations argue for the latter hypothesis : (i) whereas RNPs are tightly bound to the plasma membrane of infected cells, RNPs tend to dissociate from the virus envelope after budding (Dubois-Dalcq and Reese, 1975); (ii) there is a correlation between a defect in MV budding in a murine cell line (Vincent et al., 1999), and the poor localization of M protein in membrane rafts from infected cells (Chazal and Gerlier, 2003). The demonstration of a raft requirement for MV budding awaits further experimental evidence.

MV is released from the apical membrane domain of polarized epithelial cells, although the surface glycoproteins H and F are transported in a non-polarized fashion and to the basolateral membrane domain, respectively (Maisner et al., 1998). The expression of H and F at the basolateral membrane is required for cell-to-cell fusion and MV spreading *in vitro* and *in vivo* (Moll et al., 2004).

4. Interactions of measles virus with the immune system

4.1. Activation and subversion of innate immunity

4.1.1. Activation and subversion of interferons

Interferons (IFNs) have been named according to their potent antiviral activity. IFNs also exhibit pleiotropic functions including growth-inhibitory and immunomodulatory activity. The importance of the IFN system as a host defense against viral infection is underlined by the counteractive strategies developed by the viruses and acquired during their long co-evolution with their host. After a brief introduction on IFN, we will focus on the two mechanisms that have allowed MV to counteract the IFN response: inhibition of the initial induction of IFN-α/β and inhibition of IFN-α/β signaling in the late phase of IFN-response.

4.1.1.1. Properties of the IFN system in viral immunity

IFNs are a family of cytokines that classified into two major subtypes: alpha/beta (IFN-α/β or type 1 IFN) and gamma (IFN-γ or type 2 IFN). Whereas there are several IFN-α coded by different genes, the IFN-β and INF-γ are coded by a single gene. The type 1 IFNs are induced in response to virus infection. Most of virally infected cells are capable of synthesize IFN-α/β in cell culture. By contrast, only certain cells of the immune system including NK cells and cytotoxic $CD8^+$ T cells produce IFN-γ upon mitogenic or antigenic stimulation. IFN-α, -β and -γ are essential for the early control of viral infection since they slow down the infection before the onset of the adaptive immune response. IFNs directly limit the spread of virus by eliciting cellular resistance to viral infection in uninfected cells while promoting infected cell apoptosis (see Mogensen et al., 1999 for review). IFNs are also critical mediators of both innate and adaptive immunity (Stark et al., 1998). Briefly, IFNs activate NK cells, enhance activity and/or functions of macrophages, B lymphocytes, cytotoxic and helper T cells *in vitro* as well as *in vivo* (Le Bon et al., 2001; Le Bon and Tough, 2002).

Uninfected cells expressed a very low level of basal IFN-α/β which is critically dependant on the expression of the IFN regulatory factors IRF-3, IRF-9, and IRF-7 (Sato et al., 2000). Viral infection quickly induces high levels of IFN-α/β through multiple pathways including the activation of the transcription factors complexes IRFs and/or NF-κB. After activation, these complexes shuttle to the nucleus and bind positive regulatory domains of the IFN-α/β gene promoter regions leading to IFN-α/β transcription. The virus infection primes the IFN response by activating the so-called "virus activated kinase" made of the two IκB related kinases IKKε and TANK-binding kinase 1 (TBK-1) which phosphorylates IRF-3 (Fitzgerald et al., 2003; Sharma et al., 2003). IRF-3 plays a key role in the viral induction of IFNs since virus infected IRF-3 null mice are unable to respond by increasing IFN-β although they produced basal level of IFN-α (Sato et al., 2000). The activation of NF-κB results from the upstream activation of a growing list of kinases, including the double stranded (ds) RNA-dependent serine/threonine protein kinase (PKR). PKR is constitutively expressed in cells. Upon viral infection, PKR autophosphorylates and in turn phosphorylates substrate targets, the best characterized being the eukaryotic protein synthesis initiation factor 2 (eIF2α).

The IFNs activate also their own genes and, thus, act in both autocrine and paracrine manner. Both IFN-α and IFN-β exert their activities through binding to a common species-specific cell surface receptor. Upon binding to the secreted IFN-α/β, the interferon alpha receptor (IFNAR) induces the phosphorylation of the signal transducer and activator of transcription (STAT)-1 and -2 proteins. Phosphorylated STAT-1, STAT-2 and IRF-9 associate into active heterotrimeric IFN-stimulated gene factor 3 (ISGF3) complexes which migrate into the nucleus to activate IRF-7 transcription. IRF-7 then associates with the virus activated IRF-3 (Sato et al., 2000) to activate the IFN-α/β responsive genes. STAT-1 is a key molecule for signaling by both type 1 and type 2 IFN, since binding of IFN-γ to its receptor IFNGR also leads to STAT-1 phosphorylation. Several hundred genes are differentially regulated by IFN-α, -β or -γ (Der et al., 1998). These genes establish an antiviral state in infected as well as surrounding non infected cells. Because IFN-α and -β activate the same cellular receptor, their respective biological activities have not often been evaluated, and, whenever possible, the IFN subtype will be specified in this review.

The IFN-α/β act as important mediators for antiviral defense against MV (Leopardi et al., 1992). IFNAR null mice expressing human CD46 are highly sensitive to MV infection even after intranasal inoculation, whereas normal CD46 transgenic mice are hardly infected except after intracranial inoculation (Mrkic et al., 1998). The role of IFN in measles pathology is only partly understood.

4.1.1.2. Induction of IFN-α/β after CD46 engagement by MV-H

In vitro, macrophages from human and CD46 transgenic mice are resistant to the infection by $CD46^{using}$ MV strains, but they produce IFN-α/β and nitric oxide (NO) (Kurita-Taniguchi et al., 2000). The production of NO is further increased if the macrophages are also stimulated by IFN-γ. The inhibition of the IFN-α/β autocrine pathway by anti-IFN antibodies both alleviates the virus infection and inhibits NO production (Katayama et al., 2000). The CD46-Cytl and -Cyt2 isoforms functionally diverge in eliciting innate immune responses, and cross-linking CD46-Cytl isoform is more efficient than Cyt2 for enhancing NO production by IFN-γ-stimulated macrophages (Hirano et al., 2002). The induction of IFN-α/β requires the CD46 Cytl cytoplasmic tail. Indeed, macrophages, expressing a truncated, tail less CD46, become highly sensitive to MV infection and MV infection no longer results in enhanced NO production upon IFN-γ stimulation. MV-H binding to CD46 seems to be responsible for the IFN-α/β stimulation (Katayama et al., 2000). The enhancing effect on IFN-γ stimulated NO release can be mimicked by anti-CD46 antibodies and C3b dimers, a natural ligand of CD46 (Hirano et al., 2002). However, in coculture experiments, only H co-expressed with the F glycoprotein can mediate this effect (Hirano et al., 2002; Kurita-Taniguchi et al., 2000). Furthermore, the effect is detected only after infection with MV expressing H with high affinity for CD46. It remains to be determined if the co-expression of F contributes to maintaining its H partner in a conformation suitable for CD46 signaling and IFN-α/β induction, or if the fusion of the macrophage surface, with the opposite membrane, provides an additional stimulus.

4.1.1.3. Putative role of TLR3/7/8 in the induction of IFN-α/β by MV

a) Induction of IFN-α/β by dsRNA production during MV infection awaits evidence

MV infection induces IFN-β production by various cells including epithelial, endothelial, glial and DCs (Dhib-Jalbut and Cowan, 1993; Schneider-Schaulies et al., 1993; Vidalain et al., 2002). Neither the exact mechanism nor the factor(s) responsible for IFN-α/β secretion have been identified. Distinct pathway may be involved in the induction of IFN-β gene transcription in measles-infected cells. Thymic epithelial cell (TEC) infection by MV leads to PKR activation and to phosphorylation of eIF2α suggesting that PKR is a key factor in IFN-β secretion by MV-infected DCs (Valentin, H. unpublished). The classical component of a viral infection that stimulates PKR is double-stranded RNA (dsRNA), which accumulates during the replication of many viruses. However, viruses belonging to the *Mononegavirales* order, are unlikely to produce viral dsRNA, although the RNA dependent RNA polymerase synthesizes negative stranded genomic and positive stranded antigenomic and messenger RNAs. These RNA are unlikely to form a duplex because both genomic and antigenomic RNA templates are tightly packaged within the nucleocapsid which shields them from nuclease activity and silencing RNA (Barik, 2004). Furthermore, artificial introduction of an ambisense transcription subunit in the related Sendai virus results in artificial viral dsRNA and high sensitivity to interferon. And, this transcription subunit is quickly lost upon passage (Le Mercier et al., 2002). Thus, since the formation of viral dsRNA remains very unlikely, another pathway should lead to PKR activation.

b) Putative TLR7/8 stimulation and induction of IFN-α/β by MV

PKR is also activated upon Toll-like receptor (TLR) signaling (Barton and Medzhitov, 2003). The TLR family plays a key role in activation of the innate immune system since it mediates the recognition of many types of pathogen, including viruses. Eleven members of the TLR family have been characterized so far. Ligation of TLR on APC leads to activation of PKR and NF-κB transcription factors (Horng et al., 2001). Several *Mononegavirales* activate murine TLR7 signaling in DCs (Lund et al., 2004). Although the GU rich single strand RNAs bind to murine TLR7 and its functional human TLR8 counterpart (Heil et al., 2004), the nature of the viral ligand is unknown (Lund et al., 2004). Whichever mode of entry used by the incoming virus, fusion at neutral pH with the cell membrane or at acidic pH after internalization in the endosomal compartment, TLR7/8 mediated IFN-α/β induction requires internalization in an acidic compartment in which genomic RNA can be uncoated from N protein. Indeed, some MV is probably endocytosed because MHC-II restricted presentation of N protein is observed (Gerlier et al., 1994b). Whether MV genomic RNA become accessible for binding to TLR7/8 in the endocytic/lysosomal compartment and activates the PKR and NF-κB remains to be determined.

4.1.1.4. Subversion of interferon system by MV: role of N, V and C proteins

a) MV-N binding to IRF-3 and induction of IFN–α/β

Measles infection leads to high phosphorylation of latent cytoplasmic IRF-3 resulting in IRF-3 dimerization and cytoplasmic to nuclear translocation (Servant et al., 2001). Moreover, the isolated expression of MV-N protein induces IRF-3 hyperphosphorylation (tenOever et al., 2002) which reaches an optimum 18h post-infection (Servant et al., 2001). The IRF-3 hyperphosphorylation induced by Sendai virus, another *Paramyxoviridae* is not correlated with ISGF3 transcription but with IRF-3 proteasome-mediated degradation (Collins et al., 2004). However, the activation of IFN-α genes requires little phosphorylated IRF-3. Thus, whether MV-N binding to IRF-3 results in the activation of IFN-α genes or in the enhanced degradation of the IRF-3, and lack of activation of IFN-α genes, remains to be determined.

b) Inhibition of IFN-α/β signalling by MV-V and C proteins

To circumvent IFN induced-antiviral states, *Paramyxoviridae* have developed redundant strategies by coding proteins that block IFN activation and/or IFNAR signal transduction (Shaffer et al., 2003; Young et al., 2000). Many molecular bases for anti-IFN strategies have been established in the related Sendai virus (reviewed in Gotoh et al., 2002) and include transcriptional and/or post-transcriptional inhibition of PKR, STAT-1/2 and IRF-3 (Collins et al., 2004). The molecular basis for MV anti-IFN activity is not yet as well deciphered (Gotoh et al., 2001; 2002).

The host evasion properties of the *Paramyxoviridae* have been ascribed to the non structural C and V proteins which are encoded by the P gene. *In vitro*, C-deficient Edmonston tag, but not for V-deficient Edmonston tag strain shows reduced growth in human PBMC and in human thymus tissues xenografted into SCID mice (Escoffier et al., 1999; Valsamakis et al., 1998), whereas the former replicates efficiently in Véro cells and in murine neurons (Patterson et al., 2000; Radecke and Billeter, 1996).

In MV infected cells, C as well V proteins are found associated in complex with IFNAR, STAT-1 and the scaffolding protein RACK1, and no phosphorylation of Jak-1 or STAT-1 is observed in these conditions (Yokota et al., 2003). In 293T cells, where MV-V protein is conditionally expressed, the V protein binds to the ISGF3 ternary complex made of phosphorylated STAT-1, STAT-2 and IRF-9 and prevent its translocation into the nucleus (Palosaari et al., 2003). In HeLa cells constitutively expressing the V protein, neither STAT-1, not STAT-2 get phosphorylated (Takeuchi et al., 2003a). The reasons for this discrepancy is unknown and can reflect the use of V protein from different MV strains, different cell hosts, or to the selection of cell clones able to grow in the presence of V protein. The V protein also bind to the DNA damage binding protein (DDB1) (Lin et al., 1998a; Palosaari et al., 2003), which is part of a transcriptional complex required for both basal and IFN induced expression of the PKR (Das et al., 2004). However, the possible contribution of MV V protein to PKR expression remains to be determined. Thus, V protein could act downstream to the IFNAR stimulation by IFN-α/β. Wild-type V protein from IC-B and Edmonston B strains, but not V from the recombinant Edmonston-based tag strain, has a potent antagonistic effect on IFN-α induced signaling (Ohno et al., 2004). Both of the N- and C-terminal regions of IC-B V protein have IFN antagonistic activity. P protein from IC-B, but not from Edmonston-based tag, can also block IFN-α induced signaling likely because of the Y110H substitution in the latter. In fact, V-deficient Edmonston tag strain and its parental counterpart display equal low resistance to IFN-α/β treatment *in vitro,* (Ohno et al., 2004; Patterson et al., 2000), because of the combined Y110H and C272R substitutions in the tag V sequence (Ohno et al., 2004; Patterson et al., 2000). MV-C protein expressed alone seems to inhibit IFN-induced ISRE transcription of a reporter gene in cells stimulated with IFN-α/β (Shaffer et al., 2003), but this effect is weak (Ohno et al., 2004).

In summary, non structural V, and possibly C, proteins counteract the IFN-α/β antiviral activity through multiple molecular strategies; which may explain the variation in the ability of MV to counteract the IFN response according to the cell host and MV strain combination. *In vivo,* MV V and C proteins function as virulence factors (Patterson et al., 2000; Valsamakis et al., 1998). While the Edmonston tag strain expressing a P and V protein deficient in IFN-antagonistic activity (Ohno et al., 2004) kill CD46-transgenic mice, V-deficient Edmonston tag hardly replicates after intracranial inoculation of one-day old suckling mice, and MV-C$^-$ replicates but failed to kill the animals (Patterson et al., 2000). Accordingly, both of the V- and C-deficient Edmonston tag also display an attenuated phenotype in IFNAR null mice expressing human CD46 (Mrkic et al., 2000). Therefore, *in vivo,* MV C and V proteins encode virulence functions besides IFN antagonism and likely operate via separate mechanisms. Does the lack of the V or C protein hamper the virus dissemination through viral mRNA synthesis dysregulation?

4.1.2. Activation of proinflammatory factors

4.1.2.1. Activation of TLR2 signalling and up-regulation of CD150

Pattern recognition via TLR expressed by APC activates innate immunity. The signal transduced downstream of the TLRs engages a genetic program adapted to the nature of the pathogen. Wild-type, but not vaccine, MV strains activate cells via both human and murine TLR2, and this is a property of H protein. There is an inverse relationship between activation via TLR2 and binding to CD46: the activation via TLR2 by wild-type H protein is abolished by a single asparagine to tyrosine mutation at position 481 which converts H from low/null to high binding to CD46. TLR2 activation by MV-H protein induces the secretion of pro-inflammatory cytokines such as IL-6 in human monocytic cells and surface expression of CD150. Wild-type H protein also induce IL-6 release by TLR2 expressing macrophages but not by macrophages from TLR2 null mice. Thus, the unique property of MV wild-type strains to activate TLR2-dependent signals would contribute not only to immune activation but also to viral spread and pathogenicity by up-regulating the expression of the CD150 MV receptor on monocytes (Bieback et al., 2002).

4.1.2.2. Modulation of macrophage functions through CD46: impact on IL-12 secretion

The intracytoplasmic domains of CD46 contains the Tyr-X-X-Leu motive, related to ITAM (immunotyrosine based activatory motive), a critical shared sequence in the immune system that mediates signal transduction downstream of TCR or FcγR. The CD46 Cyt1 and Cyt2 isoforms functionally diverge in eliciting innate immune responses, but any isoform of CD46 is able to serve as a receptor for MV (Gerlier et al., 1994a; Manchester et al., 1994). Crosslinking CD46-Cyt1 isoform is more efficient than Cyt2 for enhancing NO production by IFN-γ-stimulated macrophages (Hirano et al., 2002). The cross-linking of CD46 on human macrophages down-regulates IL-12 production, a hallmark of the suppression of cell-mediated immunity (Karp et al., 1996). After six- to eight-days treatment with GM-CSF, peripheral blood monocytes become able to recruit protein-tyrosine phosphatase SHP-1 on CD46 ligation, and concomitantly to produce IL-12 p40 and NO. When stimulated with the MV KO strain, which expresses an H protein inducing early and profound CD46 down-regulation, a rapid and transient dissociation of SHP-1 from CD46 is observed. The signaling cascade leading to IL-12 down-regulation is unknown. We hypothesize an autocrine and/or paracrine role of IFN-α/β. (i) CD46 cross-linking induces IFN-α/β and its effect on NO production by macrophages is inhibited by anti-IFN-α/β antibodies (see § 4.1.1.2. above). (ii) Exogenous IFN-α induces IL-12 down-regulation and IL-10 up-regulation in human monocytes (Byrnes et al., 2001).

IL-12 is critical for the orchestration of cellular immunity since this cytokine primes Th1 differentiation of naive $CD4^+$ T cells. CD46 serves as a signal modulatory molecule and the properties of its ligands determine the suppression or activation of an innate immune response at a specific maturation stage of human macrophages (Kurita-Taniguchi et al., 2000). Thus, interactions between H and CD46 expressed on macrophages activate innate antiviral immunity via IFN-α/β and subsequent NO production, but depress adaptive cellular immunity via inhibition of IL-12 synthesis.

Many biological features of measles infection suggest that immunosuppression could result from a IL-12 defect: a Th1 to Th2 shift (Griffin and Ward, 1993), plasma IL-12 levels within the lower limit of normal range (Moss et al., 2002; Okada et al., 2001), and IL-12 secretion defects *in vitro*. In Gambian patients with measles, the stimulus-specific production of IL-12 by peripheral blood monocytes is markedly suppressed for several months (Atabani et al., 2001). Conversely, after vaccination of Rhesus macaques with a recombinant MV expressing IL-12, the Th2 bias of the immune response is altered with decrease in IL-4 secretion and increase IFN-γ secretion by stimulated $CD4^+$ T cells and the humoral neutralisation response is lowered. However, other manifestations of immunosuppression, such as the lymphopenia and low lymphoproliferation, are not improved indicating that reduced IL-12 levels are not the sole factor in MV-induced immunosuppression (Hoffman et al., 2003).

4.1.3. Pros and cons of CD46 subversion: complement activation by MV-F

The complement system is one of the potent effectors of the innate immunity which play a key role in the induction of adaptive immunity (see Aderem and Ulevitch, 2000; Fearon, 2000, for review). In 1979-80, human HeLa cells infected by MV Edmonston strain were found to activate, in the absence of antibodies, the alternative pathway of human complement, and to be sensitive to complement-mediated lysis (Sissons et al., 1980; 1979). During MV infection, the cellular receptor CD46 is down-regulated because interaction with the MV-H protein results in enhanced internalization (Krantic et al., 1995; Naniche et al., 1993b) and/or intracellular retention of CD46 (Yant et al., 1997). Since CD46 acts as a potent regulator of the alternative pathway by preventing C3b deposition on the cell surface (see § 3.3.1.1. above), this down-regulation was proposed to be responsible for the increased sensitivity of MV infected cells to complement-mediated cell lysis (Schneider-Schaulies et al., 1995; Schnorr et al., 1995). However, only a small amount of CD46 can prevent human C3b deposition (Christiansen et al., 2000a), and residual levels of CD46 after infection probably suffice to prevent the C3b deposition. Moreover, the infection of the simian B95a cells which express a SCR1 deleted form of CD46 unable to bind any MV-H, but fully able to control the complement activation, results in a high C3b deposition on the cell surface (Devaux et al., 2004). Similarly, infection of B95a cells by a MV strain, unable to bind to CD46 and to induce its down regulation, results in the lack of control of the alternative pathway. In fact, the level of C3b deposition is induced not by the expression of MV-H, but by MV-F and is proportional to the level of F expression. As a consequence, infection of cells by any MV strain will result in the activation of the Factor B dependent alternative pathway. Purified MV particles also activate the alternative pathway (Devaux et al., 2004).

How MV-F expression results in cell surface activation of the alternative pathway is unknown, although some F protein is co-immunoprecipitated with C3b β chain indicating that during exposure to human serum, some disulfide bridged C3b αβ chains become covalently linked to cell surface F protein. Whether the formation of C3b-F complex is the primary event, which engages the uncontrolled activation of the alternative pathway, can be questioned. F protein expression results in the 'escape' of C3b from its two known regulators CD46 and CD55 (known also as Decay accelerating factor). A spatial sequestration of F (and C3b) outside CD55 and CD46 rich membrane areas is possible. It is known that, while F localizes in areas rich in cholesterol and glycosphingolipids, or rafts, CD46 does not (Manie et al., 2000; Vincent et al., 2000). But, the glycosylphospatidylinositol-anchored CD55 molecule is a constituent of rafts and fully able, when expressed alone, to control the amplification loop of the C3b deposition (Christiansen et al., 2000d). The molecular mechanism of F driven activation of the alternative pathway remains to be uncovered.

When the infecting virus expresses H protein with high affinity for CD46, there is an additional modest increase in C3b deposition observed. Two mechanisms may contribute to this, the down-regulation of CD46 and a competition between H and C3b for binding to CD46. The down-regulation is unlikely to be sufficient to explain the loss in C3b deposition control since CHO cell clones expressing comparable low amounts of CD46 are fully able to prevent the C3b amplification loop (Christiansen et al., 2000a). Although the SCR2 domain of CD46 is involved in the cofactor activity (Adams et al., 1991; Manchester et al., 1995), and the binding to MV-H (Buchholz et al., 1997; 1996; Iwata et al., 1995; Manchester et al., 1995), the two functions map to distinct sites (Christiansen et al., 2000a; Liszewski et al., 2000).

What could be the role of the C3b deposited at the virus surface? IgG-depleted serum, supplemented with $MgCl_2$/EGTA, reduced the virus infectivity down to 20-25 % of that observed with heat-inactivated IgG-depleted serum suggesting that the complement activation results in partial virus inactivation. Such an inactivation occurring *in vivo* should hamper virus spreading throughout the body by free virus particles and favours a cell-to-cell virus spreading. An antigen coupled to C3d oligomers displays a 1000-10,000 fold enhancement in immunogenicity (Dempsey et al., 1996), and a DNA vaccine coding H fused to three tandem copies of the murine C3d (H-3C3d) results in increase anti-H antibodies in mice (Green et al., 2001). The underlying mechanism may involve CD21 the C3dg receptor, which, as a part of the B cell receptor (BCR) complex is a potent stimulator of B cells (Clemenza et al., 2003; Haas et al., 2002). However, C3d also functions as a molecular adjuvant in CD21/CD35 null mice (Haas et al., 2002). Then, MV coating with C3b/C3bi, and its natural C3d proteolytic derivative, likely contributes to the induction of a life-lasting virus-specific immune response in a host who is paradoxically immunodepressed by the virus infection (Fugier-Vivier et al., 1997; Hicks et al., 1977; Karp et al., 1996; Yamanouchi et al., 1981).

In summary, MV activates the alternative complement pathway with some neutralizing effect. This could play an important role, *in vivo,* to slow the viral propagation by inactivating circulating virus and eliminating MV-infected cells, and to enhance the induction of a specific immune response by opsonizing MV antigens. A critical role of C3 in the pathology induced by γ-herpesviruses has recently been demonstrated (Kapadia et al., 2002). We propose that the activation of the alternative pathway is one of the primary mechanisms of defense against MV infection, before the production of neutralizing anti-MV antibodies by the immune system. Observation of MV infection of rare cases of C3, Factor D or Factor I deficiencies not yet reported would provide valuable information.

4.2. Activation and subversion of DC linkage between innate and adaptive immunity

DCs, which play a critical role in inducing adaptive immunity, are also regulators and effectors of the innate immunity against tumoral cells and viruses. Although most of the cells are able to produce low amount of IFN-α in response to viral infection, *in vivo*, the plasmacytoid DCs are one of the major source of IFN-α/β secretion, particularly in the initial phase of virus infection (see Levy, 2002 for review). This cell population is also called natural interferon producing cells (IPCs). DCs can become cytotoxic and consequently exhibit NK function. MV was the first virus described for being able to activate DC cytotoxic functions (Vidalain et al., 2001a).

4.2.1. DCs are key targets for MV infection

Immature DCs such as monocyte-derived DCs or Langerhans cells (LC) are permissive to MV infection. The level of MV replication in immature DCs is low and comparable to the replication level obtained in LPS-activated monocytes (Fugier-Vivier et al., 1997). Even when few DCs are infected, the whole DC population undergoes maturation after engulfment of MV-infected apoptotic DCs (Servet-Delprat et al., 2000). DC maturation was demonstrated by (i) upregulation of CD40, CD80, MHC-II and MHC-I, (ii) induction of CD25, CD69, CD71, CD83, CD150, and CD86, (iii) down-regulation of CD1a and CD32 and, (iv) weak induction of IL-12, IL-1α/β and IL-6 mRNA synthesis. Furthermore, E-Cadherine, the molecule engaged in homotypic adhesion between skin DCs and keratinocytes was also down-regulated in LCs (Servet-Delprat et al., 2000). Thus, MV replicates in immature DCs and induces their maturation (Schnorr et al., 1997; Servet-Delprat et al., 2000). Viruses containing a CD150-dependent wild-type strain-derived H protein revealed a particular tropism for human DCs *in vitro*. These lymphotropic MV replicate well in secondary lymphatic tissues of cotton rats where they were also able to cause immunosuppression, as documented by an impaired proliferative response of lymphocytes *ex vivo* (Ohgimoto et al., 2001). In the cotton rat model (Sigmodon hispidus), CD150-dependent viruses spread to mediastinal lymph nodes and induced strong immune suppression. As DC maturation is associated with their migration *in vivo*, MV infection of respiratory tract DCs might be an efficient way to drive MV into the T cell area of the human draining lymph nodes.

4.2.2. Modulation of cytokine secretion by DCs

IFN-α/β induce the maturation of myeloid DCs, prime DC functions for antigen cross-priming and promote Th1-type immune responses that are characterized by the generation of $CD4^+$ Th1 cells and $CD8^+$ CTLs, which secrete high levels of IFN-γ ((Le Bon et al., 2003) and (Bogdan, 2000) for review). IFN-α/β inhibits IL-12 secretion, and induces IL-10 and IL-18 secretion by DCs and/or monocytes (Byrnes et al., 2001; Dalod et al., 2002; Dauer et al., 2003; Mohty et al., 2003). Does an initial boost of IFN-α/β secretion by plasmacytoid DCs, induced by natural measles infection, control the transient increase of IL-10 and IL-18 plasma levels observed during the rash (see Figure 1c)? Finally, IFN-α/β enhances antibody response through DC stimulation (Le Bon et al., 2001).

Together, these results strongly suggest that IFN-α/β ensures optimal immune response activation, by promoting infected cell apoptosis and DC functions. Different MV strains induce variable amounts of IFN-α/β (Naniche et al., 2000). At low multiplicity of infection, wild-type MV induces up to 80-fold less IFN-α/β secretion than do vaccine strains in activated human PBL. A low IFN-α/β secretion induced by wild-type MV would favour virus spreading in the body because of reduced activation of both innate and adaptive immune responses. An induction of high levels of IFN-α/β by ability vaccine strains would stimulate activation of both type of immunity although vaccine strains are expected to hamper strongly IFN-α/β signalling in infected cells.

4.2.3. Impairment of DC and T cell interaction

Upon arriving in the draining lymph nodes, the crosstalk between DCs and T cells usually results in DC survival, final DC differentiation including cytokine secretion and efficient antigen presentation, which induce T cell activation, proliferation and their differentiation into effector T cells. In the case of measles, the fate of MV-infected DC-T cell crosstalk is quite different. (i) They do not sustain naive T cell proliferation. (ii) They inhibit mitogen-activated T cell proliferation and develop a novel activity by becoming cytotoxic cells. (iiii) They can kill neighbouring activated T cells (Fugier-Vivier et al., 1997). (iv) MV-infected DCs do not survive more than three days with activated T cells. Between 1999 and 2004, most of the molecular mechanisms of these different observations have been unravelled.

4.2.3.1. TRAIL-mediated T cell killing by DCs

a) Activation of TRAIL-mediated cytotoxic DCs

IFN-α and IFN-β induce the maturation of myeloid DCs and prime DC functions (Bartholome et al., 1999a; 1999b; Ito et al., 2001; Luft et al., 2002; 1998; Santini et al., 2003; 2000). IFN-α/β can stimulate the expression of pro-apoptotic genes such as Fas or TRAIL, members of tumour necrosis factor family (reviewed in Chawla-Sarkar et al., 2003). TRAIL is a type II transmembrane death factor that belongs to the TNF family closely related to three other death-inducing ligands: FasL, TNF-α and TWEAK/Apo3L. TRAIL seems to be preferentially induced by IFN-β (Chawla-Sarkar et al., 2001). IFN-β, which is specifically induced after infection of immature DCs by MV, is a potent inducer of TRAIL on these cells and renders them cytotoxic for a tumour cell line (Vidalain et al., 2000; Vidalain et al., 2001b). TRAIL specifically induces apoptosis of virus-infected cells with minimal cytotoxicity towards normal tissues *in vivo*. The killing of infected cells by cytotoxic DCs may favour phagocytosis of apoptotic bodies.

b) Sensitization of uninfected T cells to TRAIL-mediated apoptosis

In vitro, the MV infection can sensitize lymphocytes to apoptosis independently from the inhibition of cell proliferation. MV induces apoptotic cell death in cells activated by phorbol ester plus ionomycin, but not those activated by the phytohemagglutinin, although the cell proliferation is suppressed in the latter case (Ito et al., 1997). In co-culture with DCs, activated peripheral blood $CD4^+$ T cells and resting CD45RA naive T cells are sensitive and resistant to apoptosis, respectively (Fugier-Vivier et al., 1997; Grosjean et al., 1997). MV Edmonston strain or dsRNA induces IFN-β production by DCs that induces TRAIL synthesis (Vidalain et al., 2000; 2001a). From these *in vitro* data and *in vivo* observations (see § 2.2. above and Figure 1), it is proposed that MV induce lymphocyte depletion by combining two actions: the induction of TRAIL synthesis by DCs that defines new killer effector cells and the sensitization of activated T cells to TRAIL-mediated apoptosis. TRAIL synthesis occurs downstream of the IFN-β secretion induced by CD46-triggering or of another mechanism yet to be described. T lymphocytes, which are targets of the DC "apoptotic kiss", would be either nearby activated T cells against antigen unrelated to MV infection, or T cells activated by MV by an unknown mechanism.

4.2.3.2. Fas-mediated DC killing by T cells

Biochemical analysis revealed that MV infection strongly modifies CD40-induced tyrosine phosphorylation in DCs (Servet-Delprat et al., 2000). Instead of being protected from Fas apoptosis, CD40L-activated MV-infected DCs are killed by Fas-expressing activated T cells. MV was shown to sensitize DCs to Fas-mediated apoptosis by up-regulating Fas, Flice/caspase-8 and FADD mRNAs (Servet-Delprat et al., 2000). As activation up-regulates Fas-ligand expressed by T cells, early apoptosis of Fas-sensitive MV-infected DCs would prevent follow-up T cell activation and could account for *in vivo* suppression of cell-mediated immunity.

4.3. Activation and subversion of adaptive immunity

For most MV victims, fatality is indirect: virus-induced transient immunosuppression predisposes the individual to opportunistic infections that, left untreated, can result in mortality. This immunosuppression is reproduced in a human CD46-transgenic mice model inoculated with MV and the bacteria Listeria monocytogenes. MV-infected mice are more susceptible to infection with Listeria than control mice, with significantly decreased numbers of macrophages and neutrophils in the spleen and substantial defects in IFN-γ production by $CD4^+$ T cells (Slifka et al., 2003). In the absence of MV replication, MV proteins could generate a systemic immunosuppression in mice through two pathways: via N and its receptor FcγR on DCs and via virus H/F and their cellular receptor, CD46. The effects comprise reduced hypersensitivity responses associated with impaired function of DCs, decreased production of IL-12, and the loss of antigen-specific T cell proliferation (Marie et al., 2001).

4.3.1. Paradoxical effects of MV-H protein: induction of specific immune response and immunosuppression

4.3.1.1. Activation of adaptive immunity by MV-H protein

H activates adaptive immunity and induces specific antibodies in infected subjects (El Mubarak et al., 2004). Indeed, the most important neutralizing and protective antibodies against MV are directed against H glycoprotein. Two distinct regions are the main target of human antibodies: the hemagglutinin noose epitope (HNE, residues 380-400) and a second epitope defined by strong neutralizing but weak hemagglutination-inhibiting activity of the antibodies. Human antibodies react more weakly with H regions that bind to the CD46 and more strongly with the putative CD150 binding site and a predicted top loop between two β-sheet of H (Ertl et al., 2003) (see below Figure 6a, right panel). The maturation of the IgG avidity is detected within 3-6 months after infection (El Mubarak et al., 2004).

In rhesus macaques, using DNA vaccines coding a secreted form of H or F proteins, H primes for type 2 cytokine responses, with suppression of IL-12 and preferential production of IL-4 after MV challenge. The fusion protein primes for a type 1 response with preferential production of IFN-γ (Polack et al., 2003b).

4.3.1.2. Immunomodulation via H binding to CD46 and CD150

Immunosuppression is partly mediated by surface contact signaling induced by viral proteins. From *in vivo* experiments in transgenic mice and *in vitro* studies on human cells, the ligation of CD46 and/or CD150 by H are involved in several immunosuppressive effects, although ligation of a third unknown receptor may be also involved.

a) CD46 is a physiological inducer of regulatory T cells (Tr1)

In addition to its complement regulator activity, CD46 is a physiological inducer of regulatory T cells (Tr1). Co-engagement of CD3 and CD46 induces T cell proliferation (Astier et al., 2000) associated with morphological changes (Zaffran et al., 2001). *In vitro* co-stimulation could be achieved with anti-CD46 MCI20.6 monoclonal antibody that inhibits MV infection by preventing H to CD46 interaction.

These data are puzzling as MV may stimulate T cell proliferation through CD46, contrary to experimental results (Astier et al., 2000; Zaffran et al., 2001). However, CD3 and CD46-stimulated T cells, in the presence of IL-2, have a cytokine profile distinct from that induced by the co-stimulation with antibodies to CD3 and CD28. They synthesize large amounts of IL-10 and intermediate amounts of TGF-β, a cytokine profile which corresponds to Tr1 cells. CD3/CD46-stimulated CD4$^+$ T cells proliferate strongly, suppress the activation of bystander CD4$^+$ T cells, and acquire a memory CD45R0$^+$ phenotype (Kemper et al., 2003). Tr1 cells are essential for maintaining peripheral tolerance and preventing autoimmunity. They may also modulate the host's immune response to many pathogens. This role of CD46 in negative regulation of cellular immune responses indicates that the use of CD46 as cellular receptor by several human pathogens, including MV, human Herpesvirus 6 (Santoro et al., 1999), Streptococcus A (Okada et al., 1995), Neisseria (Kallstrom et al., 1997) and several adenoviruses subtypes (Cattaneo, 2004) would favour their replication in immuno-competent hosts. Whether MV induces Tr1 cell activation and proliferation through H binding to CD46 remains to be demonstrated.

b) H binding to CD150 mediates the switch from Th1 to Th2 immune responses ?

The substitution of the H gene of the Edmonston vaccine MV strain with that of CD46^{nonusing} wild-type MV, changes the *in vitro* virus tropism from human peripheral lymphocytes to DCs (Ohgimoto et al., 2001). The infection of human CD150 (hCD150) transgenic mice with wild-type MV induces the down-regulation of hCD150 and inhibited the proliferation of hCD150$^+$ but not hCD150$^-$ mononuclear cells (Hahm et al., 2003). When used to infect cotton rats, CD46^{nonusing} MV spreads to mediastinal lymph nodes and induced strong immune suppression. *In vitro,* to induce the same inhibition of lymphocyte proliferation, a thousand fold less CD46^{nonusing} virus is required than CD46^{using} virus (Pfeuffer et al., 2003). However, the relevance of this observation made in the cotton rat model awaits the identification of the receptor used by MV to infect the cotton rat cells.

CD150 plays a central role at the interface of acquired and innate immune responses. It is a costimulatory receptor that mediates activation, proliferation and IFN-γ production of T cells, and that regulates signals transduced by TLR4, the major LPS receptor (Wang et al., 2004). CD150 co-localizes with the TCR following CD3 triggering in human peripheral blood T cells and, upon TCR cross-linking, it is rapidly and reversibly phosphorylated on tyrosine. The Src-like kinases Lck and Fyn phosphorylate tyrosine residues in the cytoplasmic tail of CD150. CD150-induced signal transduction is controlled by SAP/SH2D1A, AKT (Howie et al., 2002a; 2002b) and EAT-2 molecules (Sidorenko and Clark, 2003). SAP is expressed in T and NK cells and EAT-2 in DCs and macrophages (see Veilette, 2004 for review). Deletion of CD150 expression in transgenic mice results in the down-regulation of IL-4, but not IFN-γ secretion by CD4$^+$ T cells and in the inability to remove the parasite *Leishmania major* (Wang et al., 2004). MV–H down-regulates CD150 in cells either expressing or coming in contact with it (Tanaka et al., 2002). Pathogenetic consequences of CD150 down-regulation are unknown as yet but may create a favourable environment to evade viral destruction. Since high levels of CD150 were found in Th1 cells, whereas only small amounts were detectable in Th2 cells (Wang et al., 2001), down-regulation of CD150 by MV infection may play a role in the switch from Th1 type to Th2 type immune response observed *in vivo* (Moss et al., 2002). The cytokine profile during the human disease (Figure 1c) could result from strains which ligate both CD150 and CD46. The CD150 down-regulation on lymphocytes may decrease IL-2 and IFN-γ production by lack of Th1 activation. As a result, Th2 cells are less likely to be inhibited by Th1 cells, and produce high level of IL-4. Furthermore, the triggering of CD46 sustains the IL-10 production by Tr1. Thus, we propose that the concomitant ligations of CD150 and CD46 expressed by Th1 and $CD4^+$ naive T cells, respectively, deregulate the immune system skewing the primary antiviral Th1 stimulation to Th2- and Tr1-driven responses.

4.3.1.3. H and F mediated inhibition of lymphocyte proliferation

DCs are proposed to play a pivotal role in the transient immune suppression induced by MV infection. MV infection turns on the maturation program of DCs (Schnorr et al., 1997; Servet-Delprat et al., 2000) allowing migration to draining lymph nodes, where the cell surface expression of H and F glycoproteins induces potent T-cell immune suppression (Klagge et al., 2000) independently of T cell trans-infection. DC-induced suppression of T-cell proliferation is not mediated by soluble factors such as IL-10 or IFN-α/β, which are released following infection of DCs, but required contacts between MV-infected DCs and T cells. In a transgenic model, MV infected $hCD150^+$ DCs have similar lymphocyte proliferation inhibitory properties even against $CD150^-$ T lymphocytes (Hahm et al., 2004). Neutralizing anti-H and F polyclonal antibodies reverse the suppression and restore the allostimulatory ability of the DCs (Dubois et al., 2001). In the absence of MV replication, the MV-H and F glycoprotein complex seems to provide a surface contact mediated signal which directly inhibits human lymphocyte proliferation *in vitro*. Cells co-expressing H and F proteins, but not cells expressing either H or F, are inhibitory (Schlender et al., 1996). That this effect is related to the fusiogenic property of the H/F complex has not been ruled out. Indeed, the proteolytic cleavage of the MV-F protein precursor into its fusiogenic F_1 and F_2 subunits, is required for the inhibition of lymphocyte proliferation. The lack of effect of the fusion-inhibitory peptide z-D-Phe-L-Phe-Gly is not sufficient to exclude the involvement of localized microfusion between virus envelope and cell membranes. Indeed, although, z-D-Phe-L-Phe-Gly can prevent syncytia formation by inhibiting cellular fusion at the stage of hemifusion (*i.e.* it prevents the lipid mixing of the outer membrane layers) (Weidmann et al., 2000), it is unable to prevent MV entry and spreading within a cell bilayer (D. Gerlier, unpublished data; Laine et al., 2003). Moreover, z-D-Phe-L-Phe-Gly has been reported by another group to inhibit, in dose-dependent manner, the anti-proliferative activity of infectious MV (Yanagi et al., 1992). Which cell surface receptor is involved is unclear, since the effect is induced by any MV strain tested, whatever it binds to CD46 or not. Furthermore, inhibition of T cell lines not expressing CD150 is also observed (Schlender et al., 1996). Similar observations were obtained *in vitro* and *in vivo* in the cotton rat model (Niewiesk et al., 1997). Though the receptor of the H/F complex is unknown, the intracellular pathway would involve the impairment of the Akt kinase activation and the expression of a constitutively active Akt prevents the anti-proliferative effect of MV glycoproteins

(Avota et al., 2001). The molecular mechanisms underlying the H/F mediated inhibition of T cell proliferation await to be further deciphered.

4.3.2. Paradoxical effects of N protein: induction of specific immune response and immunosuppression

MV-N is an internal protein within infected cells and virus. However, some data indicate that MV-N is also released in the extracellular milieu by dying cells, and interacts with extracellular receptors. By binding to its receptors, N modulates the immune response both in a positive and a negative way.

4.3.2.1. Release of N protein in the extracellular milieu

MV-N is synthesized within the cytosol of MV-infected cells and encapsidated into the virus envelope. MV-N is also released in the extracellular milieu after apoptosis and/or secondary necrosis of MV-infected cells as demonstrated *in vitro* (Laine et al., 2003). Both MV replication and apoptosis of MV-infected cells are required for N extracellular release. MV initially replicates in the respiratory tract and then spreads to the secondary lymphoid organs where it induces syncytia or giant multinucleated WFCs (Tajima and Kudow, 1976). WFCs are observed both in the germinal centers and in the follicular areas of the draining lymph nodes (Nozawa et al., 1994). Syncytium formation is proportional to MV replication. Syncytia induce and amplify apoptosis (Scheller and Jassoy, 2001). Among mechanisms involved in the boost of MV replication, CD40 activation increases both replication, syncytia formation, and intracellular N synthesis (for a review, see Servet-Delprat et al., 2003). By interacting with N_{TAIL}, Hsp72 stimulates the transcription activity of MV and syncytium formation (Oglesbee et al., 1993; Vasconcelos et al., 1998; Zhang et al., 2002). Altogether, these combined mechanisms lead to the accumulation of high amounts of intracellular N, which are released in the extracellular compartment after apoptosis and/or secondary necrosis. In addition, N protein seems to be able to translocate from the cytosl into the late endocytic compartment from which it can be excreted when the cells express the FcγRII receptor (Marie et al., 2004). Thus, large amounts of extracellular N protein should be released within the lymphoid organs, and, after binding to cell surface receptors, act locally in B and T cell areas.

4.3.2.2. Cell-free N protein binds to several cell surface receptors

To date, MV-N binds to three different cell surface receptors: specific B cell receptor (BCR) (*i.e.* the cell surface immunoglobulin) expressed on B lymphocytes (Graves et al., 1984), low-affinity Fc receptor for IgG of type II (FcγRII, CD32) (Ravanel et al., 1997), and a yet unidentified protein receptor called nucleoprotein receptor (NR) (Laine et al., 2003).

As mammalian FcγRIIs are conserved among species, MV-N is capable of interacting with both human and murine FcγRII expressed on APC, including monocytes/macrophages, DCs and B lymphocytes, but not on T lymphocytes, epithelial cells and fibroblasts (Laine et al., 2003; Ravanel et al., 1997). Three different FcγRII isoforms are generated by alternative splicing, differing by their cytoplasmic tails: FcγRIIA, IIB and IIC (see Ravetch and Bolland, 2001; Tsubata, 1999 for review). FcγRIIB is expressed in both human and mouse, whereas FcγRIIA and IIC are expressed only in human. FcγRIIB1 is expressed only on B cells, and the FcγRIIB2 isoform is expressed mainly on myeloid cells. Human FcγRIIA favours phagocytosis and calcium mobilisation, while human and murine FcγRIIB2 molecules mediate endocytosis of their ligands. The FcγRIIA and FcγRIIB isoforms modulate the immune response via the presence in their intracytoplasmic tails of a immunoreceptor tyrosine-based activation (ITAM) and inhibitory (ITIM) motif, respectively. Recombinant MV-N binds to all FcγRII isoforms (Ravanel et al., 1997). Immune complexes, the physiological ligands of FcγRII, and MV-N may share the same or overlapping binding sites on FcγRII, because the 2.4G2 mAb blocks both of the murine FcγRII/immune complexes and FcγRII/MV-N interactions (Ravanel et al., 1997). In addition, the region of MV-N responsible for binding to human FcγRIIB maps to N_{CORE} (Laine et al., 2005). This property of N binding to human FcγRIIB is shared by the N proteins of some others *Morbillivirus* such as the canine distemper (CDV) and peste des petits ruminants virus, (PPRV), but not by the rinderpest virus (RPV) (Laine et al., 2003). A NH2-terminal region of N_{CORE} (aa 122-144) is a variable region among *Morbillivirus* N proteins, poor in hydrophobic clusters. It may therefore form a loop exposed to the solvent (Karlin et al., 2002), and it is accessible to antibodies directed against N (Buckland et al., 1989; Fournier et al., 1996). Within this loop, position 138 is occupied by a Ser residue in the N proteins of MV, CDV and PPRV, but not RPV (Diallo et al., 1994), suggesting that binding to FcγRIIB may occur through this exposed region of N_{CORE}.

A third receptor, provisionally called NR, is postulated on the basis of MV-N binding observations. NR is constitutively expressed on fibroblasts, epithelial, lymphoid and myeloid cells. NR is not detected on resting T cells, but up-regulated on human activated T cells (Laine et al., 2003). Whether NR is an ubiquitous and then unique molecule or a family of receptors is yet to be determined. MV-N binds to human as well as murine and monkey NR, suggesting that NR might be conserved among different species (Laine et al., 2003). Moreover, CDV-N, PPRV-N, and RPV-N also bind to NR. However, the region of MV-N responsible for binding to NR maps to the region 401-420 of N_{TAIL} (Laine et al., 2005), a well conserved region in the hyper-variable *Morbillivirus* N_{TAIL} domain. Although N_{TAIL} is highly sensitive to proteolysis, N_{TAIL}/NR interaction remains possible in the context of an entire nucleocapsid structure just after its release from the extracellular compartment. Since MV-N binds to FcγRIIB and NR via N_{CORE} and N_{TAIL} respectively, MV-N can simultaneously bind both to FcγRIIB and NR on the same cell type such as B cells, monocytes/macrophages and DCs. In contrast, MV-N interacts only to NR on human activated T cells (Laine et al., 2003).

4.3.2.3. Activation of adaptive immunity by N protein

During the acute phase of MV infection, both strong humoral and cellular anti-N immune responses have been documented. At the onset of the rash, the humoral response occurs after B lymphocyte stimulation by accessible and native viral proteins. Notably, the most abundant and rapidly produced antibodies are N-specific, preceding that of anti-H, anti-F and sometimes anti-M antibodies (Graves et al., 1984; Norrby and Gollmar, 1972), thus confirming that cell-free MV-N, released from MV-infected cells in the extracellular compartment, is accessible to the BCR *in vivo*. The cellular component of the immune response against MV-N involves MHC class I-restricted $CD8^+$ T cells and MHC class II-restricted $CD4^+$ T cells (Ilonen et al., 1990; Jacobson et al., 1989; van Binnendijk et al., 1989). The fact that efficient helper $CD4^+$ T cell responses against N are generated during the course of MV infection indicate that MV-N is accessible to the peptide-loading compartment after its uptake by the APC. Since infectious MV enter the host cells by neutral pH fusion at the cell membrane, the incoming N protein is delivered into the cytosol of the APC. There, it is readily available for cytosolic degradation and peptide loading onto MHC-I resulting in efficient MHC-I restricted presentation to $CD8^+$ T cells (Cardoso et al., 1995). However, instead of fusing, some viral particles are endocytosed and delivered to the MHC-II compartments which results in MHC-II presentation of N protein (Gerlier et al., 1994b). Furthermore, the cell-free exogenous N is targeted to macrophages and DCs, by binding to FcγRII, which probably likely enhances the presentation of N antigen by MHC class II to $CD4^+$ T cells. As a matter of fact, MV-N has potent immunogenic properties *in vivo*, by-passing the need of an adjuvant. Buccal immunization with recombinant MV-N protein induces *in vivo* priming of protective MHC class-I-restricted specific $CD8^+$ T cells and IFN-γ-producing N-specific effectors. Both mucosal and systemic immunization with recombinant MV-N protein induces N-specific T cell responses, a systemic anti-N antibody response and protection against MV infection challenge (Etchart et al., 2001; Olszewska et al., 2001). Those immunizations result in N transport, probably by activated DCs, via afferent lymph to local draining lymph nodes for priming of both B and T cells. However, the mechanism by which MV-N induces DC activation and/or maturation either directly via unknown receptor(s) or indirectly via the release of proinflammatory signals remains to be determined. Recent data indicate that DC activation may be induced after engagement of cell surface Hsp72 strongly suggesting that MV-N may also induce DC activation due to the ability of N_{TAIL} to bind to Hsp72 (Zhang et al., 2002). Thus, effective N-specific responses lead, with other events, to viral clearance, recovery from disease and a protective immune response against re-infection.

4.3.2.4. Immunosupressive activity of N protein

Paradoxically, exogenous recombinant MV-N is also involved in immunosuppression mechanisms after binding to FcγRIIB, NR or both receptors. MV-N suppresses hypersensitivity responses to third party antigen associated with impaired DC function, including decrease of IL-12 production and loss of antigen-specific T cell proliferation. This reduction of inflammatory reactions requires the interaction between MV-N and FcγRIIB because it is no longer observed in FcR null mice (Marie et al., 2001). Inhibition of IL-12 production has also been documented in human CD40-activated DCs *in vitro* (Servet-Delprat et al., 2000). As FcγR-dependent mechanisms of suppression of IL-12 synthesis have been demonstrated in human monocytes/macrophages (Grazia Cappiello et al., 2001), MV-N binding to FcγRII on DCs may represent a key factor for IL-12 inhibition and thus Th2 polarization observed following MV infection (Ward and Griffin, 1993). In addition, MV-N is a potent inhibitor of spontaneous and induced cell proliferation. This effect has been ascribed to FcγRIIB1 and NR. MV-binding to FcγRIIB and NR have additive inhibitory effects on the B lymphoma cell proliferation, and MV-N binding to NR

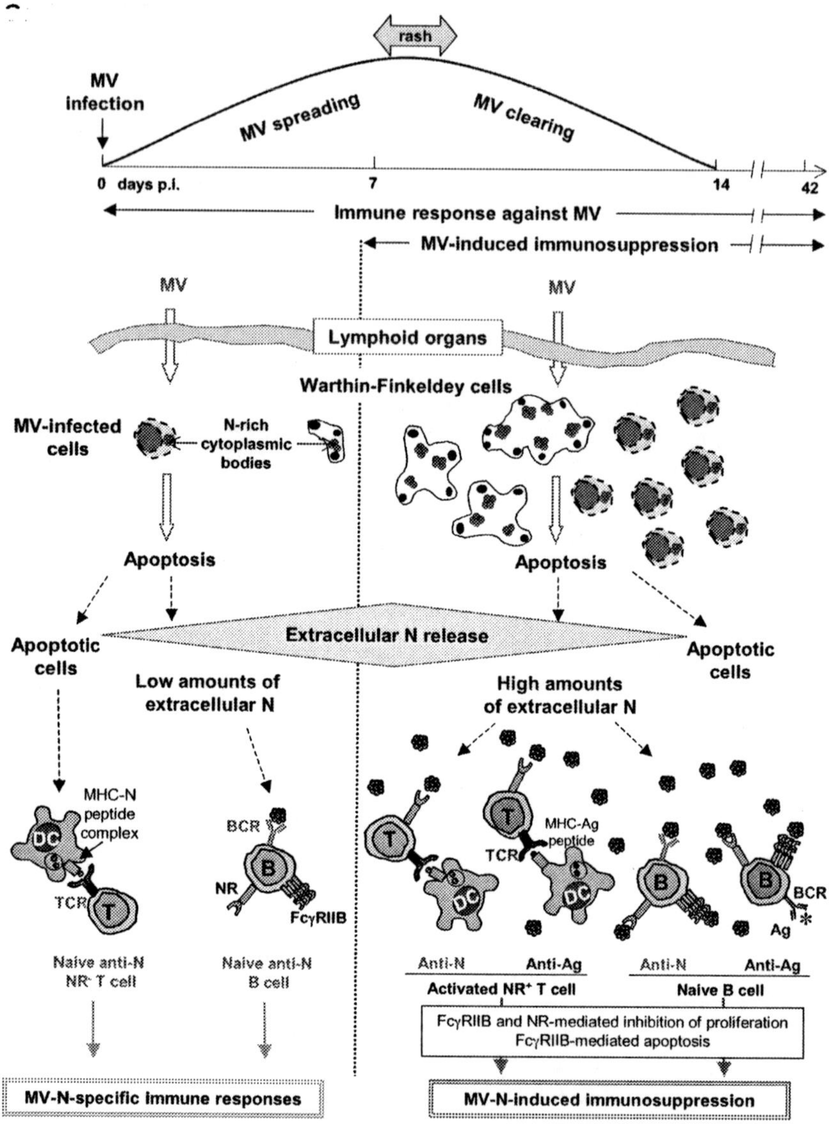

Figure 3. A two-step model of induction of early MV-N specific immune response and late generalized immunodepression by N protein. During the course of MV infection, MV-N presents paradoxical properties modulating the immune response both in a positive and a negative way. The model predicts that available increasing amounts of extracellular MV-N govern these two opposite activities. At the early stage of infection (left side of the figure), few MV particles spread and replicate into the secondary lymphoid organs. This results in the apoptosis of infected cells and as generation of few giant multinucleated cells (WFC), containing numerous intracytoplasmic inclusion bodies rich in N protein. The processing of apoptotic bodies derived from infected cells are processed by DCs leads to efficient activation of MV specific MHC restricted CD8$^+$ and CD4$^+$ T cells. Low amounts of MV-N are also released in the extracellular milieu and bind to specific and high affinity B cell receptors (BCR) resulting in efficient activation of N-specific B cells. At the onset of the rash (right side of the figure), intensive MV replication occurs in the secondary

lymphoid organs with massive infected cells apoptosis and/or secondary necrosis leading to the accumulation of high amounts of intracellular and extracellular N. Extracellular N can bind to low-affinity FcγRII and NR receptors and disrupt MV-specific and non-specific B- and T-cell-mediated responses, while N-specific effector cells are sustained.

efficiently suppresses the proliferation of epithelial and activated T cells (Laine et al., 2005; 2003). N_{CORE}/FcγRIIB and N_{TAIL}/NR interactions induce an arrest in the G_0/G_1 phases of the cell cycle, but only the former mediates an apoptotic process (Laine et al., 2005) in agreement with previous observations (Pearse et al., 1999). *In vitro*, the immunoglobulin (Ig) synthesis of CD40- or BCR-activated human B lymphocytes in the presence of IL-2 and IL-10 is strongly reduced in the presence of MV-N (Ravanel et al., 1997). Since the B cell differentiation process requires a proliferation phase prior to the Ig secretion, both FcγRIIB and NR may contribute to suppression of antibody production after MV-N binding. In conclusion, exogenous MV-N probably contributes to the panlymphopenia observed during acute MV infection (see Figure 1d) and the disruption of both B- and T-cell-mediated responses.

4.3.2.5. How can MV-N induce both specific immune response and systemic immunosuppression?

We propose a model for the paradoxical immunostimulating and immunosuppressing effects of MV-N during MV infection (Figure 3). The amount of extracellular N may affect the balance between induction versus inhibition of the immune response. Indeed, the immunostimulatory activities of N are observed using low doses of MV-N (Etchart et al., 2001; Olszewska et al., 2001), whereas high doses are required to suppress the immune response (Laine et al., 2005; 2003; Marie et al., 2001; Ravanel et al., 1997; Servet-Delprat et al., 2003). Upon entry of MV into the lymphoid organs, DCs engulf apoptotic cells, endocytose some virus particles and display N-derived peptides in complex with MHC-II molecules, and activate N-specific helper $CD4^+$ T cells. The few infected cells die by apoptosis and/or secondary necrosis, leading to the release of low amounts of extracellular MV-N (Figure 3). This low amounts of extracellular N would bind to B lymphocytes expressing high avidity N specific BCR (Obeid et al., 1995), and prime them for $CD4^+$ dependent activation (Graves et al., 1984; Norrby and Gollmar, 1972) (Figure 3). Later on, intensive replication occurs with induction of WFCs and cell apoptosis leading to the release of high amounts of N in the extracellular milieu (Figure 3). The level of extracellular MV-N reaches a sufficient level for efficient binding to its low affinity receptors FcγRIIB and NR (Ravanel et al., 1997), and triggers inhibitory signals which will prevent MV-specific and non specific B and T cell proliferation (Figure 3 left panels). MV prevents the acquisition of lymphocyte functions, but does not inhibit these functions when established (Galama et al., 1980). MV-N would be unable to down-regulate the virus-specific immune effector functions (Figure 3, right panels), thus explaining how an efficient immune response against MV can be elicited despite with the strong immunosuppressive properties of MV.

4.4. Impairment of the homeostasis of the immune system

Pan lymphopenia, monocytopenia, and neutropenia are hallmarks of MV-induced impairment of homeostasis of the immune system (Arneborn and Biberfeld, 1983; Kim et al., 2002; Okada et al., 2000). This leucopenia likely reflects (i) an impairment of cell hematopoiesis linked to the depletion of bone marrow (BM) progenitors (ii) a thymus injury and (iii) the apoptosis of infected as well as uninfected T and B cells.

4.4.1. MV-induced suppression of hematopoiesis

MV targets BM cell progenitors and suppresses hematopoiesis. There is a direct correlation between BM output, BM reconstitution and the recovery of immunosuppression (Mackall et al., 1997; McFarland et al., 2000; Okada et al., 2000). Following MV infection, BM damage is also brought about by an abnormal decrease of hematopoietic progenitors in the blood (Okada et al., 2000).

In monkey models, BM contains high titers of MV, suggesting MV replication in BM cells (Kobune et al., 1996; McChesney et al., 1997). *In vitro*, MV replication varies according to the stage of cell differentiation and maturation. Productive infection is observed in immature cells such as human $CD34^+$ committed myeloid progenitors and human granulocyte-macrophage colony-forming cells (GM-CFC) but restricted in mature monocytes/macrophages (Helin et al., 1999; Manchester et al., 2002). MV infection of BM stroma cells inhibits the hematopoietic ability of stem cells (Manchester et al., 2002). CD46 is expressed on all $CD34^+$ cells, and CD150 is expressed by only 2% of these cells. Infection of $CD34^+$ cells is not inhibited by anti-CD46 or anti-CD150 antibodies and this suggests an entry via a third cellular receptor. In contrast, BM stroma cells are infected via CD46 (Manchester et al., 2002). Thus, MV disrupts hematopoiesis because it infects BM progenitors and stroma cells. MV-mediated defects in hematopoiesis likely contributes to the long-term immunosuppression seen in measles

Osteoclasts are multinucleated cells, and, *in vitro*, they can be generated from mononuclear cells, such as monocytes and DCs (Rivollier et al., 2004). Osteoclast may participate in the immune and inflammatory responses as suggested by their expression of two ITAM molecules, DAP12 and FcγR, through which signal transduction lead to the osteoclast differenciation (Baron, 2004). *In vitro,* MV infection of BM cells from CD46 transgenic mice leads to a significant increase in size, number of nuclei, and bone-resorption ability of osteoclasts (Reddy et al., 2001). In osteoclasts from CD46-transgenic mice, MV enhances IL-6 production (Reddy et al., 2001), and increases osteoclast formation (Roodman et al., 1992). Thus, MV has also the potential ability to disturb the differentiation of osteoclast precursors.

4.4.2. Impairment of thymus functions

Measles induces a systemic disease involving both lymphoid organ and epithelial tissues. The former is the central site of the immune response, and the main site of virus growth. The thymus represents an important target for MV infection, and by this way, is probably involved into the immune suppression, particularly in infants. After a brief review of T cell development mechanisms, the putative role of thymic epithelial cells in MV-induced thymic atropy and the consequences of thymus infection on thymic output of immunocompetent T cells are discussed.

4.4.2.1. T cell development in the thymus

T cell precursors leave the bone marrow, pass into the bloodstream, enter the thymus through the subscapsular cortex to become prothymocytes. Then, they mature into T cells during their migration through the distinct thymic micro-environment in the deep cortex, the cortico-medullary junction and the medulla, and finally return to the blood. Human thymic function is essential during embryogenesis and early stages of life in the development of the T cell repertoire to ensure self-tolerance and homeostasie (Zuniga-Pflucker, 2004). To let T cells acquire immunocompetence and specificity, maturation states include signals for survival, proliferation, TCR gene rearrangements, selection, molecular and functional maturation. This maturation program is controlled by both direct cell-to-cell interactions and soluble factors between the developing T cells and the cells of the thymic micro-environment (Zuniga-Pflucker, 2004). The latter is composed of a network of stromal cells and the extracellular matrix. The major stromal cell types include TECs and hematopoietic cells, such as macrophages and DCs. TECs form a heterogeneous network within both of the cortex and medulla. The interaction of immature T cells with thymic micro-environment ensures the selection of MHC-restricted and self-tolerant T cells. In the cortex, positive selection of immature $CD4^+CD8^+$ T cells expressing functional TCR is mediated mainly by cortical TECs. In the medulla, negative selection of developing autoreactive $CD4^+CD8^+$ T cells is ensured by APCs of hematopoietic origin and, to a lesser extent, by TEC. At each stage of T cell maturation, the interactions between cells are controlled by adhesion molecules and a gradient of several growth factors including cytokines and hormones issued from autocrine, paracrine or endocrine pathways. Hassall's corpuscles, composed of different cellular micro-environment of the thymic medulla, are important functional components of the reticulo-epithelial network which provides paracrine signals to developing thymocytes (Bodey et al., 2000; Bodey and Kaiser, 1997). Thus, the thymic micro-environment supports thymocyte survival and differentiation into mature functional T cells (Figure 4).

In human, the thymus begins involution soon after birth and ends in young adults. This irreversible "physiological" involution correlates with the ability to export mature T cells into the periphery (Linton and Dorshkind, 2004). Aging involves thymic micro-environment alterations (Brelinska, 2003; Varas et al., 2003). Thymic APCs and TECs are slightly and strongly affected by age, respectively. Thymus involution is characterized by the disruption of thymocyte-TEC interactions, which results in differentiation arrest of both TECs and thymocytes. Under pathological conditions, such as virus infection, the interaction of the virus with the thymic micro-environment can lead to thymic injury, which impairs T cell development particularly in infants.

4.4.2.2. Disruption of thymus integrity following MV infection

Thymic injury is observed in young children during the acute phase of MV infection, and this may be an important factor in the impaired development of T cells (Chino et al., 1979; Ueda et al., 1982; White and Boyd, 1973). MV infects the thymus both in human and primate during the second viraemia (Kobune et al., 1996; Moench et al., 1988; Sakaguchi et al., 1986; Yamanouchi et al., 1973), and thymuses contain MV RNA and proteins (Brown et al., 1989; Moench et al., 1988). Intensive MV

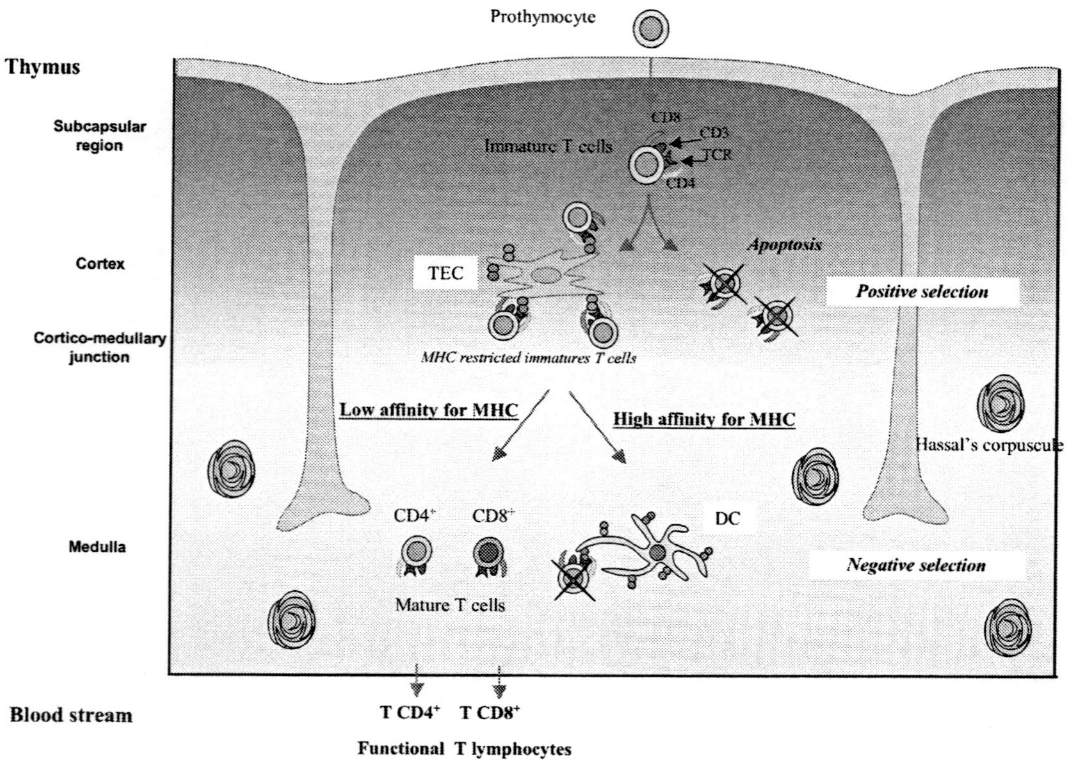

Figure 4. Thymus structure. This is a complex epithelial organ which ensures the generation of functional T cells from BM–derived hematopoietic precursors entering through the subscapsular cortex. Then, immature prothymocytes triggers T cell receptor (TCR) gene rearrangement and cells acquiring TCRα/β features and expressing CD4 and CD8 coreceptors continue to develop through the thymic microenvironment. Thymic tissue includes epithelial cells (TECs) and APC such as DCs and macrophages. Immature T cells, expressing TCR with useless or autoreactive specificities, undergo a two-step selection: (i) a positive selection of clones recognizing any peptides in context of autologous MHC molecules by cortical TECs, and (ii) a negative selection of clones recognizing autologous peptides bound to MHC molecules by APCs and, to a lesser extent, by TECs in the medulla. These selections are ensured also by soluble factors mainly provided by Hassall's corpuscles. Finally, the positively selected, non-self reactive naive T cells emigrate from the thymus into the blood stream.

replication occurs in the thymus as indicated by the presence of many multinucleated giant cells or WFC, containing numerous small hyperchromatic nuclei. These cells are strictly limited to the prodromal period of the disease and contain intranuclear and intracytoplasmic inclusion bodies and nucleocapsid protein antigens in the cytoplasm (Chino et al., 1979; Kobune et al., 1996; Nozawa et al., 1994; Sakaguchi et al., 1986; Shimizu et al., 1987; White and Boyd, 1973; Yamanouchi et al., 1973). Giant cells derive mainly from small lymphocytes or mature thymocytes in the cortex and from macrophages, DCs and TECs in the medulla (Nozawa et al., 1994). The thymic epithelium including the Hassall's corpuscles and macrophages are the major targets for MV growth. Although giant cells, expressing viral nucleocapsids, derive from both cell types, virus budding and release are detected only in the former, indicating a selective MV replication within thymic TEC (Moench et al., 1988; Sakaguchi et al., 1986). In addition, viral antigens are detected in thymus capillaries suggesting that endothelial cells are also infected (Moench et al., 1988). Thus, this primary lymphoid organ represents a peripheral reservoir of MV and may allow prolongation of viraemia, which usually ends with the onset of rash (Chino et al., 1979).

The MV-infected thymus displays in most cases degenerative and/or necrotic lesions of the cortex and/or the medulla (Chino et al., 1979; Kobune et al., 1996). As early as on the 4th day after measles rash, numerous thymocytes are degenerate and occasionally aggregate into syncytial masses similar to WFCs (White and Boyd, 1973). At a later stage (up to 64 days), a total loss of thymic cortex, with complete deletion of thymocytes, can be observed in cases of post-measles death. Recovery of the cortex has been described on the 91th and 122nd day after the rash (White and Boyd, 1973). Within the medulla of thymus from autopsies, lesions in epithelium and Hassall's corpuscles, are also observed but to a lower extend (Roberts, 1975; White and Boyd, 1973). MV replication in the thymus thus dismantles the cortico-medullary junction and the severe thymus atrophy should contribute to the systemic measles-induced cellular immune dysfunction.

Similar observations have been made in small animal models. The architecture of MV-injected human thymic implants in humanised SCID (SCID-hu) mice is entirely disrupted with eosinophilic stroma, loss of Hassall's corpuscles, and thymic cortex, while the medulla remains mostly intact (Auwaerter et al., 1996b). MV replicates primarily in cortical and medullar TECs and to a fewer extent in myelomonocytic cells, including monocytes and macrophages. DCs from other tissues are known to be susceptible to MV infection (Servet-Delprat et al., 2003), but MV infection of thymic DCs remains to be determined. In CD46 transgenic mice defective for IFN-α/β receptor, the intranasal MV infection leads to the infection of thymic macrophages, DCs and few cells with epithelial-reticular-type morphology (Mrkic et al., 2000). Thus, infection of the various thymic cell types may depend on the route of MV infection. In both models, thymic stromal cells are infected and probably contribute to the cortical and medullar thymocyte degeneration. Indeed, in MV-infected implants of SCID-hu mice, number of immature $CD4^+CD8^+$ double-positive thymocytes strongly decline whereas there is a relative increase in the $CD4^-CD8^-$ double-negative cells. The complete thymocyte loss is due to an apoptotic process as characterized by cell pyknosis, nuclear condensation, and extensive DNA fragmentation in uninfected bystander thymocytes (Auwaerter et al., 1996b). However, the intraperitonal inoculation of lck promoter driven huCD150 transgenic mice results in infection restricted to the thymocytes, but generalized to all subsets ($CD4^-CD8^-$, $CD4^+CD8^+$, $CD4^+CD8^-$ and $CD4^-CD8^+$) and without any signs of apoptosis (Hahm et al., 2003). Although a different route of MV infection was used, this very different pattern of infection, within the thymus, probably reflects the particular cellular distribution of the MV receptor in this animal model: constitutive hCD150 expression is restricted to thymocytes and T cells in huCD150 transgenic mice, whereas, in SCID-hu mice grafted with human thymus, human cells express physiological levels of CD46 and/or CD150 (Hahm et al., 2003).

The severity of thymus architecture disruption reflects the virulence of the virus. Virulent MV strains, including wild type and vaccine strains, replicate more rapidly, induce higher levels of thymocyte apoptosis and more extensive thymic injury than avirulent MV strains (Auwaerter et al., 1996b; Valsamakis et al., 1999). Change in the H, M, P/V/C, and/or L genes are associated with virulence and/or replication phenotypes (Curran and Rima, 1988; Rima et al., 1997; Takeda et al., 1998; Valsamakis et al., 1999). However, there is no correlation between the level of MV replication and damage to thymus architecture. Indeed, C-defective MV (MV-C^-) replicates more efficiently than V-defective MV (MV-V^-), whereas disruption of thymus integrity and thymocyte apoptosis is higher with the latter than the former (Valsamakis et al., 1998). Inversely, highly related vaccine strains, whose genome differ by few nucleotides, show opposite effects on virus production and thymocyte death (Valsamakis et al., 2001). Thus, the molecular basis of MV virulence and attenuation in relation with MV pathogenicity remains to be determined.

4.4.2.3. Consequences of MV-induced disruption of thymic micro-environment upon thymocyte survival signals

Combined mechanisms could account for the MV-induced thymic remodeling. In the SCID-hu mice model, local MV replication occurs in TECs within thymocyte depleted areas suggesting that thymocyte apoptosis is related to TEC infection (Auwaerter et al., 1996b). *In vitro*, human and monkey TECs are highly susceptible to MV infection: they fuse into syncytia and produce high levels of virus progeny (Numazaki et al., 1989a; 1989b; Valentin et al., 1999). In addition, MV inhibits the growth of a human cortical TECs by arresting them in the G_0/G_1 phases of the cell cycle (Valentin et al., 1999). Infected TEC become fusiform and stellate, lost their intercellular contacts and exhibit a shift from low- to high-molecular-weight keratin expression. These morphological and phenotypic changes are characteristic of TEC differentiation. This process ends by apoptosis of the fully differentiated TECs. Whether cortical and medullary TECs share a common differentiation pathway *in vitro* and *in vivo* is unknown. MV-induced differentiation of TECs is strictly dependent on the induction of IFN-β secretion and is blocked by anti-IFN-β antibodies. Moreover, recombinant human IFN-α, IFN-β, or IFN-γ is sufficient to induce the terminal differentiation of uninfected TEC (Vidalain et al., 2002). *In vivo*, injection of IFN-α induce a severe thymic atrophy and impair thymocyte development (Lin et al., 1998b), probably by inhibiting IL-7-driven survival and proliferation of $CD4^-CD8^-CD44^+CD25^+$ immature T cells (Su et al., 1995). Resident thymic plasmacytoid DCs are known to produce large amount of IFN-

α/β when stimulated with either influenza virus or HIV-1 (Bendriss-Vermare et al., 2001; Keir et al., 2002). As shown in a virus-infected mouse model (Gossmann et al., 1991), MV-specific activated T cells can infiltrate the thymus and produce locally IFN-γ (Gossmann et al., 1991). Thus, IFNs produced either by MV-infected cells or activated leukocytes could mediate damage of thymic epithelium, and/or thymocytes, leading to thymic involution. In addition, IFN-α/β produced by human thymic MV-infected DCs could generate TRAIL-mediated cytotoxic activity, suggesting a possible involvement in thymocyte deletion (see § 4.2.3.2., above), (Lichtner et al., 2004; Servet et al., 2002; Vidalain et al., 2001a). Such cytotoxic activity has been reported for thymic DCs infected by human immunodeficiency virus (Beaulieu et al., 1998). In summary, IFNs from autocrine, paracrine and/or exocrine sources, could exert their deleterious effects directly on immature T cell proliferation and indirectly by acting on thymic micro-environment.

Viruses that spread to the thymus mediate thymocyte deletion through two distinct mechanisms: replication within immature T cells leading to their apoptosis and replication within thymic stroma cells leading to the disruption of thymic micro-environment (Auwaerter et al., 1996b; Godfraind et al., 1995). Alternatively, viruses induce thymic injury without detectable local infection (Matteucci et al., 1985; Trgovcich et al., 1996). Because TECs are required for the generation of immunocompetent T cells, *in vivo* MV-induced differentiation and apoptosis would be sufficient to cause the depletion of bystander uninfected thymocyte, particularly in infants, in whom ontogenesis of the immune system requires functional thymic micro-environment. Further investigations are required to sustain this hypothesis. Can the *in vitro* destruction of thymic epithelium delivers abnormal signals to thymocytes and/or affects positive selection of thymocytes? Moreover, extracellular release of MV-N after apoptosis of MV-infected cells (Laine et al., 2003) will allow MV-N to bind to uninfected neighboring thymic epithelial cells and inhibit their proliferation. Indeed, MV-N binding NR expressed on TECs leads to signal transduction with modulation of Ca^{++} influx, and inhibits their proliferation by arresting them at the G_0/G_1 phases of the cell cycle. However, exogenous N does not induce IFN-β secretion by TECs nor their differentiation (Valentin, H., unpublished data). In this context, MV-N mediated growth arrest of uninfected TECs would affect immature T cell survival by signal deprivation. Others MV proteins could also account for thymocyte apoptosis. Thymus from $akt^{-/-}$ mice shows higher spontaneous apoptosis of thymocytes (Chen et al., 2001), and contact with MV-H and -F glycoproteins expressed on infected TECs could disrupt the Akt pathway in thymocytes as observed in mature T cells (Avota et al., 2001).

4.4.2.4. Consequences of MV-induced thymic injury upon the delivery of peripheral immunocompetent T lymphocytes

What is the contribution of the extensive thymocyte destruction on the transient T cell lymphopenia observed during the acute phase of MV infection (Figure 1d)? TCR rearrangement excision circles (TRECs) formed during rearrangement of V, D, and J genes, reflect a stable functional TCR and TREC assays are used to detect recent thymic emigrants and quantitate thymic output (Hazenberg et al., 2003). As defined by the TREC assay, the percentage of naïve $CD4^+$ and $CD8^+$

T lymphocytes from children at the onset of rash significantly decreases, with increase of activated $CD4^+$ and $CD8^+$ T lymphocytes consistent with the induction of an adaptative immune response (Permar et al., 2003b). However, the percent of naïve and of activated T cell sub-populations both return to baseline levels after 1 month. The percentage of circulating memory T cell sub-populations remains unchanged compared to control children. During both the acute and convalescent phase of measles, the $CD4^+$ and $CD8^+$ proportions are more elevated in children with measles than in control children. Finally, in the plasma, the levels of IL-7, a key cytokine for both thymic output and maintenance of T cell homeostasis, are significantly increased for several weeks (Figure 1c) (Permar et al., 2003b). These observations are interpreted as a lack of decrease of thymic output during MV infection and the transient lymphopenia would result rather from an accumulation of naïve T lymphocytes in secondary lymphoid tissue than from thymocyte deletion (Permar et al., 2003b). However, many erroneous conclusions have been drawn from TREC data as a measure of ongoing thymic output, particularly when exploring HIV-1 induced lymphopenia. Indeed, numerous biological parameters have to be taken into account to correctly interpret TRECs data. To compensate for low naïve T cell number, changes in longevity of naïve T cells, TREC dilution by division, thymic rebound, production of functional TRECs T cells at extrathymic sites, or infiltration of thymic tissue by mature T cells occur (Hazenberg et al., 2003). We favour the view that the disruption of thymic micro-environment induced by MV results in peripheral shortage of naïve T cells and/or the release of premature T cells from the thymus to the blood circulation, and, thus, contributes to the impaired functional T-lymphocyte output and possibly circulating autoreactive T cells (Permar et al., 2003b; Ryon et al., 2002). In favour of the latter, changes within TCR Vβ subsets are documented in infants following measles vaccination, independently of MV strain used (Auwaerter et al., 1996a). In conclusion, MV disrupts thymus integrity, by infecting thymic stroma cells rather than thymocytes. Our knowledge of the underlying molecular mechanisms is still in its infancy, and the biological consequences remain a matter of speculation.

4.4.3. Apoptosis-dependent B and T lymphopenia

At the onset of rash, remarkable lymphopenia (Figure 1d) has already occurred in measles patients with reduction in cell numbers of $CD4^+$ T cells, $CD8^+$ T cells, B cells, neutrophils, and monocytes. In contrast, NK cells are increased in number and activated. Apoptosis-associated molecules such as CD95 (Fas) and TRAIL-R are highly expressed on the cell surface of most lymphocytes, which show DNA fragmentation upon incubation *in vitro*. Thus, the profound lymphopenia may be due primarily to extended death of uninfected blood cells caused by apoptosis (Okada et al., 2000). *In vitro*, MV infection sensitizes lymphocytes to apoptosis which is uncoupled from the cell proliferation inhibition. MV induces apoptotic death of cells activated by phorbol ester and ionomycin, but not when they are stimulated with phytohemagglutinin, although their proliferation is suppressed (Ito et al., 1997). In DC-T cell cocultures, resting CD45RA naive T cells are resistant to apoptosis, while peripheral blood activated $CD4^+$ T cells are sensitive (Fugier-Vivier et al., 1997; Grosjean et al., 1997). MV activates IFN-β secretion by DCs (Vidalain et al., 2001a; 2001b), which in turn induces TRAIL synthesis (see §

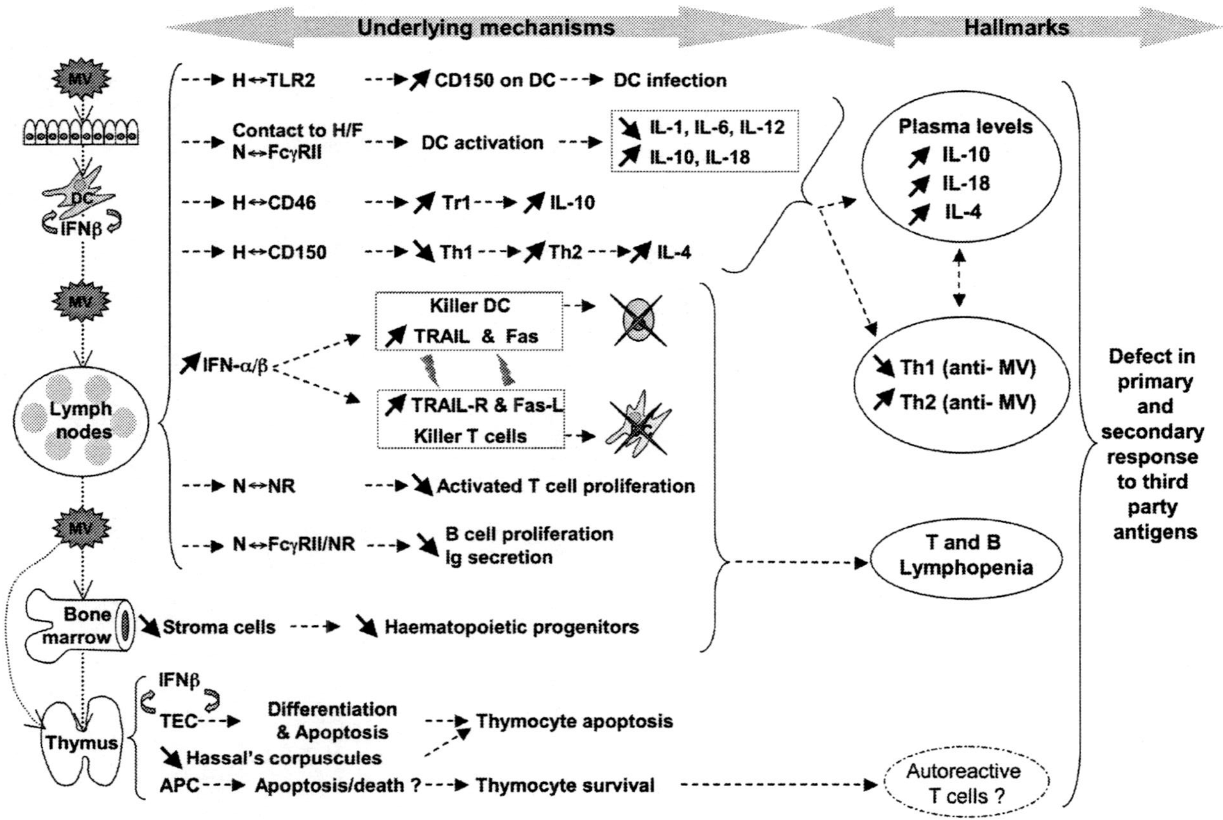

Figure 5. Model of measles virus subversion of the immune system. The model starts by MV infection of the epithelium with progressive dissemination through draining lymph nodes, thymus and bone marrow. At each step the known mechanisms are schematically indicated and their relevance to the *in vivo* immuno-modulation and suppression are indicated in the right part of the figure. Protein to protein binding is indicated as follow: protein↔protein. See also § 4.5 for a more detailed description.

4.2.3.2. above). *In vivo,* MV may induce lymphocyte depletion by both inducing TRAIL dependent cytotoxic DCs, and by sensitizing activated T cells to TRAIL-mediated apoptosis. The mechanisms underlying the induction of IFN-β secretion and the sensitization of activated lymphocytes to apoptosis are unknown.

4.5. Current model of measles virus subversion of the immune system

From our current knowledge, numerous mechanisms can account for the ability of MV to subvert the immune system. Despite being diverse, many of them focus on inducing a Th1 to Th2 switch of the cellular immune response and inducing lymphopenia. Figure 5 describes our current model of the mechanisms by which MV disturbs the immune response. MV enters the organism by infecting the epithelial barrier of the upper respiratory tract. In contact with local DCs, MV-H interacts with TRL2 and induces CD150 expression, thus enhancing MV entry into DCs. Innate antiviral immunity probably recruits and activates NK cells and plasmacytoid DC, which respond by secreting high amounts of IFN-α/β. The local high level of IFN-α/β induces TRAIL expression on DCs and renders them cytotoxic. Free MV, infected TRAIL-expressing DCs and non-infected DCs migrate to the draining lymph nodes. There, MV infects DCs, follicular DCs and activated lymphocytes. All DCs became mature, but MV-infected DCs exhibit abnormal signaling downstream CD40 and all DCs integrate signaling mediated by H and F glycoproteins binding to CD150 and CD46 and by N binding to FcγRII and/or NR. The resulting DC cytokine secretion contains high levels of IL-10 and IL-18 and low level of IL-12. TRAIL-expressing DCs are able to kill activated T cells. In turn, activated T cells expressing Fas-L become cytotoxic to Fas-expressing MV-infected DCs. H induces CD150 down-regulation in T cells, thus lowering Th1 activity and increasing Th2 activity. H-CD46 interaction induces regulatory Tr1-type cytokine secretion including high rate of IL-10. At the onset of the rash, large amount of N is released in the extracellular medium, binds to NR on activated T cells, and inhibits their proliferation. N binds also to NR and FcγRII expressed by B cells and inhibit their proliferation and their ability to secrete immunoglobulins. Secondary virus spreading to other secondary lymphoid tissues, BM and thymus results in apoptosis of the MV infected cells. As a result, there is a decrease of haematopoeitic precursors and generation of non-functional T lymphocytes. Likewise, the Th1 to Th2 switch results from the induction of IL-10 and IL-18 secretion by DC, inhibition of Th1 cells and stimulation of Tr1 cells. Thus, T and B lymphopenia may result from the defects in haematopoiesis, thymus output, B and T lymphocyte proliferations upon antigen stimulation combined with various TRAIL- and Fas-dependent mechanisms that induce lymphocyte and DC apoptosis.

5. Towards eradication and/or domestication of measles virus

Since the beginning of vaccination with attenuated live MV in the 1960s, the number of measles cases worldwide has decreased steeply. However, in 2004, measles still is endemic in South America, Africa, India and Asia (Barrero et al., 2001; El Mubarak et al., 2002; Na et al., 2001; Takahashi et al., 2000; Truong et al., 2001; Zhou et al., 2003). In Europe, Australia and North America, because of better vaccine coverage, sporadic cases and few epidemics occur. The measles strains identified from the sporadic cases and few epidemics could be allocated to five different previously well-defined sequence groups consistent with the assumption that cases of measles are due to repeated introduction from abroad rather than persistent circulation of strains in the population (Chibo et al., 2000). Recent decline in vaccine coverage of England has led to the reemergence of small epidemics (Jansen et al., 2003). In the early 1960s, both inactivated and live attenuated vaccine were tested for their efficiency in protecting young infants, but the latter has soon to abandoned, as was an inactivated vaccine against respiratory syncytial virus (RSV), because they were inducing atypical diseases after natural infection.

5.1. Historical iatrogenic subversion of the immune system: deleterious immune response induced by inactivated measles virus

In the early 1960s, inactivated MV vaccines were tested in young infants. It induced neutralizing antibodies and some protection which waned within months. Fifteen to sixty percent of the immunised children who subsequently became infected with MV developed a sever form of disease called atypical measles. The clinical and biological symptoms included prolonged high fever, extensive rash, severe pneumonitis, eosinophilia and edema (see Polack et al., 1999 and references herein). The first observation was the inability of inactivated vaccine to induce anti-F protein with neutralising activity (Norrby and Gollmar, 1972). Atypical disease was attributed to either the induction of abnormal MV-specific delayed hypersensitivity or a generalised Arthus reaction. During the same period, very similar observations were made after the infection by another *Paramyxoviridae*, RSV, of infants who have been vaccinated with inactivated RSV.

The atypical measles induced by inactivated MV can be reproduced in macaques (de Swart et al., 2002; Polack et al., 1999; 2003a; 2002), and the main biological features are summarized in Figure 6. Compared to live MV, formalin-inactivated MV does not induce any MV-specific CTL, and elicits a delayed and lower neutralising antibody (nAb) response (Figure 6, compare panels c and a). The antibody avidity is poor and, while the neutralising activity towards CD46 mediated entry is above the threshold level for efficient *in vivo* protection, the neutralising activity towards CD150-mediated entry hardly reaches it. In addition neutralising antibodies waned well below the protection level after >40 days. Upon a secondary challenge with live MV, animals previously immunised with live MV, exhibits immediate high CTL response, reduced plasma TNF-α compared to the primary infection, further sustained increase in neutralising antibodies with higher affinity and no circulating immune-complexes (Figure 6, compare panels b and d). In contrast, the challenge of the animal immunised with inactivated MV with live MV, induces a primary CTL response similar to that observed after the infection of naïve animals (Figure 6, compare panels a and d), very high transient peak of eosinophils in bronchoalveolar lavage fluids and TNFα plasma level and transient decrease in IL-12 level (Figure 6d). Furthermore, neutralising antibody towards CD46-mediated entry quickly and sharply raises well over the protective threshold, but the one towards CD150-mediated entry increases very slowly to reach the threshold protective level only three weeks p.i. Likewise, the maturation of antibody avidity is delayed by about 10 days when compared to that observed

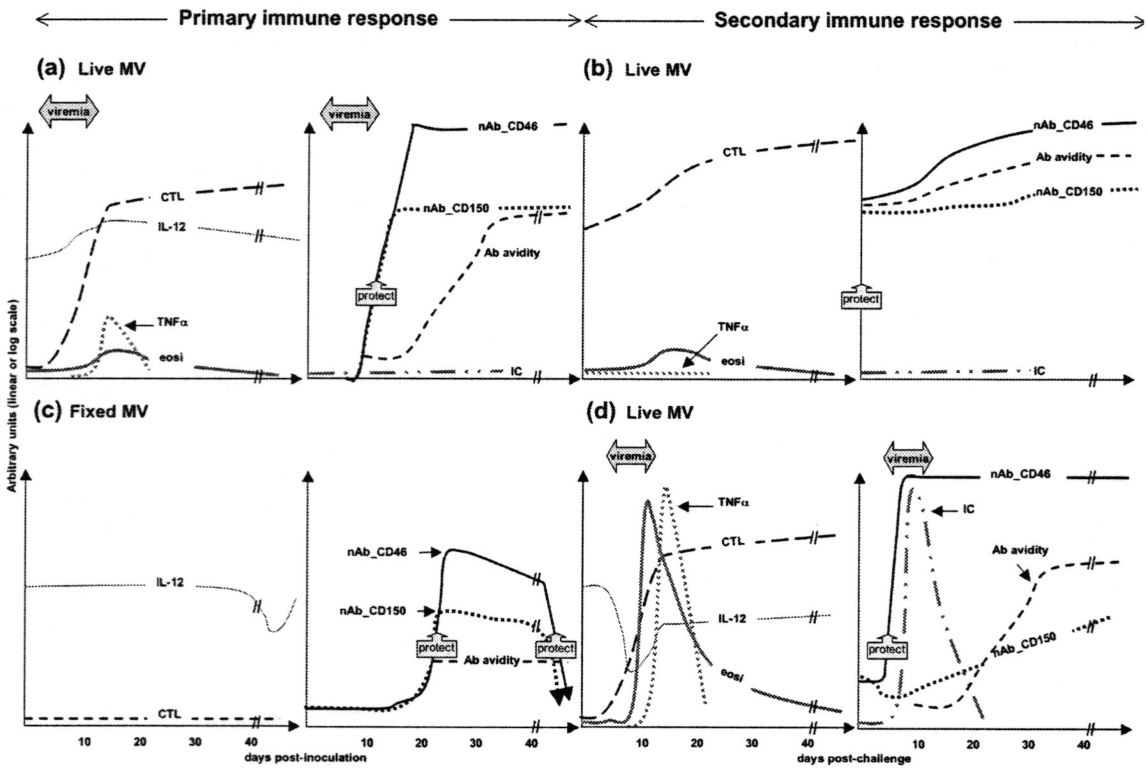

Figure 6. Model of atypical measles virus in macaques. Kinetics of cellular (left panels) and humoral (right panels) reponses after inoculation of (a), live MV, followed by (b), challenge with live MV, and, (c), immunization with fixed inactivated MV (3 times), followed by (d) challenge with live MV. Abbreviations used: CTL, cytolytic T lymphocytes; eosi, eosinophils level in bronchoalveolar lavage fluid; nAb↔CD46 and nAb↔CD150 neutralizing antibodies when infecting $CD46^{+}CD150^{-}$ Véro and $CD46^{-SCR1-}CD150^{+}$ B95a cells respectively; "protect" arrow, threshold level for full protection; IC, immune complexes. Schematic cumulative data from (Polack et al., 2003a; 2002; 2003b; 2000).

after a primary infection only to reach the same level (Figure 6, compare panels d and a). Last, but not least a sharp peak of circulating immune complexes is observed. Thus, failure of inactivated MV to induces affinity maturation results in anamnestic production of non-protective, complement fixing antibodies, immune complex deposition and atypical measles (Polack et al., 2003a). When immune complex deposition in tissues activates complement, leukocyte infiltration and tissue damage occurs (Kohl and Gessner, 1999; Buser, 1967).

What are the molecular mechanisms leading to the immune response deviation responsible of atypical measles? Formalin treatment results mainly in nucleophilic attack by NH3 and SH groups leading to CH2OH substitution of the corresponding protein residues. The random substitution of key residues of H glycoprotein for CD46 and CD150 binding may explain the lower level of elicited neutralising antibody and/or their low avidity towards native H. In addition, we predict that there is a drift in the major antigen epitopes recognised by the immune system. Since, at time of the challenge, CD150 mediated entry are likely less prone to inhibition by pre-existing antibody than CD46 mediated entry (see Figure 6c), could the resulting changes in the respective CD46 and CD150 signalling and/or type of target cells contribute to the induction of the immune disease? The fusion mediated by the interaction of H/F complexes with CD46 or CD150 receptor relies on a critical conformational change of F protein to allow exposure of the fusion peptide and dynamic apposition of the two opposite membranes (see §3.3.1.2. above). Formalin fixation inhibits these conformational changes as shown by the loss of fusion abilities of MV. Formalin-treated MV no longer lyses $CD46^{+}$ African green monkey erythrocytes (Norrby and Gollmar, 1975) and cytoplasm delivery of N protein antigen is prevented (Cardoso et al., 1995). As a result, a strong imbalance in MHC-I and MHC-II restricted presentation occurs with a complete lack of MHC-I restriction presentation of N (and likely any internal MV) protein and H (and likely F) ectodomain, a large increase of MHC-II restricted N protein antigen presentation and no change in the efficiency of MHC-II restricted presentation of H ectodomain (Cardoso et al., 1995). The lack of MHC-I restricted presentation of viral antigens from fixed MV explains the inability of this vaccine to induce any MV-specific CTL in monkeys (Figure 6c) (Polack et al., 1999).

In conclusion, formalin fixation of MV alters viral antigen processing and likely changes B and T cell epitope exposure leading to a drift in antigen recognition by the immune system. That results into a lack of $CD8^{+}$ T cells stimulation and the production of low avidity and complement fixing antibodies involved in immune complex deposition. The contributions of these immune deviations in the induction of atypical measles, as well as the role of the increased induction of TNF-α and/or yet to be identified immunoregulatory changes, remain to be established.

5.2. Measles virus vaccine provides protection

Phylogenic analysis established that the virus is monotypic and the vaccines, which are derived from a single 1954 isolate (Rota et al., 1994), still provide protection against all the virus genotypes that are circulating at present but high titre of neutralizing antibodies is required. For example, no significant difference was observed in neutralizing titres against C1, D3, D5, and H1 when sera were used at high neutralizing titres. However, some D3 and H1 strains were not completely neutralized with low levels of neutralizing antibody, and maintaining high levels of measles antibodies would be required for the control measles outbreaks since outbreaks can occur in populations with fewer than 7% seronegative individuals (Zhou et al., 2003). Consequently, a recall of vaccination is now performed in many developed countries at the age of 6 years and WHO maintains the objective to cover the entire world population in developing countries with currently available vaccines.

Comparative analysis of immunologic parameters in infants with natural disease or immunized with the attenuated vaccine reveals that vaccination is indeed less effective in inducing MV specific humoral response (Okada et al., 2001). Anti-MV IgMs are detected on the 11-14 days after the vaccination, *i.e.* with a delay close that observed after the natural infection, but rises much slower and reach lower neutralizing titers. In contrast with the natural infection, the vaccination induces little immunosuppression. It induces a mild lymphopenia, $CD4^+$, $CD8^+$ and B cell numbers change only in few cases. Accordingly, PBMC does not express significant levels of Fas, FasL or TRAIL-Rs and does not undergo *in vitro* apoptosis and only a small and transient increase in plasma Fas and FasL occurs about 10 days after the vaccination. Cytokine response in plasma remains marginal with small and transient increase in IL-18 and IL-10, but no change in IL-12 or IFN-γ. Thus, live attenuated MV is not causing apparent imunosuppression, while still holding a potential to induce apoptosis on non-infected lymphocytes.

5.3. Protective measles immunity in the presence of passively acquired antibodies

Measles causes serious morbidity in infants, with the highest risk among those who are 6 to 12 months of age. Measles vaccine is been given at 12 to 15 month age to minimize interference by passively acquired maternal antibodies and to achieve the high seroprevalence required for herd immunity. Infants from mothers with vaccine-induced immunity may lose maternal antibodies before 12 month age, leaving them susceptible to measles infection. Furthermore, humoral immunity is deficient in 6-month old infants given measles vaccine, even in the absence of detectable passively acquired neutralizing antibodies. Comparison of their responses with those of 9- and 12-month old infants indicates the need for a developmental maturation of the immune response to measles occuring during the first year of life (Gans et al., 1998).

In the 1980s, high-titer vaccine was used to immunize 6 to 9 month old babies, but this was associated, at least in some countries with a significant increase in child mortality (Garenne et al., 1991). Although the underlying mechanism is unknown (Griffin, 2001), this should keep us alerted on the very delicate balance between the induction of efficient immune response and possible deleterious immune deviation when using measles virus live vaccine. Several alternative approaches have been tested so as to successfully immunise very young infants such as DNA vaccines, synthetic peptides or recombinant vaccinia virus. DNA vaccines expressing either secreted H (Polack et al., 2000), or secreted H and F (Polack et al., 2003b), or a form of H fused to three tandem copies of the murine homologue of C3d (H-3C3d) (Green et al., 2001), illicit neutralising antibodies and T cell responses in rhesus macaques, cotton rats and mice, respectively, without any evidence of priming for atypical measles. However, the DNA vaccines are inefficient in the presence of MV-specific maternal antibodies (Schlereth et al., 2000).

Immunization with synthetic peptides from H confirms that the highly conserved sequence H379-410 of H protein is the dominant epitope. Immunization with peptides containing one or two copies of T cell epitopes and the B cell epitopes H379-400 elicit neutralizing antibodies against wild-type strains of different clades and geographic origins (El Kasmi et al., 2000). However, protection with synthetic peptides has not been evaluated in the presence of passively acquired antibodies. After two vaccinations, with a highly attenuated recombinant vaccinia virus encoding MV-F and H glycoproteins (MVA-FH), macaques developed MV-neutralizing antibodies and MV-specific T-cell responses, even in the presence of passively transferred MV-neutralizing macaque antibodies. However, no vaccinia virus-based vaccine has been approved for use in humans.

5.4. Use of MV as therapeutic vector: vaccine vector and oncolytic virus

The relative inocuity of attenuated MV strain used as anti-measles vaccine, its strong immunogenicity, its high cytopathogenicity and the available reverse genetics to build recombinant virus (Radecke et al., 1995) has led to propose its use as a vaccine carrier or tumour killing agent. Indeed, from the pioneer work showing that MV genome can accommodate cDNA encoding hepatitis B antigen (Singh et al., 1999) and be successfully used to make bivalent vaccine, numerous attempts have been made to add viral proteins from many other pathogens including HIV (Lorin et al., 2004). Multivalent vaccines based on MV as a vector look very promising because of the high immunogenicity of the parental virus. Likewise, the strong cytopathic effects of MV are thought to be exploited to kill tumour cells *in vivo*. Both MV retargeted to tumour associated antigens using single chain antibody graft onto the H protein (Hammond et al., 2001) and selective CD46 or CD150 receptor-blind virus (Vongpunsawad et al., 2004) have been successfully derived. Therefore, the building of a MV derived vector devoid of any CD46 and CD150 binding activity and having reliable targeting to a tumour associated antigen is close to be obtained. In order to prove that this recombinant virus will have a significant clinical benefit, one needs to establish its ability to escape immune neutralization owing to the intrinsically high immunogenicity of the parental virus.

6. Conclusions

The analysis of natural infection in human and experimental disease in monkeys allied to the discovery of many cellular proteins recruited by viral proteins has led to partly uncover the molecular and cellular mechanisms by which MV can simultaneously induces a strong and life-lasting antiviral immune response and a profound cellular immunodepression. Many findings, which have been obtained in this model, have contributed to enlarge our knowledge on the molecules and mechanisms involved in the establishment and/or the control of the immune response. To list a few leading discoveries, MV was the first pathogen described to use the complement regulator CD46 and the T cell costimulator CD150 as cellular receptor. New functions have been attributed to CD46, such as signal transduction, internalisation of antigen, control of IL-12 secretion, and T cell costimulation. MV-

F protein is the first protein reported to activate the alternative pathway escaping regulation by CD46 and CD55. MV-N protein is the first viral nucleoprotein described for its ability to bind to FcγRII or NR and regulates DC, B, and/or T cell functions. MV-H protein is the first envelope glycoprotein shown to bind to a member of the TLR family and induce CD150 expression. Last, but not least, MV is the first virus demonstrated to render DCs cytotoxic by inducing TRAIL expression. Despite these extensive findings, a more complete picture of all the events remains to be drawn in order to understand the immunological paradox of measles, with the hope that any additional information will prove useful in our understanding of the complex interplay between pathogens and the immune system. Selective fields could be IFN responses, NK cell responses, MV infection of plasmacytoid DCs, MV infection of follicular DCs, mechanisms of DC cross-priming to induce MV specific responses, MV infection along haematopoiesis, induction of encephalitis by MV...

Acknowledgements and Funding

The work performed by the authors and described in this review was supported in part by grants from CNRS (ATIPE, DG) and MENESR (PRFMMIP), or with financial support from the Commission of the European Communities, specific RTD programme "Quality of Life and Management of Living Resources", QLK2-CT2001-01225. It does not necessarily reflect its views and in no way anticipates the Commission's future policy in this area.

References

Adams, E. M., Brown, M. C., Nunge, M., Krych, M., and Atkinson, J. P. (1991). Contribution of the repeating domains of membrane cofactor protein (CD46) of the complement system to ligand binding and cofactor activity. J. Immunol. *147*, 3005-3011.

Aderem, A., and Ulevitch, R. J. (2000). Toll-like receptors in the induction of the innate immune response. Nature *406*, 782-787.

Arneborn, P., and Biberfeld, G. (1983). T-lymphocyte subpopulations in relation to immunosuppression in measles and varicella. Infect. Immun. *39*, 29-37.

Astier, A., Trescol-Biemont, M. C., Azocar, O., Lamouille, B., and Rabourdin-Combe, C. (2000). Cutting edge: CD46, a new costimulatory molecule for T cells, that induces p120CBL and LAT phosphorylation. J. Immunol. *164*, 6091-6095.

Atabani, S. F., Byrnes, A. A., Jaye, A., Kidd, I. M., Magnusen, A. F., Whittle, H., and Karp, C. L. (2001). Natural measles causes prolonged suppression of interleukin-12 production. J. Infect. Dis. *184*, 1-9.

Auwaerter, P. G., Hussey, G. D., Goddard, E. A., Hughes, J., Ryon, J. J., Strebel, P. M., Beatty, D., and Griffin, D. E. (1996a). Changes within T cell receptor V beta subsets in infants following measles vaccination. Clin. Immunol. Immunopathol *79*, 163-170.

Auwaerter, P. G., Kaneshima, H., McCune, J. M., Wiegand, G., and Griffin, D. E. (1996b). Measles virus infection of thymic epithelium in the SCID-hu mouse leads to thymocyte apoptosis. J. Virol. *70*, 3734-3740.

Auwaerter, P. G., Rota, P. A., Elkins, W. R., Adams, R. J., DeLozier, T., Shi, Y., Bellini, W. J., Murphy, B., and Griffin, D. E. (1999). Measles Virus Infection in Rhesus Macaques: Altered Immune Responses and Comparison of the Virulence of Six Different Virus Strains. J. Infect. Dis. *180*, 950–958.

Avota, E., Avots, A., Niewiesk, S., Kane, L. P., Bommhardt, U., ter Meulen, V., and Schneider-Schaulies, S. (2001). Disruption of Akt kinase activation is important for immunosuppression induced by measles virus. Nat. Med. *7*, 725-731.

Barik, S. (2004). Control of nonsegmented negative-strand RNA virus replication by siRNA. Virus Res. *102*, 27-35.

Baron, R. (2004). Arming the osteoclast. Nat. Med. *10*, 458-460.

Barrero, P. R., Zandomeni, R. O., and Mistchenko, A. S. (2001). Measles virus circulation in Argentina: 1991-1999. Arch. Virol. *146*, 815-823.

Bartholome, E. J., Willems, F., Crusiaux, A., Thielemans, K., Schandene, L., and Goldman, M. (1999a). IFN-beta interferes with the differentiation of dendritic cells from peripheral blood mononuclear cells: selective inhibition of CD40-dependent interleukin-12 secretion. J. Interferon Cytokine Res. *19*, 471-478.

Bartholome, E. J., Willems, F., Crusiaux, A., Thielemans, K., Schandene, L., and Goldman, M. (1999b). Interferon-beta inhibits Th1 responses at the dendritic cell level. Relevance to multiple sclerosis. Acta Neurol. Belg *99*, 44-52.

Barton, G. M., and Medzhitov, R. (2003). Linking Toll-like receptors to IFN-alpha/beta expression. Nat. Immunol. *4*, 432-433.

Beaulieu, S., Lafontaine, M., Richer, M., Courchesne, I., Cohen, E. A., and Bergeron, D. (1998). Characterization of the cytotoxic factor(s) released from thymic dendritic cells upon human immunodeficiency virus type 1 infection. Virology *241*, 285-297.

Bendriss-Vermare, N., Barthelemy, C., Durand, I., Bruand, C., Dezutter-Dambuyant, C., Moulian, N., Berrih-Aknin, S., Caux, C., Trinchieri, G., and Briere, F. (2001). Human thymus contains IFN-alpha-producing CD11c(-), myeloid CD11c(+), and mature interdigitating dendritic cells. J. Clin. Invest *107*, 835-844.

Bieback, K., Lien, E., Klagge, I. M., Avota, E., Schneider-Schaulies, J., Duprex, W. P., Wagner, H., Kirschning, C. J., Ter Meulen, V., and Schneider-Schaulies, S. (2002). Hemagglutinin protein of wild-type measles virus activates toll-like receptor 2 signaling. J. Virol. *76*, 8729-8736.

Bitko, V., and Barik, S. (2001). Phenotypic silencing of cytoplasmic genes using sequence-specific double-stranded short interfering RNA and its application in the reverse genetics of wild type negative-strand RNA viruses. BMC Microbiol. *1*, 34.

Bodey, B., Bodey, B., Jr., Siegel, S. E., and Kaiser, H. E. (2000). Novel insights into the function of the thymic Hassall's bodies. In Vivo *14*, 407-418.

Bodey, B., and Kaiser, H. E. (1997). Development of Hassall's bodies of the thymus in humans and other vertebrates (especially mammals) under physiological and pathological conditions: immunocytochemical, electronmicroscopic and *in vitro* observations. In Vivo *11*, 61-85.

Bogdan, C. (2000). The function of type I interferons in antimicrobial immunity. Curr. Opin. Immunol. *12*, 419-424.

Bohn, W., Rutter, G., Hohenberg, H., Mannweiler, K., and Nobis, P. (1986). Involvement of actin filaments in budding of measles virus: studies on cytoskeletons of infected cells. Virology *149*, 91-106.

Bojinova, V. S., Dimova, P. S., Belopitova, L. D., Mihailov, A. S., Gatcheva, N. L., Mihneva, Z. G., and Hristova, M. T. (2004). Clinical and epidemiological characteristics of subacute sclerosing panencephalitis in Bulgaria during the past 25 years (1978-2002). Eur. J. Paediatr. Neurol. *8*, 89-94.

Brelinska, R. (2003). Thymic epithelial cells in age-dependent involution. Microsc Res. Tech *62*, 488-500.

Brown, H. R., Goller, N. L., Rudelli, R. D., Dymecki, J., and Wisniewski, H. M. (1989). Postmortem detection of measles virus in non-neural tissues in subacute sclerosing panencephalitis. Ann. Neurol. *26*, 263-268.

Buchholz, C. J., Koller, D., Devaux, P., Mumenthaler, C., Schneider-Schaulies, J., Braun, W., Gerlier, D., and Cattaneo, R. (1997). Mapping of the primary binding site of measles virus to its receptor CD46. J. Biol. Chem. *272*, 22072-22079.

Buchholz, C. J., Schneider, U., Devaux, P., Gerlier, D., and Cattaneo, R. (1996). Cell entry by measles virus: long hybrid receptors uncouple binding from membrane fusion. J. Virol. *70*, 3716-3723.

Buckland, R., Giraudon, P., and Wild, F. (1989). Expression of measles virus nucleoprotein in Escherichia coli: use of deletion mutants to locate the antigenic sites. J. Gen. Virol. *70*, 435-441.

Burnet, F. M. (1968). Measles as an index of immunological function. Lancet *2*, 610-613.

Buser, F. (1967). Side reaction to measles vaccination suggesting the Arthus phenomenon. N. Engl. J. Med. *277*, 250-251.

Byrnes, A. A., Ma, X., Cuomo, P., Park, K., Wahl, L., Wolf, S. F., Zhou, H., Trinchieri, G., and Karp, C. L. (2001). Type I interferons and IL-12: convergence and cross-regulation among mediators of cellular immunity. Eur. J. Immunol. *31*, 2026-2034.

Cardoso, A. I., Beauverger, P., Gerlier, D., Wild, T. F., and Rabourdin-Combe, C. (1995). Formaldehyde inactivation of measles virus abolishes CD46-dependent presentation of nucleoprotein to murine class I-restricted CTLs but not to class II-restricted helper T cells. Virology *212*, 255-258.

Cardoso, A. I., Gerlier, D., Wild, T. F., and Rabourdin-Combe, C. (1996). The ectodomain of measles virus envelope glycoprotein does not gain access to the cytosol and MHC class I presentation pathway following virus-cell fusion. J. Gen. Virol. *77 (Pt 11)*, 2695-2699.

Casasnovas, J. M., Larvie, M., and Stehle, T. (1999). Crystal structure of two CD46 domains reveals an extended measles virus-binding surface. Embo J. *18*, 2911-2922.

Cathomen, T., Mrkic, B., Spehner, D., Drillien, R., Naef, R., Pavlovic, J., Aguzzi, A., Billeter, M. A., and Cattaneo, R. (1998a). A matrix-less measles virus is infectious and elicits extensive cell fusion: consequences for propagation in the brain. Embo J. *17*, 3899-3908.

Cathomen, T., Naim, H. Y., and Cattaneo, R. (1998b). Measles viruses with altered envelope protein cytoplasmic tails gain cell fusion competence. J. Virol. *72*, 1224-1234.

Cattaneo, R. (2004). Four viruses, two bacteria, and one receptor: membrane cofactor protein (CD46) as a pathogen magnet. J. Virol. *78*, 4385-4388.

Chawla-Sarkar, M., Leaman, D. W., and Borden, E. C. (2001). Preferential induction of apoptosis by interferon (IFN)-beta compared with IFN-alpha2: correlation with TRAIL/Apo2L induction in melanoma cell lines. Clin. Cancer Res. *7*, 1821-1831.

Chawla-Sarkar, M., Lindner, D. J., Liu, Y. F., Williams, B. R., Sen, G. C., Silverman, R. H., and Borden, E. C. (2003). Apoptosis and interferons: role of interferon-stimulated genes as mediators of apoptosis. Apoptosis *8*, 237-249.

Chazal, N., and Gerlier, D. (2003). Virus entry, assembly, budding, and membrane rafts. Microbiol. Mol. Biol. Rev. *67*, 226-237, table of contents.

Chen, M., Cortay, J. C., and Gerlier, D. (2003). Measles virus protein interactions in yeast: new findings and caveats. Virus Res. *98*, 123-129.

Chen, W. S., Xu, P. Z., Gottlob, K., Chen, M. L., Sokol, K., Shiyanova, T., Roninson, I., Weng, W., Suzuki, R., Tobe, K., et al. (2001). Growth retardation and increased apoptosis in mice with homozygous disruption of the Akt1 gene. Genes Dev. *15*, 2203-2208.

Chibo, D., Birch, C. J., Rota, P. A., and Catton, M. G. (2000). Molecular characterization of measles viruses isolated in Victoria, Australia, between 1973 and 1998. J. Gen. Virol. *81*, 2511-2518.

Chino, F., Kodama, H., Ohkawa, T., and Egashira, Y. (1979). Alterations of the thymus and peripheral lymphoid tissues in fatal measles. A review of 14 autopsy cases. Acta Pathol. Jpn *29*, 493-507.

Christiansen, D., De Sousa, E. R., Loveland, B., Kyriakou, P., Lanteri, M., Wild, F. T., and Gerlier, D. (2002). A CD46CD[55-46] chimeric receptor, eight short consensus repeats long, acts as an inhibitor of both CD46 (MCP)- and CD150 (SLAM)-mediated cell-cell fusion induced by CD46-using measles virus. J. Gen. Virol. *83*, 1147-1155.

Christiansen, D., Deléage, G., and Gerlier, D. (2000a). Evidence for distinct complement regulatory and measles virus binding sites on CD46 SCR2. Eur. J. Immunol. *30*, 3457-3462.

Christiansen, D., Devaux, P., Reveil, B., Evlashev, A., Horvat, B., Lamy, J., Rabourdin-Combe, C., Cohen, J. H., and Gerlier, D. (2000b). Octamerization enables soluble CD46 receptor to neutralize measles virus *in vitro* and *in vivo*. J. Virol. *74*, 4672-4678.

Christiansen, D., Loveland, B., Kyriakou, P., Lanteri, M., Escoffier, C., and Gerlier, D. (2000c). Interaction of CD46 with measles virus: accessory role of CD46 short consensus repeat IV. J. Gen. Virol. *81 Pt 4*, 911-917.

Christiansen, D., Loveland, B., Kyriakou, P., Lanteri, M., Rubinstein, E., and Gerlier, D. (2000d). Chimeric CD46/DAF molecules reveal a cryptic functional role for SCR1 of DAF in regulating complement activation. Mol. Immunol. *37*, 687-696.

Clemenza, L., Dieli, F., Cicardi, M., and Salerno, A. (2003). Research on complement: old issues revisited and a novel sphere of influence. Trends Immunol. *24*, 292-296.

Collins, S. E., Noyce, R. S., and Mossman, K. L. (2004). Innate cellular response to virus particle entry requires IRF3 but not virus replication. J. Virol. *78*, 1706-1717.

Crimeen-Irwin, B., Ellis, S., Christiansen, D., Ludford-Menting, M. J., Milland, J., Lanteri, M., Loveland, B. E., Gerlier, D., and Russell, S. M. (2003). Ligand binding determines whether CD46 is internalized by clathrin-coated pits or macropinocytosis. J. Biol. Chem. *278*, 46927-46937.

Curran, J. (1996). Reexamination of the Sendai virus P protein domains required for RNA synthesis: a possible supplemental role for the P protein. Virology *221*, 130-140.

Curran, J. (1998). A role for the Sendai virus P protein trimer in RNA synthesis. J. Virol. *72*, 4274-4280.

Curran, J., Marq, J. B., and Kolakofsky, D. (1995). An N-terminal domain of the Sendai paramyxovirus P protein acts as a chaperone for the NP protein during the nascent chain assembly step of genome replication. J. Virol. *69*, 849-855.

Curran, J., Pelet, T., and Kolakofsky, D. (1994). An acidic activation-like domain of the Sendai virus P protein is required for RNA synthesis and encapsidation. Virology *202*, 875-884.

Curran, M. D., and Rima, B. K. (1988). Nucleotide sequence of the gene encoding the matrix protein of a recent measles virus isolate. J. Gen. Virol. *69*, 2407-2411.

Dalod, M., Salazar-Mather, T. P., Malmgaard, L., Lewis, C., Asselin-Paturel, C., Briere, F., Trinchieri, G., and Biron, C. A. (2002). Interferon alpha/beta and interleukin 12 responses to viral infections: pathways regulating dendritic cell cytokine expression *in vivo*. J. Exp. Med. *195*, 517-528.

Das, S., Ward, S. V., Markle, D., and Samuel, C. E. (2004). DNA damage-binding proteins and heterogeneous nuclear ribonucleoprotein A1 function as constitutive KCS element components of the interferon-inducible RNA-dependent protein kinase promoter. J. Biol. Chem. *279*, 7313-7321.

Dauer, M., Pohl, K., Obermaier, B., Meskendahl, T., Robe, J., Schnurr, M., Endres, S., and Eigler, A. (2003). Interferon-alpha disables dendritic cell precursors: dendritic cells derived from interferon-alpha-treated monocytes are defective in maturation and T-cell stimulation. Immunology *110*, 38-47.

de Swart, R. L., Kuiken, T., Timmerman, H. H., van Amerongen, G., van den Hoogen, B. G., Vos, H. W., Neijens, H. J., Andeweg, A. C., and Osterhaus, A. D. M. E. (2002). Immunization of macaques with formalin-inactivated respiratory syncytial virus (RSV) induces interleukin-13-associated hypersensitivity to subsequent RSV infection. J. Virol. *76*, 11561–11569.

Dempsey, P. W., Allison, M. E. D., Akkaraju, S., Goodnow, C. C., and Fearon, D. T. (1996). C3d of complement as a molecular adjuvant: Bridging innate and acquired immunity. Science *271*, 348-350.

Der, S. D., Zhou, A., Williams, B. R., and Silverman, R. H. (1998). Identification of genes differentially regulated by interferon alpha, beta, or gamma using oligonucleotide arrays. Proc. Natl. Acad. Sci. USA. *95*, 15623-15628.

Detels, R., Brody, J. A., McNew, J., and Edgar, A. H. (1973). Further epidemiological studies of subacute sclerosing panencephalitis. Lancet *2*, 11-14.

Devaux, P., Christiansen, D., Fontaine, M., and Gerlier, D. (1999). Control of C3b and C5b deposition by CD46 (membrane cofactor protein) after alternative but not classical complement activation. Eur. J. Immunol. *29*, 815-822.

Devaux, P., Christiansen, D., Plumet, S., and Gerlier, D. (2004). Cell surface activation of the alternative complement pathway by the fusion protein of measles virus. J. Gen. Virol. *85*, 1665-1673.

Devaux, P., Loveland, B., Christiansen, D., Milland, J., and Gerlier, D. (1996). Interactions between the ectodomains of haemagglutinin and CD46 as a primary step in measles virus entry. J. Gen. Virol. *77 (Pt 7)*, 1477-1481.

Dhib-Jalbut, S. S., and Cowan, E. P. (1993). Direct evidence that interferon-beta mediates enhanced HLA-class I expression in measles virus-infected cells. J. Immunol. *151*, 6248-6258.

Diallo, A., Barrett, T., Barbron, M., Meyer, G., and Lefevre, P. C. (1994). Cloning of the nucleocapsid protein gene of peste-des-petits-ruminants virus: relationship to other morbilliviruses. J. Gen. Virol. *75*, 233-237.

Dorig, R. E., Marcil, A., Chopra, A., and Richardson, C. D. (1993). The human CD46 molecule is a receptor for measles virus (Edmonston strain). Cell *75*, 295-305.

Dubois, B., Lamy, P. J., Chemin, K., Lachaux, A., and Kaiserlian, D. (2001). Measles virus exploits dendritic cells to suppress CD4+ T-cell proliferation via expression of surface viral glycoproteins independently of T-cell trans-infection. Cell Immunol. *214*, 173-183.

Dubois-Dalcq, M., and Reese, T. S. (1975). Structural changes in the membrane of vero cells infected with a paramyxovirus. J. Cell Biol. *67*, 551-565.

El Kasmi, K. C., Fillon, S., Theisen, D. M., Hartter, H., Brons, N. H., and Muller, C. P. (2000). Neutralization of measles virus wild-type isolates after immunization with a synthetic peptide vaccine which is not recognized by neutralizing passive antibodies. J. Gen. Virol. *81*, 729-735.

El Mubarak, H. S., Ibrahim, S. A., Vos, H. W., Mukhtar, M. M., Mustafa, O. A., Wild, T. F., Osterhaus, A. D., and De Swart, R. L. (2004). Measles virus protein-specific IgM, IgA, and IgG subclass responses during the acute and convalescent phase of infection. J. Med. Virol. *72*, 290-298.

El Mubarak, H. S., van de Bildt, M. W., Mustafa, O. A., Vos, H. W., Mukhtar, M. M., Ibrahim, S. A., Andeweg, A. C., El Hassan, A. M., Osterhaus, A. D., and de Swart, R. L. (2002). Genetic characterization of wild-type measles viruses circulating in suburban Khartoum, 1997-2000. J. Gen. Virol. *83*, 1437-1443.

Erlenhoefer, C., Wurzer, W. J., Loffler, S., Schneider-Schaulies, S., ter Meulen, V., and Schneider-Schaulies, J. (2001). CD150 (SLAM) is a receptor for measles virus but is not involved in viral contact-mediated proliferation inhibition. J. Virol. *75*, 4499-4505.

Ertl, O. T., Wenz, D. C., Bouche, F. B., Berbers, G. A., and Muller, C. P. (2003). Immunodominant domains of the Measles virus hemagglutinin protein eliciting a neutralizing human B cell response. Arch. Virol. *148*, 2195-2206.

Escoffier, C., Manie, S., Vincent, S., Muller, C. P., Billeter, M., and Gerlier, D. (1999). Nonstructural C protein is required for efficient measles virus replication in human peripheral blood cells. J. Virol. *73*, 1695-1698.

Esolen, L. M., Ward, B. J., Moench, T. R., and Griffin, D. E. (1993). Infection of monocytes during measles. J. Infect. Dis. *168*, 47-52.

Etchart, N., Desmoulins, P. O., Chemin, K., Maliszewski, C., Dubois, B., Wild, F., and Kaiserlian, D. (2001). Dendritic cells recruitment and *in vivo* priming of CD8+ CTL induced by a single topical or transepithelial immunization via the buccal mucosa with measles virus nucleoprotein. J. Immunol. *167*, 384-3891.

Fearon, D. T. (2000). Innate immunity--beginning to fulfill its promise? Nat. Immunol. *1*, 102-103.

Firsching, R., Buchholz, C. J., Schneider, U., Cattaneo, R., ter Meulen, V., and Schneider-Schaulies, J. (1999). Measles virus spread by cell-cell contacts: uncoupling of contact-mediated receptor (CD46) downregulation from virus uptake. J. Virol. *73*, 5265-5273.

Fitzgerald, K. A., McWhirter, S. M., Faia, K. L., Rowe, D. C., Latz, E., Golenbock, D. T., Coyle, A. J., Liao, S. M., and Maniatis, T. (2003). IKKepsilon and TBK1 are essential components of the IRF3 signaling pathway. Nat. Immunol. *4*, 491-496.

Fournier, P., Ammerlaan, W., Ziegler, D., Giminez, C., Rabourdin-Combe, C., Fleckenstein, B. T., Wiesmuller, K. H., Jung, G., Schneider, F., and Muller, C. P. (1996). Differential activation of T cells by antibody-modulated processing of the flanking sequences of class II-restricted peptides. Int. Immunol. *8*, 1441-1451.

Fugier-Vivier, I., Servet-Delprat, C., Rivailler, P., Rissoan, M. C., Liu, Y. J., and Rabourdin-Combe, C. (1997). Measles virus suppresses cell-mediated immunity by interfering with the survival and functions of dendritic and T cells. J. Exp. Med. *186*, 813-823.

Fulginiti, V. A., Eller, J. J., Downie, A. W., and Kempe, C. H. (1967). Altered reactivity to measles virus: atypical measles in children previously immunized with inactivated measles virus vaccines. J. Amer. Med. Assoc. *202*, 1075-1080.

Galama, J. M., Ubels-Postma, J., Vos, A., and Lucas, C. J. (1980). Measles virus inhibits acquisition of lymphocyte functions but not established effector functions. Cell Immunol. *50*, 405-415.

Gans, H. A., Arvin, A. M., Galinus, J., Logan, L., DeHovitz, R., and Maldonado, Y. (1998). Deficiency of the humoral immune response to measles vaccine in infants immunized at age 6 months. Jama *280*, 527-532.

Garenne, M., Leroy, O., Beau, J. P., and Sene, I. (1991). Child mortality after high-titre measles vaccines: prospective study in Senegal. Lancet *338*, 903-907.

Gerlier, D., Loveland, B., Varior-Krishnan, G., Thorley, B., McKenzie, I. F., and Rabourdin-Combe, C. (1994a). Measles virus receptor properties are shared by several CD46 isoforms differing in extracellular regions and cytoplasmic tails. J. Gen. Virol. *75 (Pt 9)*, 2163-2171.

Gerlier, D., Trescol-Biemont, M. C., Varior-Krishnan, G., Naniche, D., Fugier-Vivier, I., and Rabourdin-Combe, C. (1994b). Efficient major histocompatibility complex class II-restricted presentation of measles virus relies on hemagglutinin-mediated targeting to its cellular receptor human CD46 expressed by murine B cells. J. Exp. Med. *179*, 353-358.

Gerlier, D., Varior-Krishnan, G., and Devaux, P. (1995). CD46-mediated measles virus entry: a first key to host-range specificity. Trends Microbiol. *3*, 338-345.

Godfraind, C., Holmes, K. V., and Coutelier, J. P. (1995). Thymus involution induced by mouse hepatitis virus A59 in BALB/c mice. J. Virol. *69*, 6541-6547.

Gossmann, J., Lohler, J., and Lehmann-Grube, F. (1991). Entry of antivirally active T lymphocytes into the thymus of virus-infected mice. J. Immunol. *146*, 293-297.

Gotoh, B., Komatsu, T., Takeuchi, K., and Yokoo, J. (2001). Paramyxovirus accessory proteins as interferon antagonists. Microbiol. Immunol. *45*, 787-800.

Gotoh, B., Komatsu, T., Takeuchi, K., and Yokoo, J. (2002). Paramyxovirus strategies for evading the interferon response. Rev. Med. Virol. *12*, 337-357.

Graves, M., Griffin, D. E., Johnson, R. T., Hirsch, R. L., de Soriano, I. L., Roedenbeck, S., and Vaisberg, A. (1984). Development of antibody to measles virus polypeptides during complicated and uncomplicated measles virus infections. J. Virol. *49*, 409-412.

Grazia Cappiello, M., Sutterwala, F. S., Trinchieri, G., Mosser, D. M., and Ma, X. (2001). Suppression of Il-12 transcription in macrophages following Fc gamma receptor ligation. J. Immunol. *166*, 4498-4506.

Green, T. D., Newton, B. R., Rota, P. A., Xu, Y., Robinson, H. L., and Ross, T. M. (2001). C3d enhancement of neutralizing antibodies to measles hemagglutinin. Vaccine *20*, 242-248.

Griffin, D. E. (1995). Immune responses during measles virus infection. Curr. Top Microbiol. Immunol. *191*, 117-134.

Griffin, D. E. (2001). Measles virus. In Fields Virology, D. M. Knipe, and P. M. Howley, eds. (Philadelphia, Lippincott-Raven), pp. 1401-1441.

Griffin, D. E., and Ward, B. J. (1993). Differential CD4 T cell activation in measles. J. Infect. Dis. *168*, 275-281.

Griffin, D. E., Ward, B. J., and Esolen, L. M. (1994). Pathogenesis of measles virus infection: an hypothesis for altered immune responses. J. Infect. Dis. *170*, S24-S31.

Griffin, D. E., Ward, B. J., Jauregui, E., Johnson, R. T., and Vaisberg, A. (1990). Immune activation during measles: interferon-gamma and neopterin in plasma and cerebrospinal fluid in complicated and uncomplicated disease. J. Infect. Dis. *161*, 449-453.

Grosjean, I., Caux, C., Bella, C., Berger, I., Wild, F., Banchereau, J., and Kaiserlian, D. (1997). Measles virus infects human dendritic cells and blocks their allostimulatory properties for CD4+ T cells. J. Exp. Med. *186*, 801-812.

Haas, K. M., Hasegawa, M., Steeber, D. A., Poe, J. C., Zabel, M. D., Bock, C. B., Karp, D. R., Briles, D. E., Weis, J. H., and Tedder, T. F. (2002). Complement receptors CD21/35 link innate and protective immunity during Streptococcus pneumoniae infection by regulating IgG3 antibody responses. Immunity *17*, 713-723.

Hahm, B., Arbour, N., Naniche, D., Homann, D., Manchester, M., and Oldstone, M. B. (2003). Measles virus infects and suppresses proliferation of T lymphocytes from transgenic mice bearing human signaling lymphocytic activation molecule. J. Virol. *77*, 3505-3515.

Hahm, B., Arbour, N., and Oldstone, M. B. (2004). Measles virus interacts with human SLAM receptor on dendritic cells to cause immunosuppression. Virology *323*, 292-302.

Hammond, A. L., Plemper, R. K., Zhang, J., Schneider, U., Russell, S. J., and Cattaneo, R. (2001). Single-chain antibody displayed on a recombinant measles virus confers entry through the tumor-associated carcinoembryonic antigen. J. Virol. *75*, 2087-2096.

Hashimoto, K., Ono, N., Tatsuo, H., Minagawa, H., Takeda, M., Takeuchi, K., and Yanagi, Y. (2002). SLAM (CD150)-independent measles virus entry as revealed by recombinant virus expressing green fluorescent protein. J. Virol. *76*, 6743-6749.

Hazenberg, M. D., Borghans, J. A., de Boer, R. J., and Miedema, F. (2003). Thymic output: a bad TREC record. Nat. Immunol. *4*, 97-99.

Heil, F., Hemmi, H., Hochrein, H., Ampenberger, F., Kirschning, C., Akira, S., Lipford, G., Wagner, H., and Bauer, S. (2004). Species-specific recognition of single-stranded RNA via toll-like receptor 7 and 8. Science *303*, 1526-1529.

Helin, E., Salmi, A. A., Vanharanta, R., and Vainionpaa (1999). Measles virus replication in cells of myelomonocytic lineage is dependent on cellular differentiation stage. Virology *253*, 35-42.

Hicks, J. T., Sullivan, J. L., and Albrecht, P. (1977). Immune responses during measles infection in immunosuppressed Rhesus monkeys. J. Immunol. *119*, 1452-1456.

Hirano, A., Kurita-Taniguchi, M., Katayama, Y., Matsumoto, M., Wong, T. C., and Seya, T. (2002). Ligation of human CD46 with purified complement C3b or F(ab')(2) of monoclonal antibodies enhances isoform-specific interferon gamma-dependent nitric oxide production in macrophages. J. Biochem. (Tokyo) *132*, 83-91.

Hirano, A., Yant, S., Iwata, K., Korte-Sarfaty, J., Seya, T., Nagasawa, S., and Wong, T. C. (1996). Human cell receptor CD46 is down regulated through recognition of a membrane-proximal region of the cytoplasmic domain in persistent measles virus infection. J. Virol. *70*, 6929-6936.

Hirsch, R. L., Griffin, D. E., Johnson, R. T., Cooper, S. J., Lindo de Soriano, I., Roedenbeck, S., and Vaisberg, A. (1984). Cellular immune responses during complicated and uncomplicated measles virus infections of man. Clin. Immunol. Immunopath *31*, 1-12.

Hoffman, S. C., Polack, F. P., Hauer, D. A., Singh, M., Billeter, M. A., Adams, R. J., and Griffin, D. E. (2003). Vaccination of rhesus macaques with a recombinant measles virus expressing interleukin-12 alters humoral and cellular immune responses. J. Infect. Dis. *188*, 1553–1561.

Horikami, S. M., Curran, J., Kolakofsky, D., and Moyer, S. A. (1992). Complexes of Sendai virus NP-P and P-L proteins are required for defective interfering particle genome replication *in vitro*. J. Virol. *66*, 4901-4908.

Horikami, S. M., and Moyer, S. A. (1995). Structure, transcription, and replication of measles virus. Curr. Top Microbiol. Immunol. *191*, 35-50.

Horikami, S. M., Smallwood, S., Bankamp, B., and Moyer, S. A. (1994). An amino-proximal domain of the L protein binds to the P protein in the measles virus RNA polymerase complex. Virology *205*, 540-545.

Horikami, S. M., Smallwood, S., and Moyer, S. A. (1996). The Sendai virus V protein interacts with the NP protein to regulate viral genome RNA replication. Virology *222*, 383-390.

Horng, T., Barton, G. M., and Medzhitov, R. (2001). TIRAP: an adapter molecule in the Toll signaling pathway. Nat. Immunol. *2*, 835-841.

Howie, D., Okamoto, S., Rietdijk, S., Clarke, K., Wang, N., Gullo, C., Bruggeman, J. P., Manning, S., Coyle, A. J., Greenfield, E., et al. (2002a). The role of SAP in murine CD150 (SLAM)-mediated T-cell proliferation and interferon γ production. Blood *100*, 2899-2907.

Howie, D., Simarro, M., Sayos, J., Guirado, M., Sancho, J., and Terhorst, C. (2002b). Molecular dissection of the signaling and costimulatory functions of CD150 (SLAM): CD150/SAP binding and CD150-mediated costimulation. Blood *99*, 957-965.

Huber, M., Cattaneo, R., Spielhofer, P., Orvell, C., Norrby, E., Messerli, M., Perriard, J. C., and Billeter, M. A. (1991). Measles virus phosphoprotein retains the nucleocapsid protein in the cytoplasm. Virology *185*, 299-308.

Ilonen, J., Makela, M. J., Ziola, B., and Salmi, A. A. (1990). Cloning of human T cells specific for measles virus haemagglutinin and nucleocapsid. Clin. Exp. Immunol. *81*, 212-217.

Ito, M., Watanabe, M., Ihara, T., Kamiya, H., and Sakurai, M. (1997). Measles virus induces apoptotic cell death in lymphocytes activated with phorbol 12-myristate 13-acetate (PMA) plus calcium ionophore. Clin. Exp. Immunol. *108*, 266-271.

Ito, T., Amakawa, R., Inaba, M., Ikehara, S., Inaba, K., and Fukuhara, S. (2001). Differential regulation of human blood dendritic cell subsets by IFNs. J. Immunol. *166*, 2961-2969.

Iwata, K., Seya, T., Yanagi, Y., Pesando, J. M., Johnson, P. M., Okabe, M., Ueda, S., Ariga, H., and Nagasawa, S. (1995). Diversity of sites for measles virus binding and for inactivation of complement C3b and C4b on membrane cofactor protein CD46. J. Biol. Chem. *270*, 15148-15152.

Jabbour, J. T., Duenas, D. A., Sever, J. L., Krebs, H. M., and Horta-Barbosa, L. (1972). Epidemiology of subacute sclerosing panencephalitis (SSPE). A report of the SSPE registry. J. Amer. Med. Assoc. *220*, 959-962.

Jacobson, S., Sekaly, R. P., Jacobson, C. L., McFarland, H. F., and Long, E. O. (1989). HLA class II-restricted presentation of cytoplasmic measles virus antigens to cytotoxic T cells. J. Virol. *63*, 1756-1762.

Jansen, V. A., Stollenwerk, N., Jensen, H. J., Ramsay, M. E., Edmunds, W. J., and Rhodes, C. J. (2003). Measles outbreaks in a population with declining vaccine uptake. Science *301*, 804.

Johansson, K., Bourhis, J. M., Campanacci, V., Cambillau, C., Canard, B., and Longhi, S. (2003). Crystal structure of the measles virus phosphoprotein domain responsible for the induced folding of the C-terminal domain of the nucleoprotein. J. Biol. Chem. *278*, 44567-44573.

Kallstrom, H., Liszewski, M. K., Atkinson, J. P., and Jonsson, A. B. (1997). Membrane cofactor protein (MCP or CD46) is a cellular pilus receptor for pathogenic Neisseria. Mol. Microbiol. *25*, 639-647.

Kapadia, S. B., Levine, B., Speck, S. H., and Virgin, H. W. t. (2002). Critical role of complement and viral evasion of complement in acute, persistent, and latent gamma-herpesvirus infection. Immunity *17*, 143-155.

Karlin, D., Ferron, F., Canard, B., and Longhi, S. (2003). Structural disorder and modular organization in Paramyxovirinae N. and P. J. Gen. Virol. *84*, 3239-3252.

Karlin, D., Longhi, S., and Canard, B. (2002). Substitution of two residues in the measles virus nucleoprotein results in an impaired self-association. Virology *302*, 420-432.

Karp, C. L., Wysocka, M., Wahl, L. M., Ahearn, J. M., Cuomo, P. J., Sherry, B., Trinchieri, G., and Griffin, D. E. (1996). Mechanism of suppression of cell-mediated immunity by measles virus. Science *273*, 228-231.

Katayama, Y., Hirano, A., and Wong, T. C. (2000). Human receptor for measles virus (CD46) enhances nitric oxide production and restricts virus replication in mouse macrophages by modulating production of alpha/beta interferon. J. Virol. *74*, 1252-1257.

Keir, M. E., Stoddart, C. A., Linquist-Stepps, V., Moreno, M. E., and McCune, J. M. (2002). IFN-alpha secretion by type 2 predendritic cells up-regulates MHC class I in the HIV-1-infected thymus. J. Immunol. *168*, 325-331.

Kemper, C., Chan, A. C., Green, J. M., Brett, K. A., Murphy, K. M., and Atkinson, J. P. (2003). Activation of human CD4+ cells with CD3 and CD46 induces a T-regulatory cell 1 phenotype. Nature *421*, 388-392.

Kim, Y. J., Kim, S. Y., Kim, Y. Y., Kim, J. W., Lee, J. H., Han, K., and Lee, W. (2002). Quantities of receptor molecules for colony stimulating factors on leukocytes in measles. Yonsei Med. J. *43*, 43-47.

Kingston, R. L., Baase, W. A., and Gay, L. S. (2004a). Characterization of nucleocapsid binding by measles and the mumps virus phosphoprotein. J. Virol. *in press*.

Kingston, R. L., Hamel, D. J., Gay, L. S., Dahlquist, F. W., and Matthews, B. W. (2004b). Structural basis for the attachment of a paramyxoviral polymerase to its template. Proc. Natl. Acad. Sci. USA. *101*, 8301-8306.

Klagge, I. M., ter Meulen, V., and Schneider-Schaulies, S. (2000). Measles virus-induced promotion of dendritic cell maturation by soluble mediators does not overcome the immunosuppressive activity of viral glycoproteins on the cell surface. Eur. J. Immunol. *30*, 2741-2750.

Kobune, F., Takahashi, H., Terao, K., Ohkawa, T., Ami, Y., Suzaki, Y., Nagata, N., Sakata, H., Yamanouchi, K., and Kai, C. (1996). Nonhuman primate models of measles. Lab. Anim Sci. *46*, 315-320.

Kohl, J., and Gessner, J. E. (1999). On the role of complement and Fc gamma-receptors in the Arthus reaction. Mol. Immunol. *36*, 893-903.

Kojima, A., Iwata, K., Seya, T., Matsumoto, M., Ariga, H., Atkinson, J. P., and Nagasawa, S. (1993). Membrane cofactor protein (CD46) protects cells predominantly from alternative complement pathway-mediated C3-fragment deposition and cytolysis. J. Immunol. *151*, 1519-1527.

Kouomou, D. W., and Wild, T. F. (2002). Adaptation of wild-type measles virus to tissue culture. J. Virol. *76*, 1505-1509.

Krantic, S., Gimenez, C., and Rabourdin-Combe, C. (1995). Cell-to-cell contact via measles virus haemagglutinin-CD46 interaction triggers CD46 downregulation. J. Gen. Virol. *76*, 2793-2800.

Kruse, M., Meinl, E., Henning, G., Kuhnt, C., Berchtold, S., Berger, T., Schuler, G., and Steinkasserer, A. (2001). Signaling lymphocytic activation molecule is expressed on mature CD83+ dendritic cells and is up-regulated by IL-1 beta. J. Immunol. *167*, 1989-1995.

Kurita-Taniguchi, M., Fukui, A., Hazeki, K., Hirano, A., Tsuji, S., Matsumoto, M., Watanabe, M., Ueda, S., and Seya, T. (2000). Functional modulation of human macrophages through CD46 (measles virus receptor): production of IL-12 p40 and nitric oxide in association with recruitment of protein-tyrosine phosphatase SHP-1 to CD46. J. Immunol. *165*, 5143-5152.

Laine, D., Bourhis, J. M., Longhi, S., Flacher, M., Cassard, L., Canard, B., Sautès-Fridman, C., Rabourdin-Combe, C., and Valentin, H. (2005). Measles virus nucleoprotein N_{CORE} and N_{TAIL} inhibits cell proliferation through cell surface FcγRII and NR respectively. J. Gen. Virol. *86*, 1771-1784.

Laine, D., Trescol-Biemont, M. C., Longhi, S., Libeau, G., Marie, J. C., Vidalain, P. O., Azocar, O., Diallo, A., Canard, B., Rabourdin-Combe, C., and Valentin, H. (2003). Measles virus (MV) nucleoprotein binds to a novel cell surface receptor distinct from FcgammaRII via its C-terminal domain: role in MV-induced immunosuppression. J. Virol. *77*, 11332-11346.

Lamb, R. A., and Kolakofsky, D. (2001). *Paramyxoviridae*: The viruses and their replication. *In* "Fields Virology" (BN Fields, DM Knipe, and PM Howley, Eds), 4th ed, pp 1305-1340 Lippincott-Raven, Philadelphia, PA.

Langedijk, J. P., Daus, F. J., and van Oirschot, J. T. (1997). Sequence and structure alignment of Paramyxoviridae attachment proteins and discovery of enzymatic activity for a morbillivirus hemagglutinin. J. Virol. *71*, 6155-6167.

Le Bon, A., Etchart, N., Rossmann, C., Ashton, M., Hou, S., Gewert, D., Borrow, P., and Tough, D. F. (2003). Cross-priming of CD8+ T cells stimulated by virus-induced type I interferon. Nat. Immunol. *4*, 1009-1015.

Le Bon, A., Schiavoni, G., D'Agostino, G., Gresser, I., Belardelli, F., and Tough, D. F. (2001). Type i interferons potently enhance humoral immunity and can promote isotype switching by stimulating dendritic cells *in vivo*. Immunity *14*, 461-470.

Le Bon, A., and Tough, D. F. (2002). Links between innate and adaptive immunity via type I interferon. Curr. Opin. Immunol. *14*, 432-436.

Le Mercier, P., Garcin, D., Haussmann, S., and Kolakofsky, D. (2002). Ambisense Sendai viruses are inherently unstable but are useful to study viral RNA synthesis. J. Virol. *76*, 5492-5502.

Lecouturier, V., Fayolle, J., Caballero, M., Carabana, J., Celma, M. L., Fernandez-Munoz, R., Wild, T. F., and Buckland, R. (1996). Identification of two amino acids in the hemagglutinin glycoprotein of measles virus (MV) that govern hemadsorption, HeLa cell fusion, and CD46 downregulation: phenotypic markers that differentiate vaccine and wild-type MV strains. J. Virol. *70*, 4200-4204.

Leopardi, R., Hyypia, T., and Vainionpaa, R. (1992). Effect of interferon-alpha on measles virus replication in human peripheral blood mononuclear cells. Apmis *100*, 125-131.

Levy, D. E. (2002). Whence interferon? Variety in the production of interferon in response to viral infection. J. Exp. Med. *195*, F15-18.

Li, L. Y., and Qi, Y. P. (1998). The Point Mutations in Hemagglutinin Gene of Measles Virus Are Responsible for Alteration in Hemadsorption. Sheng Wu Hua Xue Yu Sheng Wu Wu Li Xue Bao (Shanghai) *30*, 488-494.

Lichtner, M., Maranon, C., Vidalain, P. O., Azocar, O., Hanau, D., Lebon, P., Burgard, M., Rouzioux, C., Vullo, V., Yagita, H., et al. (2004). HIV type 1-infected dendritic cells induce apoptotic death in infected and uninfected primary CD4 T lymphocytes. AIDS Res. Hum. Retroviruses *20*, 175-182.

Lin, G. Y., Paterson, R. G., Richardson, C. D., and Lamb, R. A. (1998a). The V protein of the paramyxovirus SV5 interacts with damage-specific DNA binding protein. Virology *249*, 189-200.

Lin, Q., Dong, C., and Cooper, M. D. (1998b). Impairment of T and B cell development by treatment with a type I interferon. J. Exp. Med. *187*, 79-87.

Linton, P. J., and Dorshkind, K. (2004). Age-related changes in lymphocyte development and function. Nat. Immunol. *5*, 133-139.

Liston, P., DiFlumeri, C., and Briedis, D. J. (1995). Protein interactions entered into by the measles virus P, V, and C proteins. Virus Res. *38*, 241-259.

Liszewski, M. K., Leung, M., Cui, W., Subramanian, V. B., Parkinson, J., Barlow, P., Manchester, M., and Atkinson, J. P. (2000). Dissecting sites important for complement regulatory activity in membrane cofactor protein (MCP; CD46). J. Biol. Chem. *275*, 37692-37701.

Liszewski, M. K., Post, T. W., and Atkinson, J. P. (1991). Membrane cofactor protein (MCP or CD46): newest member of the regulators of complement activation gene cluster. Annu. Rev. Immunol. *9*, 431-455.

Longhi, S., Receveur-Brechot, V., Karlin, D., Johansson, K., Darbon, H., Bhella, D., Yeo, R., Finet, S., and Canard, B. (2003). The C-terminal domain of the measles virus nucleoprotein is intrinsically disordered and folds upon binding to the C-terminal moiety of the phosphoprotein. J. Biol. Chem. *278*, 18638-18648.

Lorin, C., Mollet, L., Delebecque, F., Combredet, C., Hurtrel, B., Charneau, P., Brahic, M., and Tangy, F. (2004). A single injection of recombinant measles virus vaccines expressing human immunodeficiency virus (HIV) type 1 clade B envelope glycoproteins induces neutralizing antibodies and cellular immune responses to HIV. J. Virol. *78*, 146-157.

Luft, T., Luetjens, P., Hochrein, H., Toy, T., Masterman, K. A., Rizkalla, M., Maliszewski, C., Shortman, K., Cebon, J., and Maraskovsky, E. (2002). IFN-alpha enhances CD40 ligand-mediated activation of immature monocyte-derived dendritic cells. Int. Immunol. *14*, 367-380.

Luft, T., Pang, K. C., Thomas, E., Hertzog, P., Hart, D. N., Trapani, J., and Cebon, J. (1998). Type I IFNs enhance the terminal differentiation of dendritic cells. J. Immunol. *161*, 1947-1953.

Lund, J. M., Alexopoulou, L., Sato, A., Karow, M., Adams, N. C., Gale, N. W., Iwasaki, A., and Flavell, R. A. (2004). Recognition of single-stranded RNA viruses by Toll-like receptor 7. Proc. Natl. Acad. Sci. USA. *101*, 5598-5603.

Mackall, C. L., Hakim, F. T., and Gress, R. E. (1997). Restoration of T-cell homeostasis after T-cell depletion. Semin Immunol. *9*, 339-346.

Maisner, A., Klenk, H., and Herrler, G. (1998). Polarized budding of measles virus is not determined by viral surface glycoproteins. J. Virol. *72*, 5276-5278.

Manchester, M., Eto, D. S., Valsamakis, A., Liton, P. B., Fernandez-Munoz, R., Rota, P. A., Bellini, W. J., Forthal, D. N., and Oldstone, M. B. (2000). Clinical isolates of measles virus use CD46 as a cellular receptor. J. Virol. *74*, 3967-3974.

Manchester, M., Liszewski, M. K., Atkinson, J. P., and Oldstone, M. B. (1994). Multiple isoforms of CD46 (membrane cofactor protein) serve as receptors for measles virus. Proc. Natl. Acad. Sci. USA. *91*, 2161-2165.

Manchester, M., and Rall, G. F. (2001). Model Systems: transgenic mouse models for measles pathogenesis. Trends Microbiol. *9*, 19-23.

Manchester, M., Smith, K. A., Eto, D. S., Perkin, H. B., and Torbett, B. E. (2002). Targeting and hematopoietic suppression of human CD34+ cells by measles virus. J. Virol. *76*, 6636-6642.

Manchester, M., Valsamakis, A., Kaufman, R., Liszewski, M. K., Alvarez, J., Atkinson, J. P., Lublin, D. M., and Oldstone, M. B. A. (1995). Measles virus and C3 binding sites are distinct on membrane cofactor protein (CD46). Proc. Natl. Acad. Sci. USA. *92*, 2303-2307.

Manie, S. N., Debreyne, S., Vincent, S., and Gerlier, D. (2000). Measles virus structural components are enriched into lipid raft microdomains: a potential cellular location for virus assembly. J. Virol. *74*, 305-311.

Marie, J. C., Kehren, J., Trescol-Biemont, M. C., Evlashev, A., Valentin, H., Walzer, T., Tedone, R., Loveland, B., Nicolas, J. F., Rabourdin-Combe, C., and Horvat, B. (2001). Mechanism of measles virus-induced suppression of inflammatory immune responses. Immunity *14*, 69-79.

Marie, J. C., Saltel, F., Escola, J. M., Jurdic, P., Wild, T. F., and Horvat, B. (2004). Cell surface delivery of the measles virus nucleoprotein: A viral strategy to induce immunosuppression. J. Virol. *78*, 11952–11961.

Markowitz, L. E., Chandler, F. W., Roldan, E. O., Saldana, M., Roach, K. C., Hutchins, S. S., Preblud, S. R., Mitchell, C. D., and Scott, G. B. (1988). Fatal measles pneumonia without rash in a child with AIDS. Journal of Infectious Diseases *158*, 480-483.

Masse, N., Ainouze, M., Néell, B., Wild, T. F., Buckland, R., and Langedijk, J. P. M. (2004). Measles Virus (MV) Hemagglutinin: Evidence that attachment sites for MV receptors SLAM and CD46 overlap on the globular head. J. Virol. *78*, 9051–9063.

Masse, N., Barrett, T., Muller, C. P., Wild, T. F., and Buckland, R. (2002). Identification of a second major site for CD46 binding in the hemagglutinin protein from a laboratory strain of measles virus (MV): potential consequences for wild-type MV infection. J. Virol. *76*, 13034-13038.

Matteucci, D., Toniolo, A., Conaldi, P. G., Basolo, F., Gori, Z., and Bendinelli, M. (1985). Systemic lymphoid atrophy in coxsackievirus B3-infected mice: effects of virus and immunopotentiating agents. J. Infect. Dis. *151*, 1100-1108.

McChesney, M. B., Miller, C. J., Rota, P. A., Zhu, Y. D., Antipa, L., Lerche, N. W., Ahmed, R., and Bellini, W. J. (1997). Experimental measles. I. Pathogenesis in the normal and the immunized host. Virology *233*, 74-84.

McFarland, R. D., Douek, D. C., Koup, R. A., and Picker, L. J. (2000). Identification of a human recent thymic emigrant phenotype. Proc. Natl. Acad. Sci. USA. *97*, 4215-4220.

Minagawa, H., Tanaka, K., Ono, N., Tatsuo, H., and Yanagi, Y. (2001). Induction of the measles virus receptor SLAM (CD150) on monocytes. J. Gen. Virol. *82*, 2913-2917.

Moench, T. R., Griffin, D. E., Obriecht, C. R., Vaisberg, A. J., and Johnson, R. T. (1988). Acute measles in patients with and without neurological involvement: distribution of measles virus antigen and RNA. J. Infect. Dis. *158*, 433-442.

Mogensen, K. E., Lewerenz, M., Reboul, J., Lutfalla, G., and Uze, G. (1999). The type I interferon receptor: structure, function, and evolution of a family business. J. Interferon Cytokine Res. *19*, 1069-1098.

Mohty, M., Vialle-Castellano, A., Nunes, J. A., Isnardon, D., Olive, D., and Gaugler, B. (2003). IFN-alpha skews monocyte differentiation into Toll-like receptor 7-expressing dendritic cells with potent functional activities. J. Immunol. *171*, 3385-3393.

Moll, M., Pfeuffer, J., Klenk, H. D., Niewiesk, S., and Maisner, A. (2004). Polarized glycoprotein targeting affects the spread of measles virus *in vitro* and *in vivo*. J. Gen. Virol. *85*, 1019-1027.

Morrison, T. G. (2003). Structure and function of a paramyxovirus fusion protein. Biochim Biophys Acta *1614*, 73-84.

Moss, W. J., Ryon, J. J., Monze, M., and Griffin, D. E. (2002). Differential regulation of interleukin IL-4, IL-5, and IL-10 during measles in Zambian children. J. Infect. Dis. *186*, 879-887.

Mrkic, B., Odermatt, B., Klein, M. A., Billeter, M. A., Pavlovic, J., and Cattaneo, R. (2000). Lymphatic dissemination and comparative pathology of recombinant measles viruses in genetically modified mice. J. Virol. *74*, 1364-1372.

Mrkic, B., Pavlovic, J., Licke, T. R., Volpe, P., Buchholz , C. J., Hourcade, D., Atkinson , J. P., AguzziI, A., and Cattanco, R. (1998). Measles virus spread and pathogenesis in genetically modified mice. J. Virol. *72*, 7420–7427.

Mumenthaler, C., Schneider, U., Buchholz, C. J., Koller, D., Braun, W., and Cattaneo, R. (1997). A 3D model for the measles virus receptor CD46 based on homology modeling, Monte Carlo simulations, and hemagglutinin binding studies. Protein Sci. *6*, 588-597.

Murakami, Y., Seya, T., Kurita, M., Fukui, A., Ueda, S., and Nagasawa, S. (1998). Molecular cloning of membrane cofactor protein (MCP; CD46) on B95a cell, an Epstein-Barr virus-transformed marmoset B cell line: B95a-MCP is susceptible to infection by the CAM, but not the Nagahata strain of the measles virus. Biochem. J. *330*, 1351-1359.

Na, B. K., Lee, J. S., Shin, G. C., Shin, J. M., Lee, J. Y., Chung, J. K., Ha, D. R., Lee, J. K., Ma, S. H., Cho, H. W., et al. (2001). Sequence analysis of hemagglutinin and nucleoprotein genes of measles viruses isolated in Korea during the 2000 epidemic. Virus Res. *81*, 143-149.

Naniche, D., Garenne, M., Rae, C., Manchester, M., Buchta, R., Brodine, S. K., and Oldstone, M. B. A. (2004). Decrease in Measles Virus–Specific CD4 T Cell Memory in Vaccinated Subjects. J. Infect. Dis. *190*, 1387-1395.

Naniche, D., Varior-Krishnan, G., Cervoni, F., Wild, T. F., Rossi, B., Rabourdin-Combe, C., and Gerlier, D. (1993a). Human membrane cofactor protein (CD46) acts as a cellular receptor for measles virus. J. Virol. *67*, 6025-6032.

Naniche, D., Wild, T. F., Rabourdin-Combe, C., and Gerlier, D. (1993b). Measles virus haemagglutinin induces down-regulation of gp57/67, a molecule involved in virus binding. J. Gen. Virol. *74*, 1073-1079.

Naniche, D., Yeh, A., Eto, D., Manchester, M., Friedman, R. M., and Oldstone, M. B. (2000). Evasion of host defenses by measles virus: wild-type measles virus infection interferes with induction of Alpha/Beta interferon production. J. Virol. *74*, 7478-7484.

Niewiesk, S., Eisenhuth, I., Fooks, A., Clegg, J. C., Schnorr, J. J., Schneider-Schaulies, S., and ter Meulen, V. (1997). Measles virus-induced immune suppression in the cotton rat (Sigmodon hispidus) model depends on viral glycoproteins. J. Virol. *71*, 7214-7219.

Norrby, E., and Gollmar, Y. (1972). Appearance and persistence of antibodies against different virus components after regular measles infections. Infect. Immun. *6*, 240-247.

Norrby, E., and Gollmar, Y. (1975). Identification of measles virus-specific hemolysis-inihibiting antibodies separate from hemagglutination-inhibiting antibodies. Infect. Immun. *11*, 231-239.

Nozawa, Y., Ono, N., Abe, M., Sakuma, H., and Wakasa, H. (1994). An immunohistochemical study of Warthin-Finkeldey cells in measles. Pathol. Int. *44*, 442-447.

Numazaki, K., Goldman, H., Bai, X. Q., Wong, I., and Wainberg, M. A. (1989a). Effects of infection by HIV-1, cytomegalovirus, and human measles virus on cultured human thymic epithelial cells. Microbiol. Immunol. *33*, 733-745.

Numazaki, K., Goldman, H., Wong, I., and Wainberg, M. A. (1989b). Replication of measles virus in cultured human thymic epithelial cells. J. Med. Virol. *27*, 52-58.

Obeid, O. E., Partidos, C. D., Howard, C. R., and Steward, M. W. (1995). Protection against morbillivirus-induced encephalitis by immunization with a rationally designed synthetic peptide vaccine containing B- and T-cell epitopes from the fusion protein of measles virus. J. Virol. *69*, 1420-1428.

Oglesbee, M. J., Kenney, H., Kenney, T., and Krakowka, S. (1993). Enhanced production of morbillivirus gene-specific RNAs following induction of the cellular stress response in stable persistent infection. Virology *192*, 556-567.

Ohgimoto, S., Ohgimoto, K., Niewiesk, S., Klagge, I. M., Pfeuffer, J., Johnston, I. C., Schneider-Schaulies, J., Weidmann, A., ter Meulen, V., and Schneider-Schaulies, S. (2001). The haemagglutinin protein is an important determinant of measles virus tropism for dendritic cells *in vitro*. J. Gen. Virol. *82*, 1835-1844.

Ohno, S., Ono, N., Takeda, M., Takeuchi, K., and Yanagi, Y. (2004). Dissection of measles virus V protein in relation to its ability to block alpha/beta interferon signal transduction. J. Gen. Virol. *85*, 2991–2999.

Okada, H., Kobune, F., Sato, T. A., Kohama, T., Takeuchi, Y., Abe, T., Takayama, N., Tsuchiya, T., and Tashiro, M. (2000). Extensive lymphopenia due to apoptosis of uninfected lymphocytes in acute measles patients. Arch. Virol. *145*, 905-920.

Okada, H., Sato, T. A., Katayama, A., Higuchi, K., Shichijo, K., Tsuchiya, T., Takayama, N., Takeuchi, Y., Abe, T., Okabe, N., and Tashiro, M. (2001). Comparative analysis of host responses related to immunosuppression between measles patients and vaccine recipients with live attenuated measles vaccines. Arch. Virol. *146*, 859-874.

Okada, N., Liszewski, M. K., Atkinson, J. P., and Caparon, M. (1995). Membrane cofactor protein (CD46) is a keratinocyte receptor for the M protein of the group A streptococcus. Proc. Natl. Acad. Sci. USA. *92*, 2489-2493.

Olszewska, W., Erume, J., Ripley, J., Steward, M. W., and Partidos, C. D. (2001). Immune responses and protection induced by mucosal and systemic immunisation with recombinant measles nucleoprotein in a mouse model of measles virus-induced encephalitis. Arch. Virol. *146*, 293-302.

Ono, N., Tatsuo, H., Tanaka, K., Minagawa, H., and Yanagi, Y. (2001). V domain of human SLAM (CDw150) is essential for its function as a measles virus receptor. J. Virol. *75*, 1594-1600.

Palosaari, H., Parisien, J. P., Rodriguez, J. J., Ulane, C. M., and Horvath, C. M. (2003). STAT protein interference and suppression of cytokine signal transduction by measles virus V protein. J. Virol. *77*, 7635-7644.

Patterson, J. B., Thomas, D., Lewicki, H., Billeter, M. A., and Oldstone, M. B. (2000). V and C proteins of measles virus function as virulence factors *in vivo*. Virology *267*, 80-89.

Pearse, R. N., Kawabe, T., Bolland, S., Guinamard, R., Kurosaki, T., and Ravetch, J. V. (1999). SHIP recruitment attenuates Fc gamma RIIB-induced B cell apoptosis. Immunity *10*, 753-760.

Permar, S. R., Klumpp, S. A., Mansfield, K. G., Kim, W. K., Gorgone, D. A., Lifton, M. A., Williams, K. C., Schmitz, J. E., Reimann, K. A., Axthelm, M. K., et al. (2003a). Role of CD8 Lymphocytes in Control and Clearance of Measles Virus Infection of Rhesus Monkeys. J. Virol. *77*, 4396–4400.

Permar, S. R., Moss, W. J., Ryon, J. J., Douek, D. C., Monze, M., and Griffin, D. E. (2003b). Increased Thymic Output during Acute Measles Virus Infection. J. Virol. *77*, 7872–7879.

Pfeuffer, J., Puschel, K., Meulen, V., Schneider-Schaulies, J., and Niewiesk, S. (2003). Extent of measles virus spread and immune suppression differentiates between wild-type and vaccine strains in the cotton rat model (Sigmodon hispidus). J. Virol. *77*, 150-158.

Plemper, R. K., Hammond, A. L., and Cattaneo, R. (2001). Characterization of a region of the measles virus hemaglutinin sufficient for its dimerization. J. Virol. *74*, 6485-6493.

Plemper, R. K., Hammond, A. L., and Cattaneo, R. (2001). Measles virus envelope glycoproteins hetero-oligomerize in the endoplasmic reticulum. J. Biol. Chem. *276*, 44239-44246.

Plemper, R. K., Hammond, A. L., Gerlier, D., Fielding, A. K., and Cattaneo, R. (2002). Strength of envelope protein interaction modulates cytopathicity of measles virus. J. Virol. *76*, 5051-5061.

Polack, F. P., Auwaerter, P. G., Lee, S. H., Nousari, H. C., Valsamakis, A., Leiferman, K. M., Diwan, A., Adams, R. J., and Griffin, D. E. (1999). Production of atypical measles in rhesus macaques: Evidence for disease mediated by immune complex formation and eosinophils in the presence of fusion-inhibiting antibody. Nature Med. *5*, 629-634.

Polack, F. P., Hoffman, S. J., Crujeiras, G., and Griffin, D. E. (2003a). A role for nonproctive complement-fixing antibodies with low avidity for measles virus in atypical measles. Nature Med. *9*, 1209-1213.

Polack, F. P., Hoffman, S. J., Moss, W. J., and Griffin, D. E. (2002). Altered Synthesis of Interleukin-12 and Type 1 and Type 2 Cytokines in Rhesus Macaques during Measles and Atypical Measles. J. Infect. Dis. *185*, 13-19.

Polack, F. P., Hoffman, S. J., Moss, W. J., and Griffin, D. E. (2003b). Differential effects of priming with DNA vaccines encoding the hemagglutinin and/or fusion proteins on cytokine responses after measles virus challenge. J. Infect. Dis. *187*, 1794-1800.

Polack, F. P., Lee, S. H., Permar, S., Manyara, E., Nousari, H. G., Jeng, Y., Mustafa, F., Valsamakis, A., Adams, R. J., Robinson, H. L., and Griffin, D. E. (2000). Successful DNA immunization against measles: neutralizing antibody against either the hemagglutinin or fusion glycoprotein protects rhesus macaques without evidence of atypical measles. Nat. Med. *6*, 776-781.

Radecke, F., and Billeter, M. A. (1996). The nonstructural C protein is not essential for multiplication of Edmonston B strain measles virus in cultured cells. Virology *217*, 418-421.

Radecke, F., Spielhofer, P., Schneider, H., Kaelin, K., Huber, M., Dotsch, C., Christiansen, G., and Billeter, M. A. (1995). Rescue of measles viruses from cloned DNA. Embo J. *14*, 5773-5784.

Rager, M., Vongpunsawad, S., Duprex, W. P., and Cattaneo, R. (2002). Polyploid measles virus with hexameric genome length. Embo J. *21*, 2364-2372.

Rall, G. F. (2003). Measles virus 1998-2002: progress and controversy. Annu. Rev. Microbiol. *57*, 343-367.

Ravanel, K., Castelle, C., Defrance, T., Wild, T. F., Charron, D., Lotteau, V., and Rabourdin-Combe, C. (1997). Measles virus nucleocapsid protein binds to FcgammaRII and inhibits human B cell antibody production. J. Exp. Med. *186*, 269-278.

Ravetch, J. V., and Bolland, S. (2001). IgG Fc receptors. Annu. Rev. Immunol. *19*, 275-290.

Reddy, S. V., Kurihara, N., Menaa, C., Landucci, G., Forthal, D., Koop, B. A., Windle, J. J., and Roodman, G. D. (2001). Osteoclasts formed by measles virus-infected osteoclast precursors from hCD46 transgenic mice express characteristics of pagetic osteoclasts. Endocrinology *142*, 2898-2905.

Reutter, G. L., Cortese-Grogan, C., Wilson, J., and Moyer, S. A. (2001). Mutations in the measles virus C protein that up regulate viral RNA synthesis. Virology *285*, 100-109.

Riedl, P., Moll, M., Klenk, H. D., and Maisner, A. (2002). Measles virus matrix protein is not cotransported with the viral glycoproteins but requires virus infection for efficient surface targeting. Virus Res. *83*, 1-12.

Rima, B. K. (2000). Measles Virus. In Nature Encyclopedia of Life Sciences. (London, Nature Publishing Group.), pp. [doi:10.1038/npg.els.0000418].

Rima, B. K., Earle, J. A., Baczko, K., ter Meulen, V., Liebert, U. G., Carstens, C., Carabana, J., Caballero, M., Celma, M. L., and Fernandez-Munoz, R. (1997). Sequence divergence of measles virus haemagglutinin during natural evolution and adaptation to cell culture. J. Gen. Virol. *78*, 97-106.

Rivollier, A., Mazzorana, M., Tebib, J., Piperno, M., Aitsiselmi, T., Rabourdin-Combe, C., Jurdic, P., and Servet-Delprat, C. (2004). Immature dendritic cell transdifferentiation into osteoclasts: a novel pathway sustained by the rheumatoid arthritis microenvironment. In revision.

Roberts, P. F. (1975). Thymic dysplasia, persistence of measles virus, and unexpected infant death. Arch. Dis. Child *50*, 401-403.

Roodman, G. D., Kurihara, N., Ohsaki, Y., Kukita, A., Hosking, D., Demulder, A., Smith, J. F., and Singer, F. R. (1992). Interleukin 6. A potential autocrine/paracrine factor in Paget's disease of bone. J. Clin. Invest *89*, 46-52.

Rota, J.S., Wang, Z.D., Rota, P.A., and Bellini, W.J. (1994). Comparison of sequences of the H, F, and N coding genes of measles virus vaccine strains. Virus Res. *31*, 317-330.

Ryon, J. J., Moss, W. J., Monze, M., and Griffin, D. E. (2002). Functional and phenotypic changes in circulating lymphocytes from hospitalized zambian children with measles. Clin. Diagn. Lab. Immunol. *9*, 994-1003.

Sakaguchi, M., Yoshikawa, Y., Yamanouchi, K., Sata, T., Nagashima, K., and Takeda, K. (1986). Growth of measles virus in epithelial and lymphoid tissues of cynomolgus monkeys. Microbiol. Immunol. *30*, 1067-1073.

Santiago, C., Bjorling, E., Stehle, T., and Casasnovas, J. M. (2002). Distinct kinetics for binding of the CD46 and SLAM receptors to overlapping sites in the measles virus hemagglutinin protein. J. Biol. Chem. *277*, 32294-32301.

Santini, S. M., Di Pucchio, T., Lapenta, C., Parlato, S., Logozzi, M., and Belardelli, F. (2003). A new type I IFN-mediated pathway for the rapid differentiation of monocytes into highly active dendritic cells. Stem Cells *21*, 357-362.

Santini, S. M., Lapenta, C., Logozzi, M., Parlato, S., Spada, M., Di Pucchio, T., and Belardelli, F. (2000). Type I interferon as a powerful adjuvant for monocyte-derived dendritic cell development and activity *in vitro* and in Hu-PBL-SCID mice. J. Exp. Med. *191*, 1777-1788.

Santoro, F., Kennedy, P. E., Locatelli, G., Malnati, M. S., Berger, E. A., and Lusso, P. (1999). CD46 is a cellular receptor for human herpesvirus 6. Cell *99*, 817-827.

Sato, M., Suemori, H., Hata, N., Asagiri, M., Ogasawara, K., Nakao, K., Nakaya, T., Katsuki, M., Noguchi, S., Tanaka, N., and Taniguchi, T. (2000). Distinct and essential roles of transcription factors IRF-3 and IRF-7 in response to viruses for IFN-alpha/beta gene induction. Immunity *13*, 539-548.

Scheller, C., and Jassoy, C. (2001). Syncytium formation amplifies apoptotic signals: a new view on apoptosis in HIV infection *in vitro*. Virology *282*, 48-55.

Schlender, J., Schnorr, J. J., Spielhoffer, P., Cathomen, T., Cattaneo, R., Billeter, M. A., ter Meulen, V., and Schneider-Schaulies, S. (1996). Interaction of measles virus glycoproteins with the surface of uninfected peripheral blood lymphocytes induces immunosuppression *in vitro*. Proc. Natl. Acad. Sci. USA. *93*, 13194-13199.

Schlereth, B., Germann, P. G., ter Meulen, V., and Niewiesk, S. (2000). DNA vaccination with both the haemagglutinin and fusion proteins but not the nucleocapsid protein protects against experimental measles virus infection. J. Gen. Virol. *81*, 1321-1325.

Schneider, U., von Messling, V., Devaux, P., and Cattaneo, R. (2002). Efficiency of measles virus entry and dissemination through different receptors. J. Virol. *76*, 7460-7467.

Schneider-Schaulies, J., Dunster, L. M., Kobune, F., Rima, B., and Ter Meulen, V. (1995). Differential downregulation of CD46 by measles virus strains. J. Virol. *69*, 7257-7259.

Schneider-Schaulies, J., Schneider-Schaulies, S., and Ter Meulen, V. (1993). Differential induction of cytokines by primary and persistent measles virus infections in human glial cells. Virology *195*, 219-228.

Schnorr, J. J., Dunster, L. M., Nanan, R., Schneider-Schaulies, J., Schneider-Schaulies, S., and ter Meulen, V. (1995). Measles virus-induced down-regulation of CD46 is associated with enhanced sensitivity to complement-mediated lysis of infected cells. Eur. J. Immunol. *25*, 976-984.

Schnorr, J. J., Xanthakos, S., Keikavoussi, P., Kampgen, E., ter Meulen, V., and Schneider-Schaulies, S. (1997). Induction of maturation of human blood dendritic cell precursors by measles virus is associated with immunosuppression. Proc. Natl. Acad. Sci. USA. *94*, 5326-5331.

Servant, M. J., ten Oever, B., LePage, C., Conti, L., Gessani, S., Julkunen, I., Lin, R., and Hiscott, J. (2001). Identification of distinct signaling pathways leading to the phosphorylation of interferon regulatory factor 3. J. Biol. Chem. *276*, 355-363.

Servet, C., Zitvogel, L., and Hosmalin, A. (2002). Dendritic cells in innate immune responses against HIV. Curr. Mol. Med. *2*, 739-756.

Servet-Delprat, C., Vidalain, P. O., Bausinger, H., Manie, S., Le Deist, F., Azocar, O., Hanau, D., Fischer, A., and Rabourdin-Combe, C. (2000). Measles virus induces abnormal differentiation of CD40 ligand-activated human dendritic cells. J. Immunol. *164*, 1753-1760.

Servet-Delprat, C., Vidalain, P. O., Valentin, H., and Rabourdin-Combe, C. (2003). Measles virus and dendritic cell functions: how specific response cohabits with immunosuppression. Curr. Top Microbiol. Immunol. *276*, 103-123.

Shaffer, J. A., Bellini, W. J., and Rota, P. A. (2003). The C protein of measles virus inhibits the type I interferon response. Virology *315*, 389-397.

Sharma, S., tenOever, B. R., Grandvaux, N., Zhou, G. P., Lin, R., and Hiscott, J. (2003). Triggering the interferon antiviral response through an IKK-related pathway. Science *300*, 1148-1151.

Shimizu, T., Fukushima, N., Ebihara, Y., Sata, T., and Aoyama, Y. (1987). Giant cell pneumonia in Letterer-Siwe disease. Acta Pathol. Jpn *37*, 493-501.

Sidorenko, S. P., and Clark, E. A. (2003). The dual-function CD150 receptor subfamily: the viral attraction. Nat. Immunol. *4*, 19-24.

Singh, M., Cattaneo, R., and Billeter, M. A. (1999). A recombinant measles virus expressing hepatitis B virus surface antigen induces humoral immune responses in genetically modified mice. J. Virol. *73*, 4823-4828.

Sissons, J. G. P., Oldstone, M. B. A., and Schreiber, R. D. (1980). Antibody-independent activation of the alternative complement pathway by measles virus-infected cells. Proc. Natl. Acad. Sci. USA. *77*, 559-562.

Sissons, J. G. P., Schreiber, R. D., Perrin, L. H., Cooper, N. R., Muller-Eberhard, H. J., and Oldstone, M. B. A. (1979). Lysis of measles virus-infected cells by the purified cytolytic alternative complement pathway and antibody. J. Exp. Med. *150*, 415-454.

Slifka, M. K., Homann, D., Tishon, A., Pagarigan, R., and Oldstone, M. B. (2003). Measles virus infection results in suppression of both innate and adaptive immune responses to secondary bacterial infection. J. Clin. Invest *111*, 805-810.

Smallwood, S., Cevik, B., and Moyer, S. A. (2002). Intragenic Complementation and oligomerization of the L subunit of the Sendai virus RNA polymerase. Virology *304*, 235-245.

Smith, A. E., and Helenius, A. (2004). How viruses enter animal cells. Science *304*, 237-242.

Spann, K. M., Collins, P. L., and Teng, M. N. (2003). Genetic recombination during coinfection of two mutants of human respiratory syncytial virus. J. Virol. *77*, 11201-11211.

Spehner, D., Drillien, R., and Howley, P. M. (1997). The assembly of the measles virus nucleoprotein into nucleocapsid-like particles is modulated by the phosphoprotein. Virology *232*, 260-268.

Spielhofer, P., Bachi, T., Fehr, T., Christiansen, G., Cattaneo, R., Kaelin, K., Billeter, M. A., and Naim, H. Y. (1998). Chimeric measles viruses with a foreign envelope. J. Virol. *72*, 2150-2159.

Stark, G. R., Kerr, I. M., Williams, B. R., Silverman, R. H., and Schreiber, R. D. (1998). How cells respond to interferons. Annu. Rev. Biochem. *67*, 227-264.

Stein, C. E., Birmingham, M., Kurian, M., Duclos, P., and Strebel, P. (2003). The global burden of measles in the year 2000--a model that uses country-specific indicators. J. Infect. Dis. *187*, S8-14.

Su, L., Kaneshima, H., Bonyhadi, M., Salimi, S., Kraft, D., Rabin, L., and McCune, J. M. (1995). HIV-1-induced thymocyte depletion is associated with indirect cytopathogenicity and infection of progenitor cells *in vivo*. Immunity 2, 25-36.

Tajima, M., and Kudow, S. (1976). Morphology of the Warthin-Finkeldey giant cells in monkeys with experimentally induced measles. Acta Pathol. Jpn *26*, 367-380.

Takahashi, M., Nakayama, T., Kashiwagi, Y., Takami, T., Sonoda, S., Yamanaka, T., Ochiai, H., Ihara, T., and Tajima, T. (2000). Single genotype of measles virus is dominant whereas several genotypes of mumps virus are co-circulating. J. Med. Virol. *62*, 278-285.

Takasu, T., Mgone, J. M., Mgone, C. S., Miki, K., Komase, K., Namae, H., Saito, Y., Kokubun, Y., Nishimura, T., Kawanishi, R., et al. (2003). A continuing high incidence of subacute sclerosing panencephalitis (SSPE) in the Eastern Highlands of Papua New Guinea. Epidemiol. Infect. *131*, 887-898.

Takeda, M., Kato, A., Kobune, F., Sakata, H., Li, Y., Shioda, T., Sakai, Y., Asakawa, M., and Nagai, Y. (1998). Measles virus attenuation associated with transcriptional impediment and a few amino acid changes in the polymerase and accessory proteins. J. Virol. *72*, 8690-8696.

Takeuchi, K., Kadota, S. I., Takeda, M., Miyajima, N., and Nagata, K. (2003a). Measles virus V protein blocks interferon (IFN)-alpha/beta but not IFN-gamma signaling by inhibiting STAT1 and STAT2 phosphorylation. FEBS Lett. *545*, 177-182.

Takeuchi, K., Miyajima, N., Nagata, N., Takeda, M., and Tashiro, M. (2003b). Wild-type measles virus induces large syncytium formation in primary human small airway epithelial cells by a SLAM(CD150)-independent mechanism. Virus Res. *94*, 11-16.

Tanaka, K., Minagawa, H., Xie, M. F., and Yanagi, Y. (2002). The measles virus hemagglutinin downregulates the cellular receptor SLAM (CD150). Arch. Virol. *147*, 195-203.

Tatsuo, H., Ono, N., Tanaka, K., and Yanagi, Y. (2000). SLAM (CDw150) is a cellular receptor for measles virus. Nature *406*, 893-897.

Tatsuo, H., and Yanagi, Y. (2002). The morbillivirus receptor SLAM (CD150). Microbiol. Immunol. *46*, 135-142.

tenOever, B. R., Servant, M. J., Grandvaux, N., Lin, R., and Hiscott, J. (2002). Recognition of the measles virus nucleocapsid as a mechanism of IRF-3 activation. J. Virol. *76*, 3659-3669.

Tober, C., Seufert, M., Schneider, H., Billeter, M. A., Johnston, I. C., Niewiesk, S., ter Meulen, V., and Schneider-Schaulies, S. (1998). Expression of measles virus V protein is associated with pathogenicity and control of viral RNA synthesis. J. Virol. *72*, 8124-8132.

Trgovcich, J., Aronson, J. F., and Johnston, R. E. (1996). Fatal Sindbis virus infection of neonatal mice in the absence of encephalitis. Virology *224*, 73-83.

Truong, A. T., Mulders, M. N., Gautam, D. C., Ammerlaan, W., de Swart, R. L., King, C. C., Osterhaus, A. D., and Muller, C. P. (2001). Genetic analysis of Asian measles virus strains--new endemic genotype in Nepal. Virus Res. *76*, 71-78.

Tsubata, T. (1999). Co-receptors on B lymphocytes. Curr. Opin. Immunol. *11*, 249-255.

Ueda, K., Imamura, T., Kawaguchi, T., Tamari, H., and Kambara, T. (1982). Subacute sclerosing panencephalitis. An autopsy case with impaired cellular immunity. Acta Pathol. Jpn *32*, 1103-1110.

Valentin, H., Azocar, O., Horvat, B., Williems, R., Garrone, R., Evlashev, A., Toribio, M. L., and Rabourdin-Combe, C. (1999). Measles virus infection induces terminal differentiation of human thymic epithelial cells. J. Virol. *73*, 2212-2221.

Valsamakis, A., Auwaerter, P. G., Rima, B. K., Kaneshima, H., and Griffin, D. E. (1999). Altered virulence of vaccine strains of measles virus after prolonged replication in human tissue. J. Virol. *73*, 8791-8797.

Valsamakis, A., Kaneshima, H., and Griffin, D. E. (2001). Strains of measles vaccine differ in their ability to replicate in an damage human thymus. J. Infect. Dis. *183*, 498-502.

Valsamakis, A., Schneider, H., Auwaerter, P. G., Kaneshima, H., Billeter, M. A., and Griffin, D. E. (1998). Recombinant measles viruses with mutations in the C, V, or F gene have altered growth phenotypes *in vivo*. J. Virol. *72*, 7754-7761.

van Binnendijk, R. S., Poelen, M. C., de Vries, P., Voorma, H. O., Osterhaus, A. D., and Uytdehaag, F. G. (1989). Measles virus-specific human T cell clones. Characterization of specificity and function of CD4+ helper/cytotoxic and CD8+ cytotoxic T cell clones. J. Immunol. *142*, 2847-2854.

van Binnendijk, R. S., van Baalen, C. A., Poelen, M. C., de Vries, P., Boes, J., Cerundolo, V., Osterhaus, A. D., and UytdeHaag, F. G. (1992). Measles virus transmembrane fusion protein synthesized de novo or presented in immunostimulating complexes is endogenously processed for HLA class I- and class II-restricted cytotoxic T cell recognition. J. Exp. Med. *176*, 119-128.

van Binnendijk, R. S., van der Heijden, R. W., and Osterhaus, A. D. (1995). Monkeys in measles research. Curr. Top Microbiol. Immunol. *191*, 135-148.

Varas, A., Sacedon, R., Hernandez-Lopez, C., Jimenez, E., Garcia-Ceca, J., Arias-Diaz, J., Zapata, A. G., and Vicente, A. (2003). Age-dependent changes in thymic macrophages and dendritic cells. Microsc Res. Tech *62*, 501-507.

Vasconcelos, D. Y., Cai, X. H., and Oglesbee, M. J. (1998). Constitutive overexpression of the major inducible 70 kDa heat shock protein mediates large plaque formation by measles virus. J. Gen. Virol. *79*, 2239-2247.

Veilette, A. (2004). SLAM family receptors regulate immunity with and without SAP-related adaptors. J. Exp. Med. *199*, 1175-1178.

Vidalain, P. O., Azocar, O., Lamouille, B., Astier, A., Rabourdin-Combe, C., and Servet-Delprat, C. (2000). Measles virus induces functional TRAIL production by human dendritic cells. J. Virol. *74*, 556-559.

Vidalain, P. O., Azocar, O., Rabourdin-Combe, C., and Servet-Delprat, C. (2001a). Measle virus-infected dendritic cells develop immunosuppressive and cytotoxic activities. Immunobiology *204*, 629-638.

Vidalain, P. O., Azocar, O., Yagita, H., Rabourdin-Combe, C., and Servet-Delprat, C. (2001b). Cytotoxic activity of human dendritic cells is differentially regulated by double-stranded RNA and CD40 ligand. J. Immunol. *167*, 3765-3772.

Vidalain, P. O., Laine, D., Zaffran, Y., Azocar, O., Servet-Delprat, C., Wild, T. F., Rabourdin-Combe, C., and Valentin, H. (2002). Interferons mediate terminal differentiation of human cortical thymic epithelial cells. J. Virol. *76*, 6415-6424.

Vincent, S., Gerlier, D., and Manie, S. N. (2000). Measles virus assembly within membrane rafts. J. Virol. *74*, 9911-9915.

Vincent, S., Spehner, D., Manie, S., Delorme, R., Drillien, R., and Gerlier, D. (1999). Inefficient measles virus budding in murine L.CD46 fibroblasts. Virology *265*, 185-195.

von Messling, V., Milosevic, D., Devaux, P., and Cattaneo, R. (2004). Canine distemper virus and measles virus fusion glycoprotein trimers: partial membrane-proximal ectodomain cleavage enhances function. J. Virol. *78*, 7894-7903.

Von Pirquet, C. (1908). Das verhalten der kutanen tuberculinreaktion wahrend der masern. Deutche Midizinische Wochenschrift *34*, 1297.

Vongpunsawad, S., Oezgun, N., Braun, W., and Cattaneo, R. (2004). Selectively receptor-blind measles viruses: Identification of residues necessary for SLAM- or CD46-induced fusion and their localization on a new hemagglutinin structural model. J. Virol. *78*, 302-313.

Wang, N., Morra, M., Wu, C., Gullo, C., Howie, D., Coyle, T., Engel, P., and Terhorst, C. (2001). CD150 is a member of a family of genes that encode glycoproteins on the surface of hematopoietic cells. Immunogenetics *53*, 382-394.

Wang, N., Satoskar, A., Faubion, W., Howie, D., Okamoto, S., Feske, S., Gullo, C., Clarke, K., Rodrigues Sosa, M., Sharpe, A. H., and Terhorst, C. (2004). The cell surface receptor SLAM controls T cell and macrophage functions. J. Exp. Med. *199*, 1255-1264.

Ward, B. J., and Griffin, D. E. (1993). Changes in cytokine production after measles virus vaccination: predominant production of IL-4 suggests induction of a Th2 response. Clin. Immunol. Immunopathol *67*, 171-177.

Weidmann, A., Fischer, C., Ohgimoto, S., Ruth, C., ter Meulen, V., and Schneider-Schaulies, S. (2000). Measles virus-induced immunosuppression *in vitro* is independent of complex glycosylation of viral glycoproteins and of hemifusion. J. Virol. *74*, 7548-7553.

Welstead, G. G., Hsu, E. C., Iorio, C., Bolotin, S., and Richardson, C. D. (2004). Mechanism of CD150 (SLAM) Down Regulation from the Host Cell Surface by Measles Virus Hemagglutinin Protein. J. Virol. *78*, 9666–9674.

White, R. G., and Boyd, J. F. (1973). The effect of measles on the thymus and other lymphoid tissues. Clin. Exp. Immunol. *13*, 343-357.

Whittle, H. C., Dossetor, J., Oduloju, A., Bryceson, A. D., and Greenwood, B. M. (1978). Cell-mediated immunity during natural measles infection. J. Clin. Invest *62*, 678-684.

World Health Organization (1998). Standardization of the nomenclature for describing the genetic characteristics of wild-type measles viruses. In Weekly epidemiological report (Geneva), pp. 265-272.

Xu, C., Mao, D., Holers, V. M., Palanca, B., Cheng, A. M., and Molina, H. (2000). A critical role for murine complement regulator crry in fetomaternal tolerance. Science *287*, 498-501.

Yamanouchi, K., Chino, F., Kobune, F., Kodama, H., and Tsuruhara, T. (1973). Growth of measles virus in the lymphoid tissues of monkeys. J. Infect. Dis. *128*, 795-799.

Yamanouchi, K., Ohta, H., Kataoka, T., Kobune, F., Yoshikawa, Y., and Tokunaga, T. (1981). Suppression of delayed hypersensitivity by measles virus infection in guinea pigs. Jpn J. Med. Sci. Biol. *34*, 81-94.

Yanagi, Y., Cubitt, B. A., and Oldstone, M. B. (1992). Measles virus inhibits mitogen-induced T cell proliferation but does not directly perturb the T cell activation process inside the cell. Virology *187*, 280-289.

Yant, S., Hirano, A., and Wong, T. C. (1997). Identification of a cytoplasmic Tyr-X-X-Leu motif essential for down regulation of the human cell receptor CD46 in persistent measles virus infection. J. Virol. *71*, 766-770.

Yokota, S., Saito, H., Kubota, T., Yokosawa, N., Amano, K., and Fujii, N. (2003). Measles virus suppresses interferon-alpha signaling pathway: suppression of Jak1 phosphorylation and association of viral accessory proteins, C and V, with interferon-alpha receptor complex. Virology *306*, 135-146.

Young, D. F., Didcock, L., Goodbourn, S., and Randall, R. E. (2000). Paramyxoviridae use distinct virus-specific mechanisms to circumvent the interferon response. Virology *269*, 383-390.

Zaffran, Y., Destaing, O., Roux, A., Ory, S., Nheu, T., Jurdic, P., Rabourdin-Combe, C., and Astier, A. L. (2001). CD46/CD3 costimulation induces morphological changes of human T cells and activation of Vav, Rac, and extracellular signal-regulated kinase mitogen-activated protein kinase. J. Immunol. *167*, 6780-6785.

Zhang, X., Glendening, C., Linke, H., Parks, C. L., Brooks, C., Udem, S. A., and Oglesbee, M. (2002). Identification and characterization of a regulatory domain on the carboxyl terminus of the measles virus nucleocapsid protein. J. Virol. *76*, 8737-8746.

Zhou, J., Fujino, M., Inou, Y., Kumada, A., Aoki, Y., Iwata, S., and Nakayama, T. (2003). H1 genotype of measles virus was detected in outbreaks in Japan after 2000. J. Med. Virol. *70*, 642-648.

Zhu, Y. D., Heath, J., Collins, J., Greene, T., Antipa, L., Rota, P., Bellini, W., and McChesney, M. (1997). Experimental measles. II. Infection and immunity in the rhesus macaque. Virology *233*, 85-92.

Zuniga-Pflucker, J. C. (2004). T-cell development made simple. Nat. Rev. Immunol. *4*, 67-72.

Appendix 1

Abbreviations

APC	Antigen presenting cells
BCR	B cell receptor
BM	Bone marrow
CTL	Cytotoxic T lymphocyte
Cyt	Cytoplasmic tail
DC	Dendritic cells
DDB1	DNA damage binding protein
dsRNA	Double strand RNA
F	Fusion
Fas-L	Fas ligand
FcR	Fc receptor
FcγRII	Immunoglobulin G Fc receptor II
GM-CSF	Granulocyte monocyte colony stimulating factor
H	Hemagglutinin
HSP	Heath shock protein
IFN	Interferon
IFNAR	interferon alpha receptor
Ig	Immunoglobulin
IL	Interleukin
IPC	Interferon producing cells
IRF	Interferon regulatory factor
ISGF	Interferon stimulated gene factor
ISRE	Interferon specific responsive element
ITAM	Immunotyrosine based activatory motive
Jak	Jun activating kinase
L	Large (protein)
LC	Langherans Cell
LPS	Lipopolysaccharide
M	Matrix
MAC	membrane attack complex
MCP	Membrane cofactor protein
MHC	Major histocompatibility complex
MIBE	Measles inclusion bodies encephalitis
MV	measles virus
N	Nucleoprotein
NC	Nucleocapsid
N_{CORE}	nucleoprotein core (N-terminus)
NK	natural killer
NO	Nitric oxid
NR	Nucleoprotein receptor
N_{TAIL}	nucleoprotein tail (C-terminus)
P	Phosphprotein
PBL	Peripheral blood lymphocytes
PBMC	Peripheral blood monocytes
PCT	Phosphprotein C-terminus
PKR	RNA-dependent serine/threonine protein kinase
PNT	Phosphoprotein N-terminus
PPRV	Peste des Petits Ruminants Virus
RCA	Regulator of complement activation
RdRp	RNA dependent RNA polymerase
RNA	Ribonucleic acid
RNAi	RNA interfering
RNP	Ribonucleoparticle
RPV	Rinderpest Virus
RSV	Respiratory Syncytial Virus
SAEC	Small airway epithelial cells
SCID	Severe combined immunodeficiency
SCR	Short consensus repeat
SHP	Protein-tyrosine phosphatase
sIL2R	Soluble IL-2 receptor
SLAM	Signaling lymphocyte activating molecule
SSPE	Sub-ACUTE sclerosis pan-encephalitis
STAT	Signal transducer and activator of transcription
STP	Serine threonine proline
TEC	Thymic epithelial cell
TCR	T cell receptor
TEC	Thymic epithelial cell
TGF	Tumor growth factor
TGN	Trans-Golgi network
Th	T helper
TLR	Toll like receptor
TNF	Tumor necrosis factor
Tr	T regulatory
TRAIL	TNF-related apoptosis-inducing ligand
TRAIL-R	TRAIL Receptor
TREC	TCR rearrangement excision circles
WFC	Warthin-Finkeldey cells
WHO	World health organisation
CDV	Canine Distemper virus
HIV	Human Immunodeficiency Virus
ITIM	Immunotyrosine based inhibitory motif

Printed in the United Kingdom
by Lightning Source UK Ltd.
108968UKS00001BB/3-28

9 781904 455059